MARKETING TECHNOLOGY PLATFORMS

LICENSE, DISCLAIMER OF LIABILITY, AND LIMITED WARRANTY

By purchasing or using this book and companion files (the "Work"), you agree that this license grants permission to use the contents contained herein, including the disc, but does not give you the right of ownership to any of the textual content in the book / disc or ownership to any of the information or products contained in it. *This license does not permit uploading of the Work onto the Internet or on a network (of any kind) without the written consent of the Publisher.* Duplication or dissemination of any text, code, simulations, images, etc. contained herein is limited to and subject to licensing terms for the respective products, and permission must be obtained from the Publisher or the owner of the content, etc., in order to reproduce or network any portion of the textual material (in any media) that is contained in the Work.

MERCURY LEARNING AND INFORMATION ("MLI" or "the Publisher") and anyone involved in the creation, writing, or production of the companion disc, accompanying algorithms, code, or computer programs ("the software"), and any accompanying Web site or software of the Work, cannot and do not warrant the performance or results that might be obtained by using the contents of the Work. The author, developers, and the Publisher have used their best efforts to ensure the accuracy and functionality of the textual material and/or programs contained in this package; we, however, make no warranty of any kind, express or implied, regarding the performance of these contents or programs. The Work is sold "as is" without warranty (except for defective materials used in manufacturing the book or due to faulty workmanship).

The author, developers, and the publisher of any accompanying content, and anyone involved in the composition, production, and manufacturing of this work will not be liable for damages of any kind arising out of the use of (or the inability to use) the algorithms, source code, computer programs, or textual material contained in this publication. This includes, but is not limited to, loss of revenue or profit, or other incidental, physical, or consequential damages arising out of the use of this Work.

The sole remedy in the event of a claim of any kind is expressly limited to replacement of the book and/or disc, and only at the discretion of the Publisher. The use of "implied warranty" and certain "exclusions" varies from state to state and might not apply to the purchaser of this product.

Marketing Technology Platforms

Strategy, Evaluation, and Implementation

Greg Kihlstrom

MERCURY LEARNING AND INFORMATION
Boston, Massachusetts

Copyright ©2025 by Mercury Learning And Information.
An Imprint of DeGruyter Inc. All rights reserved.

This publication, portions of it, or any accompanying software may not be reproduced in any way, stored in a retrieval system of any type, or transmitted by any means, media, electronic display, or mechanical display, including, but not limited to, photocopy, recording, Internet postings, or scanning, without prior permission in writing from the publisher.

Mercury Learning And Information
121 High Street, 3rd Floor
Boston, MA 02110
info@merclearning.com

G. Kihlstrom. *Marketing Technology Platforms: Strategy, Evaluation, and Implementation.*
ISBN: 978-1-5015-2423-3

The publisher recognizes and respects all marks used by companies, manufacturers, and developers as a means to distinguish their products. All brand names and product names mentioned in this book are trademarks or service marks of their respective companies. Any omission or misuse (of any kind) of service marks or trademarks, etc. is not an attempt to infringe on the property of others.

Library of Congress Control Number: 2025937554

242526321   This book is printed on acid-free paper in the United States of America.

Our titles are available for adoption, license, or bulk purchase by institutions, corporations, etc.

All of our titles are available in digital format at various digital vendors.

For Lindsey,
My partner in agility and in constantly staying curious about what's next.
I'm looking forward to many more discoveries ahead.

Contents

Acknowledgments		*xxxiii*
Introduction		*xxxv*
Part 1:	**An Overview of Marketing Technology**	**1**
Chapter 1:	**Defining Marketing Technology**	**3**
	Categories of MarTech Platforms	3
	Artificial Intelligence and MarTech Platforms	4
	Customer Data	5
	Example Platforms	6
	Content, Campaign, and Multichannel Delivery	6
	Example Platforms	6
	Workflow and Automation	7
	Example Platforms	7
	Analytics and Reporting	7
	Example Platforms	8
	Related Platforms	8
	VoltAge Begins Its MarTech Journey	9
	Customer Data: Understanding the Customer Journey	9
	Creation, Workflow, and Operations: Streamlining Campaigns	9

	Content, Campaign, and Multi-Channel Delivery: Reaching Customers Everywhere	10
	Measurement, Reporting, and Analysis: Closing the Feedback Loop	10
	Building the Roadmap	10
	Conclusion	11
	Notes	11
Chapter 2:	**A Brief History of MarTech**	**13**
	Early MarTech (Pre-1960s)	14
	The Pre-Digital Era	14
	The Birth of Database Marketing (1960s-1980s)	14
	The Rise of the Internet (1990-2001)	16
	World Wide Web	16
	Search Engines and Search Engine Optimization (SEO)	16
	The Birth of Digital Advertising	17
	The Social Media Era (2002-2009)	18
	Web 2.0 and User-Generated Content	18
	Early Social Platforms	19
	Social Media Marketing	19
	Paid Social Advertising	20
	The Mobile Emergence (2007-2011)	20
	SMS Marketing	20
	Smartphones	21
	The Era of Big Data and Automation (2010s)	22
	Data Explosion and Advanced Analytics	22
	Marketing Automation	23
	Growth of CRM and CDP	23
	Privacy and Consent Management	24
	The AI and Personalization Era (2020s)	25
	The Rise of AI in Marketing	25
	Focus on First-Party Data Strategies	26

	Hyper-Personalization	26
	Growing Data Privacy Focus	27
	Conclusion	28
	Notes	28
Chapter 3:	**MarTech and Marketing Teams**	**33**
	Marketing Leadership	34
	How They Use MarTech	34
	Types of MarTech Platforms Used	35
	Additional Considerations	35
	Customer Acquisition	36
	How They Use MarTech	36
	Platforms Utilized	36
	Digital Marketing Campaigns	37
	How They Use MarTech	37
	Platforms Utilized	38
	Content Marketing	39
	How They Use MarTech	39
	Web Site Marketing	40
	How They Use MarTech	40
	Additional Considerations	41
	Social Media Marketing	42
	How They Use MarTech	42
	Additional Considerations	43
	Digital Advertising	43
	Email Marketing	45
	How They Use MarTech	45
	Platforms Utilized	45
	Events Marketing	46
	How They Use MarTech	46
	Search Marketing	48
	How They Use MarTech	48

	Platforms Utilized	48
	Additional Considerations	49
E-commerce		49
	How They Use MarTech	49
	Platforms Utilized	50
	Additional Considerations	50
Analytics Teams		51
	How They Use MarTech	51
	Platforms Utilized	51
	Additional Considerations	52
CRM Team		52
	How They Use MarTech	52
	Platforms Utilized	53
	Additional Considerations	53
Marketing Operations		54
	How They Use MarTech	54
	Platforms Utilized	54
	Additional Considerations	55
Conclusion		55

Chapter 4: How Other Teams in an Organization Utilize MarTech — **57**

Non-Marketing Executives	57
How They Use MarTech	58
Platforms Utilized	58
Additional Considerations	59
Sales Teams	59
How They Use MarTech	60
Platforms Utilized	60
Additional Considerations	60
Data Teams	61
How They Use MarTech	61

	Platforms Utilized	62
	Additional Considerations	62
	Technology Teams	63
	How They Use MarTech	63
	Platforms Utilized	63
	Additional Considerations	64
	Customer Service Teams	64
	How They Use MarTech	64
	Platforms Utilized	65
	Additional Considerations	65
	Creative Teams	66
	How They Use MarTech	66
	Platforms Utilized	67
	Additional Considerations	67
	Customer Experience	68
	How They Use MarTech	68
	Platforms Utilized	68
	Additional Considerations	69
	Conclusion	69
	Notes	70
Chapter 5:	**MarTech and the Evolving Customer Relationship**	**71**
	Greater Adoption of Digital Channels	71
	What This Means for Brands	72
	How This Affects MarTech	72
	Social Media Mindset	73
	What This Means for Brands	73
	How This Affects MarTech	73
	Mobile-First Behavior	74
	What This Means for Brands	74
	How This Affects MarTech	75

Increased Importance of the Omnichannel Customer Experience ... 75
   What This Means for Brands ... 75
   How This Affects MarTech ... 76
Concern About Consumer Data Privacy ... 76
   What This Means for Brands ... 76
   How This Affects MarTech ... 77
Expectation of Personalization ... 77
   What This Means for Brands ... 77
   How This Affects MarTech ... 77
Desire for Self-Serve Options ... 78
   What This Means for Brands ... 78
   How This Affects MarTech ... 78
Conclusion ... 79
Notes ... 79

Chapter 6: The Business Case for MarTech Investments ... 83
Evaluating Where MarTech Can Make the Biggest Impact ... 83
   Ensuring Marketing and Communication Effectiveness ... 84
   Closing the Customer Experience Gap Between Customer Expectations and Product/Service Delivery ... 86
   Increasing Competitive Advantage ... 88
   Creating Greater Internal Efficiency and Collaboration ... 89
   Enabling More Data-Driven Decisions ... 91
Introducing a MarTech Maturity Model ... 93
Conclusion ... 94
Notes ... 95

Chapter 7:	**Methods of MarTech Implementation**	**97**
	All-in-one, Best-of-Breed, and Composable	97
	All-in-one	98
	Best-of-Breed	99
	Composable	100
	Open Source and Proprietary	101
	Proprietary Software	102
	Open Source Software	103
	Hybrid Scenarios	104
	Conclusion	105
Chapter 8:	**AI and MarTech**	**107**
	Generative AI	107
	Hypothetical Use Cases	108
	Examples of Generative AI Platforms	109
	Predictive Analytics	109
	Hypothetical Use Cases	110
	Examples of Platforms That Incorporate Predictive Analytics	110
	Workflow Automation	111
	Hypothetical Use Cases	112
	Example Software Platforms That Utilize Workflow Automation	112
	Personalization	112
	Hypothetical Use Cases	113
	Example Software Platforms That Utilize Personalization	114
	Synthetic Personas and Research	114
	Hypothetical Use Cases	115
	Example Synthetic Research Platforms	116

	AI-Powered Customer Support	116
	Hypothetical Use Cases	117
	Example Platforms	118
	Content Optimization and Analysis	118
	Hypothetical Use Cases	119
	Examples of Platforms That Support Content Optimization	119
	Audience Targeting and Segmentation	120
	Hypothetical Use Cases	120
	Example Platforms	121
	Advertising Optimization and Media Buying	121
	Hypothetical Use Cases	122
	Example Platforms	122
	Brand Monitoring and Sentiment Analysis	123
	Hypothetical Use Cases	124
	Example Platforms	124
	Cross-Channel Attribution	124
	Hypothetical Use Cases	125
	Example Platforms	126
	Visual Recognition and Image Processing	126
	Hypothetical Use Cases	127
	Example Platforms	127
	Customer Journey Orchestration	128
	Hypothetical Use Cases	128
	Example Platforms	129
	Conclusion	129
	Notes	130
Part 2:	**Customer Data**	**131**
Chapter 9:	**Key Customer Data Considerations**	**133**
	Understanding Data Ownership	134
	Types of Customer Data	134

	Benefits of a First-Party Data Strategy	137
	Factors Affecting Customer Data Quality	139
	Recency and Relevance	139
	Data Hygiene	140
	Consistency Across Channels	140
	Data Enrichment	140
	Data Source Reliability	141
	Data Storage and Access	141
	Case Study: VoltAge Motors and the Customer Data Dilemma	143
	The Absence of a First-Party Data Strategy	143
	Data Siloes Between Marketing and Customer Retention Teams	144
	Taking a Strategic Turn	144
	Looking Ahead	145
	Conclusion	145
	Note	145
Chapter 10:	**CRM, CDP, and DMP**	**147**
	Challenges of Managing Customer Data	147
	Platforms That Can Address These Challenges	148
	How The Platforms Work Together	150
	Combined Use Case	151
	Customer Relationship Management (CRM)	152
	CRMs and Their Usage by Marketers	152
	Key Components of a CRM	152
	How to Evaluate a CRM	156
	Contact and Relationship Management	156
	Sales and Pipeline Management	157
	Integration and Customization	157
	Automation and Workflow Efficiency	158
	Analytics, Reporting, and Performance Tracking	158
	Customer Data Platform (CDP)	159

Brief History and Usage by Marketers	159
Key Features of a CDP	160
How to Evaluate a CDP	164
Data Collection and Integration	164
Identity Resolution and Data Unification	164
Audience Segmentation and Activation	165
Compliance, Security, and Data Governance	166
Analytics and AI Capabilities	166
Key Differences Between a CDP and a CRM	167
Purpose and Focus	167
Data Collection and Scope	167
Data Types	168
Real-Time Capabilities	168
Intended Users	168
Data Integration	168
Scalability	168
Analytics and Insights	169
Personalization	169
Data Accessibility	169
Data Management Platform (DMP)	169
Key Features of a DMP	170
Using a DMP	170
DMPs and a First-Party Data Strategy	171
The Future of DMPs	171
Case Study: VoltAge Motors and the Intersection of CDPs, CRMs, and DMPs	172
How VoltAge Uses a CDP, CRM, and DMP	172
Challenges in Differentiating CDP and CRM Usage	173
Areas for Improvement	173
Looking Ahead	173
Conclusion	174
Notes	174

Chapter 11:	**Privacy and Consent**	**175**
	The Trust Gap Between Consumers and Brands	175
	The Need for Customer Data Remains	176
	Data Privacy Regulations	176
	Data Minimization	181
	Cross-Border Data Transfers	181
	AI and Privacy	182
	Consent Management Platforms (CMPs)	182
	Customer Data Platforms (CDPs)	183
	Identity Resolution Platforms	183
	Case Study: VoltAge Motors and the Challenge of GDPR Compliance	183
	The Current State	184
	The Path Forward	185
	Conclusion	186
	Notes	186
Part 3:	**Creation, Workflow, and Operations**	**189**
Chapter 12:	**Key Considerations with Creation, Workflow, and Operations**	**191**
	Strategy and the Creative Process	191
	Content Planning and Ideation Tools	192
	Asset Management	192
	Aligning Creative Efforts with Marketing Objectives	192
	Multi-Channel Needs	193
	Channel-Specific Optimization	193
	Repurposing Content and Assets	193
	Automation and AI-Driven Recommendations	194
	Efficiency	194
	Project and Task Management Tools	194

	Workflow Automation	194
	Version Control and Approval Processes	195
	Collaboration	195
	Content Collaboration Tools	195
	Cross-Team Coordination	195
	Remote and Global Teams	196
	AI's Growing Role in Content Creation	196
	Case Study: VoltAge Motors' Quest to Manage Content Creation	197
	A Disconnected Set of Tools	197
	The Solution	197
	Next Steps, Results, and Lessons Learned	197
	Conclusion	198
Chapter 13:	**Content Marketing Platforms**	**199**
	How a Content Marketing Platform Helps Marketers	199
	Enhancing Personalization Efforts	200
	Optimizing Content Distribution	201
	Who Are the Users of a CMP?	201
	Content Strategists and Marketing Managers	201
	Writers, Editors, and Content Creators	202
	SEO Specialists and Content Optimizers	202
	Social Media and Email Teams	203
	Project Managers	203
	Analytics and Performance Teams	203
	Key Features	204
	Content Planning and Strategy	204
	Content Creation Tools	205
	Collaboration Tools	205
	Content Distribution	205
	Analytics and Performance Tracking	206

	Asset Management	206
	Integration Capabilities	206
	Brand Management	207
	How to Evaluate a CMP	207
	Usability and User Experience	207
	Integration Capabilities	208
	Content Management and Workflow Automation	208
	Analytics and Performance Tracking	209
	Scalability and Future-Proofing	209
	Related Platforms	210
	Content Management Systems (CMSs) and Digital Experience Platforms (DXPs)	210
	Focused Content Creation and Collaboration Tools	210
	Content Curation Platforms	211
	Video Content Marketing Platforms	211
	Influencer Marketing Platforms	211
	Conclusion	211
Chapter 14:	**Digital Asset Management (DAM) Systems**	**213**
	How a DAM Helps Marketers	214
	Who Are the Users of a DAM?	214
	Marketing Teams	214
	Creative Teams (Designers, Videographers, and Copywriters)	214
	Sales Teams	215
	Product Management Teams	215
	Legal and Compliance Teams	216
	IT and Data Teams	216
	External Agencies and Partners	217
	Key Features of a DAM Platform	217
	How to Evaluate a DAM	221

	Asset Organization and Searchability	222
	Integration with Existing Systems	222
	User Access, Security, and Compliance	223
	Collaboration and Workflow Automation	223
	Scalability, Performance, and Future-Proofing	224
	Conclusion	224
Chapter 15:	**Project Management Platforms**	**225**
	How Does Project Management Software Help Marketers?	226
	Resource Allocation and Time Management	227
	Visibility and Transparency	227
	Scalability and Adaptability	227
	Who Are the Users of Project Management Software for Marketing?	228
	Marketing Leadership	228
	Campaign and Content Teams	229
	Creative and Design Teams	229
	Marketing Operations and Analytics Teams	230
	Cross-Functional Teams (Sales, Product, IT, and Legal)	230
	Varying Needs Depending on Teams	231
	Small Marketing Teams and Startups	231
	Mid-Sized Marketing Teams	232
	Enterprise-Level Marketing Departments	233
	Key Features of Project Management Platforms	233
	Task and Workflow Management	234
	Campaign Planning and Calendar Management	234
	Team Collaboration and Communication Tools	235
	Resource Allocation and Time Tracking	235
	Integration with Other Marketing Tools	236
	Approval Workflows and Version Control	236

	Analytics and Reporting	237
	How to Evaluate Project Management Software for Marketing Teams	237
	Ease of Use and Adoption	238
	Customization and Scalability	238
	Collaboration and Communication Features	239
	Automation and Workflow Efficiency	239
	Reporting and Performance Tracking	240
	Conclusion	240
Part 4:	**Content, Campaign, and Multi-Channel Delivery**	**243**
Chapter 16:	**Key Considerations for Content, Campaign, and Multichannel Delivery**	**245**
	The Marketing Lifecycle and Its Impact on Platform Selection	246
	Aligning Platforms with the Marketing Lifecycle	246
	Platform Needs Across the Customer Journey	247
	Balancing Long-Term Brand-Building and Short-Term Performance Marketing	247
	The Challenges of Planning, Creation, Management, and Delivery	248
	Fragmentation Across Teams and Workflows	248
	Maintaining Agility Without Compromising Strategy	249
	A Growing Number of Marketing Channels	250
	The Proliferation of Digital Touchpoints	250
	Adapting Content Across Channels	251
	Customer Channel Switching and Evolving Preferences	253
	The Expectation of Seamless Transitions Between Channels	253
	Changing Consumer Behavior and Platform Selection	254

	Balancing Channel-Specific Expertise with Omnichannel Excellence	255
	Complexity of Managing Multiple Channels Effectively	255
	Training and Adoption Challenges	256
	Integration and Coordination Between Channels	257
	Providing a Cohesive Omnichannel Experience	258
	Personalizing Across Multiple Channels	258
	Creating Personalized Content at Scale	259
	VoltAge Motors' Multichannel Marketing Challenges	260
	Conclusion	262
Chapter 17:	**DXPs and CMS**	**263**
	CMSs	264
	How a CMS Empowers Non-Technical Users	264
	Examples of CMS Platforms	265
	DXPs	266
	The Shift from Traditional CMS to DXPs	266
	DXPs and a Multi-channel Focus	267
	Examples of DXPs	268
	The Differences Between a DXP and a CMS	269
	Scope and Capabilities	269
	Integration and Extensibility	270
	Personalization and AI Capabilities	270
	Target Users and Use Cases	271
	Primary Users of DXPs and CMSs	272
	Marketers	272
	Content Creators and Editors	273
	Developers, Engineers, and IT Teams	273
	UX and Design Teams	274
	Evaluating DXP and CMS Platforms	274
	DXPs	275

	CMSs	276
	Differing Approaches to Managing Content	278
	Headless CMS	279
	Conclusion	282
	Note	282
Chapter 18:	**Marketing Automation, Multichannel Personalization, and Journey Orchestration**	**283**
	Comparing the Types of Platforms	284
	When to Use Each Type of Platform	285
	How Do These Platforms Help Marketers?	285
	Who Uses These Platforms?	286
	Marketing Teams	287
	Customer Experience Teams	287
	Sales Teams	287
	IT and Data Teams	288
	MAPs	288
	Strengths	289
	Weaknesses	290
	How to Evaluate MAPs	290
	Multichannel Personalization Platforms	290
	Key Features	291
	Example Multichannel Personalization Platforms	291
	Strengths	292
	Weaknesses	292
	How to Evaluate Multichannel Personalization Platforms	292
	CJO Platforms	293
	Key Features	293
	Example CJO Platforms	294
	Strengths	294

	Weaknesses	294
	How to Evaluate CJO Platforms	295
	Conclusion	296
	Notes	296
Chapter 19:	**Single-Channel Marketing Platforms**	**297**
	How Do Single-Channel Marketing Platforms Help Marketers?	298
	Examples of Single-Channel Platforms	298
	Email Marketing Platforms	299
	Social Media Management	302
	Digital Advertising Management	304
	Search Engine Optimization Platforms	307
	Conclusion	310
	Notes	310
Chapter 20:	**Conversational Marketing and Real-time Communication Platforms**	**311**
	What Falls Under This Category?	312
	Example Platforms	312
	How Do They Help Marketers?	315
	Instant Engagement and Lead Conversion	315
	Personalized Customer Experiences	315
	24/7 Customer Support and Automation	316
	Omni-Channel Communication	316
	Data Collection and Behavioral Insights	316
	Who Are the Users of These Platforms?	317
	Marketing Teams	317
	Sales Teams	317
	Customer Support and Customer Service Teams	317
	E-Commerce and Retail Teams	318
	Technology Teams	318
	Key Features	318

	How to Evaluate	319
	AI and NLP Capabilities	320
	Omni-Channel Functionality	320
	Integration Capabilities	320
	Scalability and Automation	321
	Customization and Personalization Features	321
	Conclusion	322
	Notes	322
Chapter 21:	**Multi-Function Platforms and Suites**	**323**
	Adobe Experience Platform (AEP)	323
	Path to Growth	324
	Optimizely ONE	325
	Current Market Focus	325
	Key Acquisitions and Feature Expansions	326
	Current State of Optimizely ONE	326
	Sitecore DXP	327
	Current Market Focus	327
	Evolution of Sitecore DXP	328
	Acquia DXP	328
	Current Market Focus	328
	Key Acquisitions and Feature Expansions	329
	Current State of the Acquia DXP	329
	Hubspot Marketing Cloud	329
	Origins and Growth of HubSpot Marketing Hub	329
	Current Market Focus	330
	Current State of HubSpot Marketing Hub	330
	Salesforce Marketing Cloud	330
	Current Market Focus	331
	Strengths of Multi-Function Platforms	332
	Weaknesses of Multi-Function Platforms	333
	Lack of Robust Features in Specific Areas	333
	Lack of Interoperability	333

	How to Evaluate	333
	Existing Features and Functionality	334
	Customer Base	334
	Product Roadmap	334
	Conclusion	334
	Notes	335
Part 5:	**Measurement and Reporting**	**339**
Chapter 22:	**Key Considerations for Measurement, Reporting, and Analysis**	**341**
	Defining Objectives and KPIs	341
	Data Collection and Data Integrity	342
	Privacy and Compliance Considerations	343
	Budget and Resource Constraints	343
	Illustrating the Considerations: VoltAge Motors Example	344
	Conclusion	345
Chapter 23:	**Single Platform, In-App Measurement, and Multi-Channel Measurement**	**347**
	Single-Channel Measurement Tools	347
	In-App Analytics	349
	Multichannel Measurement Tools	352
	Case Example: VoltAge Motors and Its Multilayered Analytics Strategy	355
	Conclusion	356
Chapter 24:	**Data Visualization and Analysis Tools**	**357**
	Key Features of Data Visualization Platforms	358
	Interactive Dashboards and Workflow Management	358
	Advanced Visualization Options	358
	Customizable Reporting	359
	Integration and Automated Data Ingestion	359
	Real-Time and Near-Real-Time Data Capabilities	359

	Resource Allocation and Scalability	360
	Security and Compliance	360
	Optional Features	361
	AI-Driven Insights	361
	Mobile Access	361
	Embedded Analytics	361
	Storytelling and Presentation Tools	362
	Natural Language Processing (NLP)	362
	Advanced Data Modeling and Preparation	362
	Community and Marketplace Resources	362
	How to Evaluate Data Visualization Tools	363
	Flexibility with Visualization	363
	Compatibility with Data Sources	363
	User-Friendliness vs. Advanced Features	364
	Pricing Structures	364
	VoltAge Motors' Transformation with Tableau	364
	Conclusion	365
Part 6:	**Evaluation of Your MarTech Stack and Platforms**	**367**
Chapter 25:	**Define Your Goals**	**369**
	MarTech Infrastructure Goals Framework	369
	Recommended Strategic Goals	370
	Example KPIs to Measure Success	373
	Example Steps to Achieve Each Strategic Goal (Technology-Focused)	374
	Key Takeaways	377
	Revisiting the MarTech Maturity Model	377
	An Overview of Each Stage in the MarTech Maturity Model	380
	Stage 1: Foundational	380
	Stage 2: Integrated	380
	Stage 3: Advanced	381
	Stage 4: Innovative/Next-Level	381

	Involving the Right Teams and Unique Roles	382
	Case Study: VoltAge Motors' Maturity Journey	383
	Conclusion	384
	Note	384
Chapter 26:	**A MarTech Stack Evaluation Framework**	**385**
	Start by Visualizing	386
	Requirements for a MarTech Infrastructure Map	387
	Who to Involve	388
	Evaluating Your MarTech Stack	388
	Step 1: Define the Scope and Purpose	391
	Step 2: Assess the Current Maturity Level	393
	Step 3: Evaluate Core Infrastructure and Architecture	395
	Step 4: Review Data Management and Accessibility	397
	Step 5: Examine System Integration and Interoperability	399
	Step 6: Analyze Automation and Personalization Capabilities	402
	Step 7: Evaluate Governance, Compliance, and Security	404
	Step 8: Assess Vendor and Platform Roadmaps	405
	Step 9: Examine Organizational Readiness and Skills Alignment	407
	Step 10: Quantify Performance and ROI	409
	Step 11: Identify Gaps and Prioritize Next Steps	411
	Step 12: Evaluation Summary	413
	Other Considerations	415
	Identify Overarching Gaps	415
	Identify Specific Gaps	415
	Conclusion	417

Chapter 27:	**Evaluation of Individual MarTech Platforms**	**419**
	When Should You Evaluate a Single Platform in Your MarTech Stack?	420
	An Evaluation Framework	421
	Key Features	421
	Utilization	422
	Growth	422
	Risk Mitigation	422
	Investment	423
	Prioritize the Categories	423
	Benefits of the Evaluation Process	423
	Vendor Evaluation	424
	When Do You Replace a Platform?	424
	From Big Picture to Small Picture	425
	What to Do with Weak Links in Your MarTech Chain	425
	Procuring a New Platform	426
	VoltAge: Evaluating Its Web Site CMS	428
	The Evaluation Process	428
	Key Features	428
	Utilization and Growth Potential	429
	Risk Mitigation and Investment Considerations	429
	The Decision: Moving to a DXP	429
	Results and Next Steps	429
	Conclusion	430
	Note	430
Chapter 28:	**Building a MarTech Roadmap**	**431**
	Your MarTech Roadmap	431
	Why a Strategic Roadmap Is Necessary for Your MarTech Infrastructure	432
	Steps to Create a Comprehensive Roadmap	432
	One: Start with a Static Infrastructure Map	434

Two: Determine the Best Format for
Your Roadmap 434
Three: Socialize Your Roadmap 435
Four: Analyze Gaps Between the Current
and Desired State of Technology 436
Five: Align KPIs with Your Roadmap 436
Six: Identify and Prioritize Current Marketing
Technologies 436
Ensure Your Roadmap Stays up to Date 437
VoltAge Motors: Using a MarTech Roadmap to
Align DXP and Customer Data Initiatives 437
Establishing the Roadmap: From Static Map
to Dynamic Strategy 438
How the Roadmap Brought Teams Together 438
Results and Impact: A Clear Path to Personalization
and Efficiency 439
Conclusion 439

Chapter 29: Measuring Success 441
Overall Return on Investment (ROI) 441
TCO 442
Speed to Delivery 443
Speed to Insights 443
Customer Satisfaction (Internal and External) 444
Adoption Rate 446
Cross-Team Adoption 446
Level of Usage of Specific Products 446
Level of Usage of the Entire MarTech Ecosystem 446
Level of Integration 446
Benefit to the Business 447
Case Study: VoltAge Motors Aligns MarTech
Adoption with Business Benefits 449

VoltAge Motors Searched for Reasons Why Its Business Results Didn't Align with How Its Teams Used MarTech Platforms	449
Diagnosis: The Root Causes of Limited Adoption	450
The Turning Point: VoltAge Motors Developed a Practical Adoption and Alignment Roadmap to Address Its Issues	450
Results: Aligning Adoption with Business Benefits	451
Conclusion	451
Epilogue	**453**
Define Your Goals with Continuous Improvement in Mind	453
Keep up with the Changing Technology Landscape	454
Anticipate Future Needs and Trends	454
Integrate Your Existing Systems	455
Maintain Flexibility	455
Regularly Revisit and Reconcile Your Strategic Goals	456
About the Author	**457**
Index	**459**

Acknowledgments

As with any book, countless people have a hand in the thoughts and ideas that are contained within. I will endeavour to thank many of them, but a full list would take up its own book, so please excuse this abbreviated list.

Thanks to the many people that have influenced this book and my knowledge of Marketing Technology. I'll list a few here, though there are many throughout my career. Brian Browning, Carlos Manalo, Chris Bach, Adam Chen, Chad Solomonson, Tommi Marsans, Peter Fogelsanger, Robin Ross, Kevin Li, Bennett Boucher, Adeline Ashley, Don Schuermann, Abby See, Brian Marzullo, Wiselin Jayajos, and others.

Thank you also to the Marketing Technology platform companies that have participated in my podcast and invited me to their conferences over the years. Meeting and discussing in person or over an interview has provided a greater depth of insight and understanding, and I am forever thankful for that. Some of my guests include: Gary Vaynerchuk, Alex Atzberger, Dave O'Flanagan, Heidi Bullock, Bernadette Nixon, Fabrice Martin, Pasquale DeMaio, Arianna Vogel, Katherine Lehman, Dani Yogatama, Jen Rapp, Mitsu Kikuchi, Noah Zamansky, and others. These include executives and leaders from platforms like Optimizely, Sitecore, Hubspot, Acquia, Pega, EagleAI, Braze, Bloomreach, Tealium, Treasure Data, Salesforce, Databricks, ActionIQ, Redpoint Global, Medallia, Qualtrics, Netlify, Persado, Constructor, Algolia, Hootsuite, Siteimprove, Microsoft, Inferos, Fizz, Kount, SendBird, Topsort, Constellation, Znode, Cogito, Churnkey, TrueDialog, Bazaarvoice, The Trade Desk, Radial, arrivia, and many more.

I am forever grateful to my wife Lindsey, who is always supportive of me, no matter how many books I write during the course of a year. She is forever an inspiration, and I'm thankful to have such a great partner in all things.

Finally, thanks to everyone reading this book and anyone who has listened to my podcast The Agile Brand with Greg Kihlström, read an article, and supported me in any way over the last several years. I hope that the thoughts and ideas shared by myself and others have been helpful in your work.

Let's move forward and create great things together!

INTRODUCTION

The topic of this book is one that I deal with almost every day to one degree or another. In my consulting work, I help clients make big investment decisions in marketing technology (MarTech) platforms. I know full well that the implications of making the right—or wrong—choice can be quite dramatic and last many years. Many times in my career, I have worked with clients suffering from the effects of poor platform choices, misguided integration recommendations, and short-sighted roadmaps. In some cases, I was able to help them make the most of these inherited decisions and investments, and in others, I created a roadmap and strategic plan to get their MarTech infrastructure back on the right path.

Anyone familiar with the Chiefmartec Marketing Technology Landscape Supergraphic knows this to be true: the MarTech landscape is not getting any smaller or more narrowly defined any time soon[1]. The number of platforms that consider themselves part of the MarTech landscape increased from about 1,867 in 2015 to nearly 10,000 as of mid-2022[2], and a whopping 14,106 as of 2024[3]. With so many choices, it can become nearly impossible to make good decisions or even to know where to start narrowing down your choices.

Additionally, there is the sheer size and scale of the issue of evaluating the right platforms to use. This is particularly difficult for large organizations, which use an average of 367 different software tools (many of which fall under the "MarTech" umbrella)[4], and which often have multiple "flavors" of a single type of platform. For instance, a single organization may have a different Web site content management system (CMS) or digital experience platform (DXP) running each of the following: their marketing Web site, their e-commerce platform, a customer or partner portal, a customer support site, and more, and that is just one category of platform!

Additionally, according to Clevertouch's State of MarTech 2023 report, 44% of marketers reported that they had several MarTech platforms that have largely gone unused[5]. This means the arduous steps of evaluation, procurement, and implementation are undergone only to have a platform gathering dust months later.

Anyone reading this who has worked at an enterprise can probably relate to those scenarios described above, yet solving them is never simple. So many factors go into choosing the right platforms, coordinating with other internal (and often external) teams managing legacy systems, or working on parallel efforts.

Thus, there is never a single, perfect answer to any of the tough questions that marketing teams face when building their MarTech stack or deciding on the best platform. By utilizing a consistent approach to planning and decision-making, however, you can avoid the many missteps other organizations have taken. After all, according to that same Clevertouch report, 36% of marketers are planning to spend the largest proportion (35%) of their budget on software as opposed to other areas, such as campaigns (19.7%), services (19.3%), or people (9.6%).

So, it is important to be able to make good decisions about the software investments that you *do* make and to find ways to look at your entire infrastructure as a whole, as well as the components. That's what this guide is designed to do.

While this book is intended for those with some experience with MarTech and at least some decision-making authority in selecting the platforms and directions in which to proceed, I will, from time to time, step back and give a little background for those who may be new to all of this.

WHAT IS A MARTECH INFRASTRUCTURE?

Those of you with decades of experience in the MarTech space may know the answer already, but with the degree of specialization in so many marketing roles these days, it is also important to ensure we thoroughly understand exactly what we mean when we say MarTech infrastructure.

A MarTech stack is a collection of software tools and platforms that marketers use to plan, execute, measure, and optimize their marketing activities. A well-designed MarTech stack can help marketers achieve their goals more efficiently and effectively by automating tasks, integrating data, personalizing messages, and providing insights.

Building a MarTech stack is not as simple as buying and installing a bunch of software and then getting back to the work of marketing, however! Marketing teams need to consider several factors when choosing and configuring their tools, such as the following:

- The tools that will best fulfill the objectives and strategies of their marketing campaigns
- The needs, preferences, and rising expectations of their target audiences, and the shifting competitive landscape
- The compatibility and interoperability of different tools within the MarTech stack
- The budget and resources available for investing in technology, both in the short- and long-term
- The skills and expertise required to use and manage the tools, as well as to support them in the future

A MarTech stack can be divided into different categories or layers, depending on the functions they perform, and within this book, we will be looking at a framework to help evaluate and understand where gaps may exist in your MarTech stack.

MarTech stacks can vary in size and complexity, depending on the scale and scope of the business. Some businesses may have only a few essential tools in their stack, while others may have dozens or even hundreds of tools. The quality of a MarTech stack, however, is not determined by how many tools it has within it but rather by how well the tools within the MarTech stack work together to deliver great customer experience and value to the business.

Therefore, marketers should not focus on acquiring more tools but on aligning their tools with their objectives, audience needs, and best practices. They should also regularly evaluate their existing tools to see if they are still relevant, effective, and efficient.

Additionally, an organization's MarTech stack is not a static entity but a dynamic one that evolves over time as the business grows and changes and markets and customer expectations evolve. Marketers should always be on the lookout for new opportunities and challenges in the market and adapt their technology accordingly, while strategically balancing investments and guarding against chasing "shiny objects" that may have little to no material impact on the bottom line. By building a robust and flexible MarTech stack, marketers can enhance their capabilities and competitiveness in today's omnichannel world, where customer experience is a key point of competition and differentiation among brands.

THIS IS BASED ON RESEARCH AND EXPERIENCE

My work continually informs my writing, and this book is an example of that. I have been privileged to work with several organizations of varying sizes (from Fortune 50 to 1000) and assisted with creating strategies, finding solutions, and delivering many initiatives that the approaches described in this book can help teams achieve. I am committed to being both a writer-researcher and a practitioner; I want my insights to be more than purely theoretical. I hope this makes the concepts on the page more actionable, insightful, and beneficial to you, the reader.

The subject matter in this guide is near and dear to me, as I have helped many organizations with their high-level MarTech infrastructure strategy, as well as evaluate, procure, and implement individual elements. Having done this dozens of times in the last few years alone, I can most certainly say I know what I speak of (or write, as the case may be)!

I've been living in the world of MarTech since the early 2000s, and anyone on the same ride (or longer) has seen the rise of many technologies, some of which have continued to evolve and others that have fallen by the wayside. Either way, though, there has been tremendous growth.

You've probably seen the infographic that Chiefmartec has been creating for years that has mapped the growth of the MarTech landscape. If you haven't, it started with about 150 platforms in 2011, hit over 3,500 in 2016, and reached nearly 10,000 in 2022[6]. That's over 6,500% growth in less than a dozen years.

This growth does not seem to be slowing any time soon, either, with MarketWatch predicting that the global MarTech market size will grow from an estimated valuation in 2021 of $168 billion to nearly $370 billion by 2027[7]. Thus, investments in MarTech need to be made with growth in mind. In this book, we will talk about how to make the most strategic investments and clearly define your approach to building a winning MarTech infrastructure.

WHO THIS BOOK IS FOR

This book is for marketing executives and professionals who want to understand how to evaluate and understand the current state of your MarTech infrastructure, as well as how to take steps to improve it, including the evaluation of new platforms.

For the purposes of this book, we will focus on the needs of large organizations and have an enterprise mindset, although if you work at a smaller company, you can still gain plenty of insights.

WHAT THIS BOOK IS NOT

While I lay out a framework for categorizing MarTech platforms and provide tools to help evaluate them, this book is not a guide to what the best MarTech platforms are, nor is it intended to tell you what types of platforms you need based on your own organization. There are simply too many variables to give any specific recommendations. In other words, what works for a hospital system will not necessarily work for an apparel company or a financial services company, and so on.

Instead, consider this guide a toolset you can use for both strategic and tactical evaluation and implementation. You'll have several evaluation tools you can use and a solid understanding of the available tools and methods of implementation to be successful.

WHAT WE WILL COVER

I'm excited to share my knowledge and experience of creating strategies for MarTech infrastructure and then evaluating, implementing, and optimizing the individual tools and the MarTech platform as a whole. We'll touch on all of this in this book. This book is divided into four main sections:

Part 1: An Overview of MarTech

This part of the book starts by defining what is included under the umbrella of MarTech, then provides a history of MarTech, discusses the teams that utilize it, and looks at the business case and methods of implementation commonly used.

The categories of MarTech platforms will also be introduced, which will be used to explore the breadth of these tools in Parts 2–6.

Part 2: Customer Data

The full range of MarTech platforms related to customer data and how the data is stored, shared, and utilized are covered in this section, including customer data platforms (CDPs), customer relationship management systems (CRMs), and more.

Part 3: Creation, Workflow, and Operations

MarTech platforms responsible for the creation of content and campaigns, as well as those that are integrally involved in managing their creation, are discussed in this part, including digital asset management (DAM) systems, content marketing platforms, and project management systems.

Part 4: Content, Campaign, and Multi-Channel Delivery

This part of the book is focused on MarTech platforms that enable marketing teams to deliver and serve content, campaigns, advertising, and more across many channels to reach customers in a variety of methods. These include Web site CMSs, DXPs, and more.

Part 5: Measurement, Reporting, and Analysis

Successful analysis and reporting of marketing efforts require a solid set of tooling to support it. This part of the book focuses on the measurement tools that allow marketers to analyze one or more marketing channels, as well as to visualize and analyze the results of their work.

Part 6: Evaluation of Your MarTech Stack and Platforms

The last part of the book focuses on tools and methods to evaluate the current state of an organization's MarTech infrastructure as well as ways to create roadmaps to make improvements that return business value and customer value.

INTRODUCING VOLTAGE MOTORS

Throughout this book, a fictional company will offer the opportunity to explore the role of MarTech in the growth and success of an organization. While the company used may not always have direct applicability to all readers of this book, the use cases and examples have been chosen to be as broadly applicable as possible.

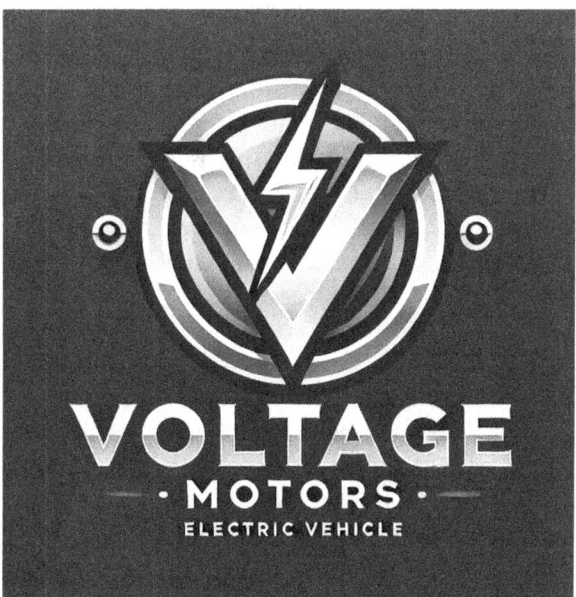

FIGURE 0.1 VoltAge Motors Logo. Generated by Dall-E 2

VoltAge Motors, headquartered in Reading, Pennsylvania in the United States—a suburb of Philadelphia—is a dynamic startup in the electric vehicle (EV) industry, challenging established giants with its commitment to innovation, sustainability, and customer centricity. Founded with a mission to accelerate the adoption of electric mobility, VoltAge Motors combines cutting-edge technology with bold design to deliver vehicles that are as efficient as they are exciting to drive. Positioned as a disruptor, the company is laser-focused on creating a seamless and engaging car-buying experience that aligns with the needs of modern, tech-savvy consumers.

VoltAge Motors' Challenges and Opportunities

As a startup competing with industry heavyweights, VoltAge Motors faces the dual challenge of building brand recognition while delivering a customer experience that rivals more mature companies. This requires the adoption of a sophisticated MarTech stack to effectively reach and engage target audiences. Key priorities include developing compelling content that tells the VoltAge story, educating potential buyers on the advantages of EVs, and nurturing leads throughout a long and considered purchase journey.

VoltAge also integrates its marketing efforts with an innovative e-commerce platform, which allows customers to explore, customize, and purchase vehicles entirely online. The platform extends beyond vehicle purchases to include financing, warranties, and subscription services, creating a holistic digital ecosystem. For VoltAge to maximize the potential of this platform, its MarTech stack must seamlessly connect customer data, content delivery, and analytics to create a personalized and frictionless experience.

Setting the Stage for Exploration

In the pages to come, we will explore VoltAge Motors' journey to building a mature MarTech stack, examining the challenges and opportunities it encounters at each stage. From developing a robust customer data platform to implementing marketing automation, predictive analytics, and personalized content strategies, we'll uncover how VoltAge transforms its marketing capabilities to support growth and compete effectively in the EV market. Through VoltAge's story, you'll gain actionable insights into the practical application of MarTech solutions and learn how to navigate similar challenges in your own organization.

BEFORE BEGINNING

Keep in mind that the concepts and categories introduced early in the book will enable you to best utilize the evaluation tools and maturity model that is more thoroughly explored in Part 6 of the book.

Additionally, in some industries, there can be specific tools and even communication methods that are more prevalent than others. While the exact tools that your organization uses may not be included in some of the descriptions, it is highly likely that they will fit into one of the categories described.

Finally, consider the maturity model and evaluation framework a starting point that can be used in your own organization. They have been provided to be used, and therefore have been made as easy as possible to adopt. Your feedback and use cases of how they have been adopted within your organization are always welcome as well so that we can continue to improve them in the future.

NOTES

1. Chiefmartec. "Marketing Technology Landscape 2022." Retrieved April 16, 2023 from https://chiefmartec.com/2022/05/marketing-technology-landscape-2022-search-9932-solutions-on-martechmap-com/

2. Statista. "Number of MarTech Solutions." Retrieved April 16, 2023 from https://www.statista.com/statistics/1131436/number-martech-solutions/

3. Brinker, S. (2024, May 6). 2024 marketing technology landscape supergraphic — 14,106 martech products (27.8% growth YoY). ChiefMartec. https://chiefmartec.com/2024/05/2024-marketing-technology-landscape-supergraphic-14106-martech-products-27-8-growth-yoy/

4. Forrester Research and Airtable. "Report: Software is Fracturing Your Organization." Retrived April 16, 2023 from https://www.airtable.com/lp/resources/reports/crisis-of-the-fractured-organization

5. Clevertouch and University of Southampton. The State of Martech 2023 Report. Retrieved April 16, 2023 from https://clever-touch.com/love-martech/

6. Chiefmartec. "Marketing Technology Landscape 2022." Retrieved April 16, 2023 from https://chiefmartec.com/2022/05/marketing-technology-landscape-2022-search-9932-solutions-on-martechmap-com/

7. Research Reports World. "Marketing Technology Market Growth, Consumer Demand, Technologyical Advancements and Industry Trends." Retrieved April 26, 2023 from: https://www.marketwatch.com/press-release/marketing-technology-market-2023-growth-consumer-demand-technological-advancements-and-industry-trends-2023-04-21

PART 1

AN OVERVIEW OF MARKETING TECHNOLOGY

As digital channels and experiences continue to play a more prevalent role in how brands engage with their customers, the software tools and platforms that marketers use to do their work, reach their customers, and analyze their results become increasingly important—and increasingly digital.

Yet, with over 14,000 marketing technology (MarTech) platforms available today (Statista, 2024)[1], it can be overwhelming for marketers to identify the right tools to meet their needs. Additionally, the MarTech landscape is evolving rapidly, driven by fast-growing consumer expectations and the increasing digitization of businesses, as well as the rise of AI-based tools and solutions. Companies must navigate an ever-changing environment while delivering seamless customer experiences and improving operational efficiency all the while.

This exploration of MarTech will begin with a history of MarTech and a definition of the term, which includes what falls within the realm of MarTech, as well as what falls outside but is still relevant.

In this first part of the book, we are going to look at the following:

- An overview of MarTech
- A history of MarTech from the early twentieth century to today
- How the customer experience has evolved to the highly digital, omnichannel journey it is today

- The evolution of the digital business
- The categories of MarTech platforms
- AI and MarTech
- Methods of implementing MarTech

These chapters will set the stage for a deeper exploration of the individual categories of MarTech platforms that will follow in Part 2.

CHAPTER 1

DEFINING MARKETING TECHNOLOGY

Marketing technology, or MarTech, refers to the software and technology tools and platforms that marketers use to plan, execute, and measure their marketing campaigns.

MarTech can include many different types of tools, such as content management systems (CMS), customer relationship management systems (CRMs), social media management tools, email marketing solutions, Web site analytics platforms, and much more. Additionally, MarTech enables marketers to create, deliver, and measure personalized, relevant, and effective marketing experiences that drive customer satisfaction, loyalty, and growth to one or more customer bases.

According to a 2022 study of B2B marketers by Dun and Bradstreet, a company has an average of 10 tools in their MarTech stacks, though many enterprise organizations have considerably more[2].

These MarTech platforms also help internal teams streamline their workflow and output, enabling them to increase their efficiency to meet the demands of customers who increasingly want personalized content, offers, and experiences. Thus, MarTech is essential for modern marketing as it helps marketers adapt to the changing customer needs and expectations in a digital world.

CATEGORIES OF MARTECH PLATFORMS

There are several methods of cataloging and categorizing the thousands of platforms that comprise the MarTech category. While there is no single way to do this, the framework in this book can be used to do so.

Customer Data
Helping brands understand their customers

Content, Campaign, and Multichannel Delivery
Serving customers with content, offers, and experiences across the journey

Measurement, Reporting, and Analysis
Enabling tracking and insights for marketing efforts

Creation, Workflow, and Operations
Empowering teams to be more efficient in content and campaign creation

FIGURE 1.1 The four categories of MarTech platforms

This book will explore the vast array of MarTech platforms by grouping them into four main categories, as shown in Figure 1.1:

- *Customer Data*, which helps businesses to better understand their customers
- *Creation, Workflow, and Operations*, which enables teams to more efficiently create content and campaigns
- *Content, Campaign, and Multi-Channel Delivery*, which serves customers with content, offers, and experiences including Web site, email, social media, and advertising
- *Measurement, Reporting, and Analysis*, which allows tracking and analysis of marketing efforts, including channel-specific and multi-channel measurement.

Artificial Intelligence and MarTech Platforms

You might have noticed that there is not a specific category for AI platforms in this list, and there is a good reason why. Instead of being treated as separate and distinct, AI is treated as a feature within platforms that enables—and in many cases augments—marketers' ability to do existing tasks. For instance, AI that is introduced to a MarTech platform could play one of the following roles:

- Creating multiple variations of content to utilize in A/B testing
- Generate different image sizes of social media images to be distributed across brand channels
- Calculate the propensity of a customer to purchase a specific product
- Automate a workflow between marketing teams to streamline complex approval and revision processes

Of course, these are just a few examples, but, as demonstrated with the list above, AI is a component of a broader application or set of processes in each. Thus, we will primarily look at AI in this way, rather than as a separate category of platform.

CUSTOMER DATA

The first category is the set of platforms within a MarTech stack that collect and share customer information for prospective, current, and lapsed customers. There are two primary types of platforms that comprise this category and help businesses manage their customer data and interactions: customer relationship managers (CRMs) and customer data platforms (CDPs).

A CRM is a system that tracks and organizes the sales, marketing, and service activities of a company with its customers.

A CDP, or customer data platform, is a system that collects and unifies customer data from various sources, such as Web sites, apps, social media, and offline channels. CDPs complement CRMs because they are more suitable for collecting information about customers on channels outside of the domain of a company, such as external social networks, advertising, and other marketing activities.

The purpose of all customer information tools is to provide a comprehensive and accurate view of the customer journey, preferences, and behavior, and to enable personalized and effective communication and engagement with the customers.

Example Platforms

Some types of platforms that fall into this category are as follows:

- *CRMs* such as Salesforce, Microsoft Dynamics, and Hubspot
- *CDPs* such as Tealium, Amperity, and Treasure Data
- *Data management platforms (DMPs)* such as Adobe Audience Manager, BluKai, Nielsen DMP, and The Trade Desk

CONTENT, CAMPAIGN, AND MULTICHANNEL DELIVERY

Next, we have MarTech platforms that allow brands and their marketing teams to launch and deliver their content, offers, and campaigns to their intended audiences. The platforms in this category span a wide range of functionality and features, as well as the channels they support.

For instance, there may be one or more platforms that support each of the following channels:

- Web site and e-commerce
- Mobile app
- Email marketing
- SMS messaging
- Social media
- Digital advertising

Your brand probably serves other channels as well. Ensuring that cross-channel content and campaign delivery is critical to your MarTech stack. This is why in many cases an organization will choose an all-in-one approach to consolidate management of several marketing channels on a single platform.

Example Platforms

Some types of platforms that fall into this category are as follows:

- *Digital experience platforms (DXPs) and Web site CMSs* such as Adobe, Optimizely, Sitecore, and Drupal
- *Marketing automation and email marketing platforms* such as Eloqua, Marketo, Pardot, and Constant Contact

- *Testing and optimization platforms* such as Adobe Target, Optimizely, and Omniconvert
- *Customer journey orchestration (CJO) platforms* such as CSG Xponent, Medallia Thunderhead, and Genesys Pointillize
- *Social media management platforms* such as Hootsuite, Buffer, and Sprout Social
- *Enterprise search* platforms such as Algolia, Coveo, and Solr

WORKFLOW AND AUTOMATION

Workflow and task automation tools are software applications that can help marketers streamline and optimize their workflows, such as creating and executing campaigns, managing leads, and analyzing results. Workflow automation tools can reduce manual work, improve efficiency, and enhance collaboration among marketing teams.

Workflow automation tools can help marketers save time and money, eliminate errors, and boost productivity. By automating repetitive and tedious tasks, marketers can focus on the creative and strategic aspects of their work and deliver better results for their clients and customers.

Example Platforms

Some types of platforms that fall into this category are as follows:

- *Project management and workflow management* such as Adobe Workfront and Atlassian Jira
- *Content workflow management* such as Optimizely Orchestrate

ANALYTICS AND REPORTING

Analytics and reporting tools are essential for enterprise businesses to gain insights from their data and make informed decisions. These tools can help enterprises collect, analyze, visualize, and share data across different departments, functions, and locations. Some of the benefits of using analytics and reporting tools for enterprise businesses are as follows:

- They can improve operational efficiency and performance by helping teams identify bottlenecks, optimizing processes, and reducing costs.

- They can enhance customer satisfaction and loyalty by enabling marketing teams to gain a better understanding of customer behavior, preferences, and feedback.
- They can enable marketing teams to find ways to increase revenue and profitability by discovering new opportunities, creating personalized offers, and predicting future trends.
- They can foster innovation and collaboration in marketing teams by enabling data-driven experimentation, learning, and problem-solving.

Example Platforms

Some types of platforms that fall into this category are as follows:

- Web and mobile app analytics such as Google Analytics or Adobe Analytics
- Mobile app analytics such as Amplitude
- Customer journey analytics such as SAS
- Reporting tools such as Tableau and Microsoft PowerBI

RELATED PLATFORMS

In a sense, any platform that a marketing team touches could be considered MarTech (or at least part of the ecosystem), but aside from acknowledging them here, this book will not address them in depth.

These related platforms include the following:

- General productivity software is used by marketers and many others in business. Platforms that fall into this category include Microsoft Office and Google Workspace, as well as visualization tools such as Miro and communication tools such as Slack, Teams, and Zoom.
- E-commerce is closely related to many marketers' work, yet there are components that fall outside the scope of MarTech. These tools could be inventory management, product information management, fulfillment and logistics, and others.
- Customer service, like e-commerce, is often closely related to the work marketers do, yet many of the tools that customer service teams use are not considered part of a MarTech stack. These include ticketing applications such as ZenDesk and call center applications.

- Finally, data infrastructure, including data warehouses and other data storage often relates to marketers' work but falls outside the scope of MarTech. This might include platforms such as Snowflake and AWS.

VOLTAGE BEGINS ITS MARTECH JOURNEY

VoltAge Motors, a rapidly growing electric vehicle (EV) startup based in Reading, Pennsylvania, faced a critical juncture as it sought to scale up its operations and marketing efforts. With increasing competition from established players and ambitious growth goals, the marketing leadership team decided to evaluate whether their existing MarTech stack was equipped to support their trajectory. The evaluation revealed significant gaps across all four key categories of their MarTech stack:

- Customer Data
- Creation and Workflow
- Content, Campaign, and Multi-Channel Delivery
- Measurement

To sustain its growth and remain competitive, VoltAge concluded that a complete overhaul and the development of a roadmap were necessary.

Customer Data: Understanding the Customer Journey

VoltAge's first area of focus was its customer data capabilities. Despite having a CRM system in place, the team discovered that their tools were not fully integrated, leading to fragmented customer profiles. This lack of a unified customer view made it challenging to deliver personalized experiences and measure the impact of their campaigns effectively. For instance, while they could track initial leads from Web site forms, connecting this data to broader insights such as purchase history or engagement with post-sale services was difficult. The team realized that to provide the seamless, tailored experiences that their tech-savvy EV customers demanded, they needed to invest in a robust CDP that could consolidate and centralize data from all touchpoints.

Creation, Workflow, and Operations: Streamlining Campaigns

Next, the VoltAge team assessed their processes for creating and managing marketing campaigns. While they used a mix of project management tools and standalone content creation software, these tools often operated in silos, resulting in inefficiency and delays. Teams reported difficulty in tracking project timelines, managing creative assets, and ensuring consistent branding across campaigns. This lack of cohesion became increasingly problematic as their marketing efforts expanded to include multi-channel campaigns across digital, social, and offline platforms. VoltAge concluded that a comprehensive content management and workflow automation solution was essential to streamline their processes, improve collaboration, and enable quicker go-to-market timelines.

Content, Campaign, and Multi-Channel Delivery: Reaching Customers Everywhere

VoltAge's ability to deliver content and campaigns across channels was another area in need of improvement. Its current tools allowed basic email campaigns and social media scheduling but lacked the sophistication to execute and coordinate complex multi-channel campaigns. The marketing team struggled to connect campaigns across email, social media, digital ads, and their e-commerce platform, leading to inconsistent customer experiences. Additionally, their Web site lacked dynamic personalization capabilities, which limited their ability to tailor content to different customer segments. They recognized that investing in a more advanced multi-channel delivery platform would allow them to provide cohesive experiences and scale their efforts as their audience grew.

Measurement, Reporting, and Analysis: Closing the Feedback Loop

The final area of evaluation focused on measurement and reporting. VoltAge relied heavily on channel-specific analytics tools, which provided limited visibility into the effectiveness of its overall marketing efforts. The team lacked the ability to perform multi-channel attribution and struggled to connect marketing performance data to business outcomes, such as vehicle sales or customer retention. Without a clear feedback loop, it was difficult to identify which strategies were driving the most value. The VoltAge team determined that a unified analytics platform capable of both granular and multi-channel insights was critical to inform their decision-making and optimize campaign performance.

Building the Roadmap

After this comprehensive evaluation, VoltAge's marketing leadership reached a clear conclusion: their current MarTech stack was not equipped to support their ambitious goals. To address the gaps and build a more future-ready infrastructure, they decided to re-evaluate all four categories of platforms and create a detailed roadmap for implementation. This roadmap would not only prioritize immediate needs, such as integrating customer data and streamlining workflow, but also ensure scalability to support the company's transition from startup to industry leader.

The company's commitment to a robust, scalable MarTech strategy underscored its dedication to providing exceptional customer experiences while maintaining operational efficiency and staying ahead in the competitive EV market.

In the pages that follow, VoltAge's journey to greater MarTech maturity will be followed.

CONCLUSION

With this categorization explained and a few examples for each category given, a vocabulary has now been established to describe the components of a MarTech stack and a categorization scheme to group the potentially many platforms that comprise our MarTech infrastructure.

The next chapter will explore a brief history of MarTech, from the early days of mass media communication, through our current era dominated by AI and a focus on personalized customer content and experiences.

NOTES

1. https://www.statista.com/statistics/1131436/number-martech-solutions/
2. Dun & Bradstreet. (n.d.). *8th annual B2B sales and marketing data report*. Dun & Bradstreet. https://www.dnb.com/perspectives/marketing-sales/8th-annual-b2b-sales-and-marketing-data-report.html

CHAPTER 2

A Brief History of MarTech

The current state of MarTech would not exist without consistent effort over many decades to improve the way that brands capture information about their customers and communicate with those same customers in more sophisticated ways.

Thus, to give a better perspective on the current state of MarTech, it is helpful to explore its evolution over the last several decades. As shown in Figure 2.0, starting in the 1950s, its history can be separated into seven eras, each of which will be explored in more detail in the pages that follow.

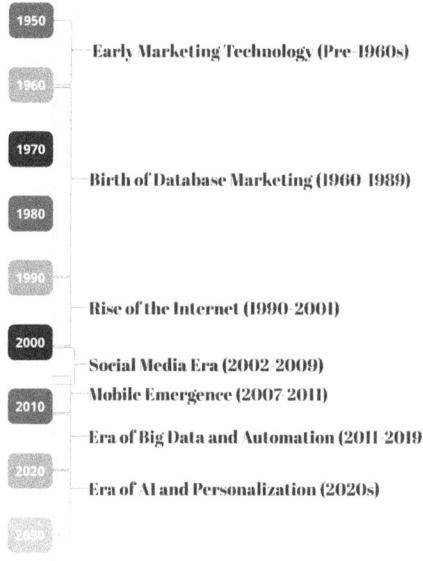

FIGURE 2.0 History of MarTech's eras

EARLY MARTECH (PRE-1960s)

While it could be traced back to before the 1950s, when computers were not machines but rather humans that *computed*, this review of the history of MarTech will start at the beginning of mass communication.

The Pre-Digital Era

The first part of the Early MarTech era can be considered when early mass communication tools such as print media, radio, and television come into use as marketing channels.

For instance, radio telegraphy, while proposed much earlier, was first made usable by the Italian inventor Guglielmo Marconi in the late 1890s, and was first established as a method of nautical communication[1]. Only later would it become a method of communication for marketers and advertisers.

While each of these mediums had their own unique purposes for which they were originally created, over time, companies found ways to reach consumers through them.

Additionally, as mass communication grew, methods of more direct marketing to individual consumers, such as direct mail and telemarketing, were introduced during this time period as early forms of what is now considered personalized marketing.

> *Key Milestones of Early MarTech*
> - *July 1, 1941: First TV Ad*
>
>   The first television advertisement ran on NBC's WNBT station in New York City. It was a 10-second commercial for the watch and jewelry company Bulova that ran before a baseball game between the Brooklyn Dodgers and the Philadelphia Phillies[2].

THE BIRTH OF DATABASE MARKETING (1960s-1980s)

With mass communication such as television growing from an $85 million market in the United States in the early 1950s to a billion-dollar market by the 1960s, it had established itself firmly in marketing budgets for companies of all sizes[3].

The concept of mass communication led to a growing desire to reach individuals with specific product and service needs. This was facilitated by the rise of computers and data storage systems that enabled companies to track more granular details about current and potential customers.

These databases include early customer relationship management (CRM) systems that facilitated storing and organizing customer data for targeted marketing and replaced more rudimentary analog Rolodex systems (invented in 1956[4]) or custom-built mainframe computer customer data digitization instances.

This era also saw the beginning of the use of mainframe computers for managing mailing lists and tracking customer interactions.

Key Milestones of the Birth of Database Marketing

- *1960s: Introduction of Electronic Data Interchange (EDI)*

    EDI allowed businesses to conduct electronic transactions through primitive computer networks, enabling the sharing of invoices, order forms, and shipping confirmations[5]

- *1971: Invention of Email*

    Ray Tomlinson sent the first network email through ARPANET, laying the foundation for future email marketing[6]

- *1978: First Spam Email*

    Gary Thuerk sent the first commercial email message promoting DEC computers to 400 recipients, marking the beginning of email marketing (and spam)[7]

- *1987: ACT! CRM Is Launched*

    Mike Muhney and Pat Sullivan launched what is largely considered the first contact management software, making it the first CRM to be in wide distribution beyond some one-off cases of more rudimentary mainframe customer data digitization.[8]

THE RISE OF THE INTERNET (1990-2001)

This era is arguably the biggest turning point in the history of MarTech because it began a shift of focus from what would soon become known as more "traditional" media (e.g., print, radio, and television) to online consumer behavior. That said, this was not an overnight transition, as the Web started much like other communication channels: intended for non-marketing purposes.

World Wide Web

Originally conceived as a tool for scientists to collaborate, and invented by Tim Berners-Lee at CERN in 1989[9], the World Wide Web (WWW) has, in subsequent years, grown in its use and depth well beyond that originally imagined.

- Emergence of Web sites as a new marketing channel
- Email marketing: One of the first digital marketing tactics

> *Key Milestone*
>
> - *1990-91: Creation of the World Wide Web (WWW)*
>
>   Tim Berners-Lee built the first web server and web browser, releasing the World Wide Web protocol in August 1991, leading to the development of URL, HTML, and HTTP[10].

Search Engines and Search Engine Optimization (SEO)

The concept of a search engine pre-dates the advent of the World Wide Web, with Elizabeth Feinler and her team at WHOis creating the WHOIS directory in the early 1970s[11]. That said, the first web-based search engine is believed to be W3Catalog, released September 2, 1993[12]. It distinguished itself from earlier tools by being specifically designed for the World Wide Web.

Additionally, search engines were not the only method of finding information and new Web sites in the early days of the Web. Yahoo! launched in 1994 as a web directory that was browsed via category, and gained popularity in that way before launching its own search engine later[13].

Early search engines include a host of platforms such as Lycos (1994), Yahoo, which added search functionality to its directory (1995)[14], AltaVista (1995), Ask Jeeves (1997), and later, Google (1998)[15]. While many of these search engines have not stood the test of time, others, such as Google, have grown beyond their roots as search engines and into other areas.

With growing search engine usage and the resulting user traffic for Web sites that are positioned favorably in those results, a new domain called search engine optimization (SEO)appeared to drive web traffic from search results.

This led to the rise of SEO marketing firms as well as SEO marketing roles within organizations, all aimed at driving traffic from organic search results.

> *Key Milestone*
>
> - *1998: Launch of Google Search Engine*
>
>   Google's search engine revolutionized how people find information online, eventually leading to the development of SEO and search engine marketing (SEM). Its unique PageRank algorithm took into account page popularity and other aspects that made it arguably more effective than its competition.[16]
>
>   Within 4 years, the term "googling" entered the common vocabulary, with one of the first recorded usages being in an episode of "Buffy the Vampire Slayer," and the American Dialect Society selecting "google" as the most "useful new word" of 2002.[17]

The Birth of Digital Advertising

From the introduction of the first banner advertisements to the growth of digital advertising as an effective and profitable marketing tactic, this era saw the introduction of many new inventions, such as banner ads themselves to dynamic ad servers, as well as the standardization of banner sizes and measurements.

Additionally, Google, while still early in its growth as a company, introduced Google AdWords, which would become a dominant presence in search marketing and digital advertising more broadly in the years to come.

> *Key Milestone*
>
> - *1994: AT&T Banner Ad on HotWired*
>
>   HotWired sold one of the first clickable banner ads to several advertisers, marking the beginning of digital display advertising, with the very first being an ad for AT&T Corp placed on October 27, 1994.[18]
>
> - *1994: Time Warner's Pathfinder Is Launched with Banner Ads*
>
>   Within days of HotWired's launch of the AT&T banner ad, the Pathfinder Web site was launched, complete with banner ads[19]. There is some disagreement about which was launched first.
>
> - *2000: Introduction of Google AdWords*
>
>   Google launched AdWords on October 23, 2000, allowing advertisers to promote products within search results and pioneering pay-per-click advertising[20].

THE SOCIAL MEDIA ERA (2002-2009)

With the Internet maturing as a way for brands to reach consumers, the use of Web-based sites and platforms as a marketing channel grew in prominence, even as some brands and businesses struggled to attach a return on investment (ROI) to some of their efforts.

Web 2.0 and User-Generated Content

While the very first Web sites were focused on a small number of authors providing content to readers (or Web 1.0), the rise of Web 2.0 marked a shift in how people interacted online, emphasizing user-generated content and dynamic engagement. Blogs, forums, and review sites such as Blogger (1999[21]) and Yelp (2004[22]) empowered individuals to share opinions, experiences, and knowledge publicly. This era redefined consumer behavior, as people began relying more on peer-generated reviews and community forums rather than solely trusting traditional brand messaging. Companies recognized the influence of these platforms and started to monitor and engage in these discussions to protect and enhance their reputations.

This surge in user-generated content also laid the foundation for the democratization of content creation, where individuals could shape public discourse. With consumers increasingly turning to blogs and forums for product information, businesses had to adapt by embracing authenticity and transparency to connect with their audiences.

Early Social Platforms

The early 2000s witnessed the emergence of the first social media platforms that would define a new era of connectivity and interaction. Friendster (2002) and MySpace (2003) gained massive popularity as social hubs for connecting with friends, while Facebook (2004) began as a university-exclusive network before expanding to the public. The launch of YouTube (2005) introduced a new visual medium, empowering creators and viewers to share and engage with video content. Twitter, launched in 2006, further transformed social interactions with its concise and fast-paced communication style.

These platforms revolutionized how individuals connected and shared content, creating new spaces for community building and discussion. For businesses, the rise of social networks opened up a direct communication channel with consumers, offering new opportunities for engagement and feedback.

While there remained steep competition among several social media platforms during this time, Facebook became a winner in terms of audience and traffic, surpassing 1 billion users as of late 2011[23].

Social Media Marketing

As social platforms gained traction, businesses quickly recognized their potential for building brand awareness and fostering engagement. Companies began creating dedicated profiles on platforms such as MySpace and Facebook, using them to share updates, promote products, and interact with followers. Social media marketing in this era was largely organic, focusing on authentic communication and community engagement rather than overt advertising.

This period marked a shift in marketing strategies, as brands learned to navigate the conversational nature of social media. Early adopters that embraced these platforms found new ways to humanize their brand and develop loyal online communities, setting the stage for the strategic social media marketing practices we see today.

Paid Social Advertising

By the mid-2000s, social platforms began experimenting with advertising models, introducing businesses to paid social advertising opportunities. MySpace was one of the first platforms to explore banner ads[24], while Facebook introduced its "Flyers" ad system in 2004[25], which later evolved into targeted ads based on user demographics and behaviors. These early efforts allowed businesses to reach highly specific audiences, marking the beginning of personalized digital advertising.

Paid social advertising quickly gained popularity as businesses realized its potential for driving measurable results. The ability to target specific users and track campaign performance made social media advertising an essential component of marketing strategies, laying the foundation for the sophisticated ad ecosystems that dominate platforms today.

> *Key Milestones*
> - *2004*: Facebook is launched as TheFacebook.com, and later that year begins its first ad revenue project[26].
> - *2006*: Myspace temporarily surpasses Google as the most-visited Web site in the United States[27].
> - *2009*: Facebook surpasses MySpace in total US visitors for the first time[28].
> - *2010*: Instagram is launched on October 6, 2010, later being sold to Facebook for $1 billion in cash and stock[29].

THE MOBILE EMERGENCE (2007-2011)

With consumers increasingly comfortable with using the Web in daily communications, for shopping, and as an information source, this era saw a shift towards taking the Internet anywhere and everywhere, with an ever-increasing focus on mobile technologies.

SMS Marketing

Marketers were soon to follow this mobile growth, with businesses realizing the potential of SMS as a marketing channel, many of which were run by local businesses such as pizza shops and dentist offices[30]. This was facilitated by the introduction of SMS short codes in 2003, 5 or 6-digit codes used by companies such as Pontiac and Nike to launch large-scale consumer marketing campaigns.[31]

Smartphones

The origins of smartphones trace back to the late 1990s and early 2000s, when devices such as thePalm Pilot and BlackBerry began merging the functionalities of personal digital assistants (PDAs) with basic mobile communication. Palm Pilots, introduced in 1996[32], were handheld PDAs that allowed users to manage calendars, contacts, and notes. A few years later, on January 19, 1999, BlackBerry introduced mobile devices with email capabilities, capturing the attention of business professionals who relied on instant communication[33]. These early innovations demonstrated the potential of handheld devices to go beyond simple communication, incorporating organizational tools that were crucial in a professional context.

The concept of a fully integrated smartphone began to take shape with devices such as the Nokia Communicator series, which combined a mobile phone with internet access and productivity tools. It wasn't until the introduction of the iPhone in 2007, however, that the modern smartphone era truly began. Apple's iPhone revolutionized the market by integrating a sleek touchscreen design, a robust operating system, and access to third-party applications through its App Store.

The iPhone set the standard for what a smartphone could achieve, catalyzing rapid advancements in mobile technology. Following its debut, competitors such as Android-powered devices quickly entered the market, creating a new era of innovation. By blending communication, entertainment, and computing power into a single device, the advent of the iPhone turned smartphones into indispensable tools for both personal and professional use, forever changing how people interact with technology, each other, and with brands that saw the opportunity in marketing to consumers via this channel.

> *Key Milestones*
>
> - *2003: Introduction of SMS Short Codes*
>
>   Allows marketers to more easily connect with consumers via text messages.[34]
>
> - *2006: AdMob Mobile Advertising Network Is Launched*
>
>   Even before the iPhone and App Store, mobile advertising emerges as a viable new channel[35].
>
> - *2007: Introduction of the iPhone*
>
>   The iPhone revolutionizes mobile Internet usage, leading to the development of mobile marketing strategies and app-based advertising[36].
>
> - *2008: Apple App Store Is Launched*
>
>   With the introduction of the App Store, Apple improves consumers' ability to gain new functionality on their smartphones, while fundamentally changing the way that software is acquired[37].

THE ERA OF BIG DATA AND AUTOMATION (2010s)

By the 2010s, MarTech as a practice was maturing, and the rapid rise of consumers' usage of digital channels such as the Web and mobile devices only cause that to accelerate.

Data Explosion and Advanced Analytics

The 2010s marked the rise of big data, as organizations began leveraging vast amounts of information generated from online activities, social media, Internet of Things (IoT) devices, and more. This data explosion provided unprecedented opportunities for brands to gain deeper insights into consumer behavior, preferences, and trends. By analyzing large datasets, marketers could move beyond traditional segmentation and adopt a more granular approach, identifying patterns and targeting specific customer needs with precision. The ability to harness this data became the cornerstone of strategic decision-making, as companies sought to understand their audiences in increasingly sophisticated ways.

Advanced analytics tools emerged during this era, enabling marketers to process data in real time and gain actionable insights. Solutions such as Google Analytics, Tableau, and Adobe Analytics helped track campaign performance, optimize customer journeys, and attribute marketing efforts to measurable business outcomes. Real-time reporting capabilities allowed brands to pivot quickly, improving campaign effectiveness and maximizing ROI. The integration of predictive analytics further empowered marketers to anticipate future behaviors and trends, driving proactive strategies rather than reactive responses.

Marketing Automation

The introduction of marketing automation tools such as HubSpot, Marketo, and Pardot revolutionized how companies managed campaigns and engaged with customers. These platforms enabled businesses to streamline repetitive tasks, such as email marketing, social media posting, and lead nurturing, freeing up time for more strategic initiatives. Automated email campaigns became a hallmark of this era, delivering personalized messages based on user behavior, preferences, and lifecycle stages. With features such as triggered emails and dynamic content, marketers could engage with customers at the right time with relevant information.

Another significant advancement was lead scoring and segmentation, which allowed marketing and sales teams to prioritize prospects more effectively. By assigning scores based on factors such as Web site visits, content downloads, and email interactions, businesses could focus on high-value leads, improving conversion rates. Customer segmentation became more dynamic, enabling brands to tailor campaigns to specific audiences with unprecedented accuracy. These innovations transformed marketing from a labor-intensive process into an efficient, data-driven discipline.

Growth of CRM and CDP

CRM platforms such as Salesforce and Microsoft Dynamics redefined how sales and marketing teams aligned their efforts, enabling seamless communication and collaboration. These systems provided a centralized repository for customer data, making it easier to track interactions, manage leads, and analyze performance metrics. CRM platforms became essential tools for maintaining long-term customer relationships, enhancing the ability to nurture prospects and retain loyal clients. By integrating with marketing automation tools, CRMs created a unified workflow that bridged the gap between lead generation and conversion.

The emergence of CDPs further enhanced marketing capabilities by providing a comprehensive view of individual customers. Unlike CRMs, which focus on managing relationships, CDPs collect and unify data from multiple sources to build detailed customer profiles. These profiles allowed marketers to deliver highly personalized experiences across channels, meeting consumer expectations for relevance and consistency. As data became the backbone of modern marketing, CRM and CDP technologies set the foundation for more sophisticated and customer-centric strategies.

Privacy and Consent Management

The 2010s also brought significant challenges regarding data privacy and consent management. High-profile data breaches and growing concerns over how personal information was used led to increased scrutiny of data practices. In response, governments enacted regulations such as the General Data Protection Regulation (GDPR) in the European Union and the California Consumer Privacy Act (CCPA) in the United States. These laws gave consumers greater control over their data, requiring companies to be more transparent about collection practices and to obtain explicit consent for data usage.

For marketers, these regulations introduced a new layer of complexity. Brands needed to ensure compliance by implementing secure data storage practices, updating privacy policies, and providing opt-out options. MarTech platforms adapted by introducing features for consent management, such as cookie banners and preference centers. While these changes initially posed challenges, they also encouraged companies to build trust with their audiences, reinforcing the importance of ethical and transparent data practices in modern marketing.

> *Key Milestones*
>
> - *2015: More Email Opens on Mobile Devices Than Desktops*
>
>   In 2015, emails are opened on mobile devices more frequently than on desktops, signaling a shift towards mobile-first marketing strategies[38].
>
> - *2018: GDPR Is Introduced in the EU*
>
>   This sets the stage for several other similar sets of consumer data privacy regulations in the United States and other countries, and demonstrates the need for more robust first-party data strategies[39].
>
> - *2019: Digital Advertising Spend Surpasses Traditional Media*
>
>   For the first time, the advertising spend on digital media surpassed traditional media such as TV, radio, and print[40].

THE AI AND PERSONALIZATION ERA (2020s)

At the time of writing, we are still within this era, which is a mix of optimistic usage of AI by marketers and others in business, with consumers' concerns for data privacy mixed with a desire for more tailored content and experiences that match their preferences and activity. The balance between a need for privacy and the desire for personalization characterizes an era where there are conflicting forces at work to provide an optimal customer experience.

The Rise of AI in Marketing

The 2020s ushered in a transformative era for marketing, driven by the rapid adoption of AI. Predictive analytics became the cornerstone of marketing strategies, enabling brands to anticipate customer behavior and optimize campaigns accordingly. Recommendation engines, pioneered by companies such as Amazon and Netflix, gained prominence, helping brands deliver relevant content and product suggestions to individual users. Chatbots, powered by AI, revolutionized customer service by offering instant, personalized assistance at scale, improving both efficiency and customer satisfaction.

The rise of generative AI tools, such as ChatGPT and other large language models (LLMs), marked a new frontier in marketing innovation. These tools enabled brands to create high-quality content, design campaigns, and generate marketing copy in seconds, significantly reducing production timelines. AI-driven ideation tools also empowered marketers to explore creative strategies, simulate customer interactions, and test messaging variations with unprecedented speed. This explosion of generative AI reshaped the creative process, making advanced marketing capabilities accessible to organizations of all sizes.

Focus on First-Party Data Strategies

In response to evolving consumer data privacy regulations, brands in the 2020s shifted their focus to first-party data strategies. First-party data, collected directly from customers with their consent, became the foundation of many marketing efforts, offering both compliance with privacy laws and deeper insights into customer preferences. By investing in first-party data, brands could maintain consumer trust while developing more accurate and reliable customer profiles.

This emphasis on first-party data also unlocked new levels of personalization, as brands leveraged data from direct interactions to tailor content, offers, and experiences. CDPs saw significant growth during this period, enabling companies to unify and analyze data from various touchpoints. These platforms allowed marketers to create seamless, personalized customer journeys across channels while respecting user privacy, making first-party data an indispensable asset in modern marketing.

Hyper-Personalization

Hyper-personalization became a defining characteristic of the 2020s, with brands leveraging AI to deliver individualized experiences in real time. Web sites dynamically adjusted content based on user behavior, emails featured products tailored to specific interests, and ads were targeted with precision using advanced data analytics. This level of personalization created deeper engagement, as customers felt understood and valued at every stage of their journey.

AI played a pivotal role in scaling hyper-personalization by analyzing vast amounts of data and predicting customer needs. Machine learning algorithms helped brands anticipate what customers wanted before they asked, offering proactive solutions and recommendations. By integrating AI into their marketing stacks, organizations could enhance customer experiences across channels, driving loyalty and higher conversion rates in an increasingly competitive market.

Growing Data Privacy Focus

As data privacy concerns heightened in the 2020s, brands adopted privacy-first approaches to align with consumer expectations and regulatory requirements. Tools and techniques such as cookie-less tracking emerged as alternatives to third-party cookies, ensuring user data could be gathered ethically and transparently. Privacy-first marketing emphasized building trust, giving consumers more control over how their data was used.

This growing focus on privacy reshaped marketing practices, encouraging brands to adopt consent-driven strategies and invest in secure data infrastructure. Privacy-compliant platforms and processes became essential, not just for meeting legal obligations but for maintaining customer loyalty in an age of increasing skepticism. By prioritizing transparency and ethical data use, brands demonstrated their commitment to protecting consumer rights, fostering stronger relationships with their audiences.

Key Milestone

- *2022: ChatGPT Exposes Millions to the Power of Generative AI*

    The release of ChatGPT on November 30, 2022, by OpenAI sparked the widespread adoption of AI in marketing automation tools, leading to advancements in predictive analytics and generative content creation[41].

- *2024: Global MarTech Spending Estimated at $148 Billion*

    According to Forrester's Global MarTech Software Forecast, the number is expected to surpass $215 billion by 2027, implying a 13.3% annual growth rate (AGR)[42].

- *2024: 56% of US Consumers Bought on TikTok*

    A survey revealed the prevalence of social commerce amongst consumers[43].

CONCLUSION

Of course, this was only a high-level exploration of the history of this complex and ever-growing space. Understanding the evolution helps to put the changes into perspective and also highlights the current opportunities and limitations.

While marketers and consumers alike take many of these accomplishments for granted in their daily work, it is important to understand the dynamic nature of MarTech and acknowledge that the evolution in the profession that started prior to the 1960s is a continual work in progress. With the rise of AI and increased automation, there will certainly be further changes to come.

For these reasons, it can be said with reasonable certainty that MarTech has not stopped evolving, and in the pages that follow, we will explore some of the new areas that marketers are integrating into their MarTech infrastructure, including composable approaches, AR/VR, and predictive modeling.

The next chapter will explore the relationship between MarTech and the marketing teams that use them.

NOTES

1. Klooster, John W. Icons of Invention [2 Volumes]: The Makers of the Modern World from Gutenberg to Gates. Bloomsbury Academic, 2009. ISBN 0313347433, 9780313347436
2. https://www.strategus.com/blog/the-history-of-commercials-and-tv-advertising
3. Media Culture. (2020, April 6). *Expert interview: The evolution of TV advertising*. Media Culture. Retrieved December 5, 2024, from https://www.mediaculture.com/insights/expert-interview-the-evolution-of-tv-advertising
4. Caballero, B. (2023, June 9). *The history of CRM from the 1950s to today*. Fit Small Business. Retrieved December 5, 2024, from https://fitsmallbusiness.com/history-of-crm/
5. https://www.shipmonk.com/blog/the-history-of-ecommerce-from-the-1960s-to-the-2020s
6. https://www.discosloth.com/visual-timeline-of-digital-marketing/

7. https://community.cisco.com/t5/the-break-room/may-3rd-1978-the-world-s-first-spam-email-sent/td-p/4827862

8. Caballero, B. (2023, June 9). *The history of CRM from the 1950s to today*. Fit Small Business. Retrieved December 5, 2024, from https://fitsmallbusiness.com/history-of-crm/

9. CERN. (n.d.). *A short history of the web*. CERN. https://home.cern/science/computing/birth-web/short-history-web

10. McPherson, Stephanie Sammartino (2009). Tim Berners-Lee: Inventor of the World Wide Web. Twenty-First Century Books. ISBN 978-0-8225-7273-2.

11. Evans, Claire L. (2018). Broad Band: The Untold Story of the Women Who Made the Internet. New York}: Portfolio/Penguin. p. 116. ISBN 978-0-7352-1175-9.

12. WordStream. (n.d.). *Search engine marketing: The basics and beyond*. WordStream. https://www.wordstream.com/search-engine-marketing

13. Warren, C. (2014, September 27). *Yahoo Directory, once the center of a web empire, will shut down*. The Verge. https://www.theverge.com/2014/9/27/6854139/yahoo-directory-once-the-center-of-a-web-empire-will-shut-down

14. Oppitz, Marcus; Tomsu, Peter (2017). Inventing the Cloud Century: How Cloudiness Keeps Changing Our Life, Economy and Technology. Springer. p. 238. ISBN 978-3-319-61161-7.

15. Search Engine History. (n.d.). *History of search engines: From 1945 to Google today*. Search Engine History. http://www.searchenginehistory.com/

16. Mirror Review. (n.d.). *History of Google – Tracing the evolution of a tech titan*. Mirror Review. Retrieved December 5, 2024, from https://www.mirrorreview.com/history-of-google/

17. Heffernan, V. (2017, November 15). *Just Google it: A short history of a newfound verb*. WIRED. Retrieved December 5, 2024, from https://www.wired.com/story/just-google-it-a-short-history-of-a-newfound-verb/

18. LaFrance, Adrienne (20 April 2017). "The First-Ever Banner Ad on the Web: It was an advertisement for AT&T in 1994, and people clicked on it like crazy". *The Atlantic*.

19. Carmody, Deirdre (October 24, 1994). "THE MEDIA BUSINESS; Time Inc. Raises Its Multimedia Profile With an Internet Test". *The New York Times*

20. Google. (2000, October 23). *Google launches self-service advertising program: Google AdWords offers cost-per-thousand impression pricing, basic ad management tools.* Google Press Center. https://googlepress.blogspot.com/2000/10/google-launches-self-service.html

21. Notre Dame of Maryland University. (2018, March 22). *History of blogging: A blog timeline.* NDMU Online. Retrieved December 5, 2024, from https://online.ndm.edu/news/communication/history-of-blogging/

22. Bowman, J. (2017, December 5). *Building Yelp: A history lesson on how to launch a major tech company.* Medium. Retrieved December 5, 2024, from https://medium.com/swlh/building-yelp-bc4e62c4db3b

23. Ortiz-Ospina, E. (2019, September 18). *The rise of social media.* Our World in Data. Retrieved December 5, 2024, from https://ourworldindata.org/rise-of-social-media

24. Baer, J. (2010, November 22). *6 lessons learned from the demise of MySpace.* Convince & Convert. Retrieved December 5, 2024, from https://www.convinceandconvert.com/social-media/6-lessons-learned-from-the-demise-of-myspace/

25. Flanagan, J. (2017, March 8). *Ad evolution: The history of Facebook.* Adtaxi. Retrieved December 5, 2024, from https://www.adtaxi.com/blog/ad-evolution-history-facebook/

26. Flanagan, J. (2017, March 8). *Ad evolution: The history of Facebook.* Adtaxi. Retrieved December 5, 2024, from https://www.adtaxi.com/blog/ad-evolution-history-facebook/

27. Cashmore, P. (2006, July 11). *MySpace: America's number one.* Mashable. Retrieved December 5, 2024, from https://mashable.com/archive/myspace-americas-number-one

28. Kramer, S. (2009, June 16). *More Americans go to Facebook than MySpace.* PCMag. Retrieved December 5, 2024, from https://www.pcmag.com/archive/more-americans-go-to-facebook-than-myspace-241432

29. Blystone, D. (2024, July 9). *Instagram: What it is, its history, and how the popular app works.* Investopedia. Retrieved December 5, 2024, from

https://www.investopedia.com/articles/investing/102615/story-instagram-rise-1-photo0sharing-app.asp

30. Weiss, E. (2019, October 15). *The history and future of SMS marketing*. LinkedIn. Retrieved December 5, 2024, from https://www.linkedin.com/pulse/history-future-sms-marketing-eli-weiss/

31. Tatango. (2022, December 2). *30 years of SMS and the history of mobile fundraising*. Tatango. Retrieved December 5, 2024, from https://www.tatango.com/blog/history-of-text-message-marketing/

32. Team illumy. (2024, May 2). *What happened to Palm Pilot?* illumy. Retrieved December 5, 2024, from https://www.illumy.com/what-happened-to-palm-pilot/

33. History.com Editors. (1999, January 19). *First BlackBerry device hits the market*. HISTORY. Retrieved December 5, 2024, from https://www.history.com/this-day-in-history/first-blackberry-device-850-released

34. Tatango. (2022, December 2). *30 years of SMS and the history of mobile fundraising*. Tatango. Retrieved December 5, 2024, from https://www.tatango.com/blog/history-of-text-message-marketing/

35. Clearcode. (n.d.). *The history of digital advertising technology*. The AdTech Book. Retrieved December 8, 2024, from https://adtechbook.clearcode.cc/history-advertising-technology/

36. The Editors of Encyclopaedia Britannica. (2024, December 8). *iPhone*. Encyclopaedia Britannica. Retrieved December 8, 2024, from https://www.britannica.com/technology/iPhone

37. AppleInsider Staff. (2008, July 10). *Apple's App Store launches with more than 500 apps*. AppleInsider. Retrieved December 8, 2024, from https://appleinsider.com/articles/08/07/10/apples_app_store_launches_with_more_than_500_apps

38. Abdo, R. (2020, January 22). *Do you remember these unforgettable marketing trends of the 2010s?* Campaign Monitor. Retrieved December 8, 2024, from https://www.campaignmonitor.com/blog/email-marketing/do-you-remember-these-unforgettable-marketing-trends-of-the-2010s/

39. Council of the European Union. (n.d.). *General data protection regulation (GDPR)*. Council of the European Union. Retrieved December 8, 2024, from https://www.consilium.europa.eu/en/policies/data-protection/data-protection-regulation/

40. Miller, H. (2019, February 20). *Digital advertising to surpass print and TV for the first time, report says*. The Washington Post. Retrieved December 8, 2024, from https://www.washingtonpost.com/technology/2019/02/20/digital-advertising-surpass-print-tv-first-time-report-says/

41. Marr, B. (2023, May 19). *A short history of ChatGPT: How we got to where we are today*. Forbes. Retrieved December 8, 2024, from https://www.forbes.com/sites/bernardmarr/2023/05/19/a-short-history-of-chat-gpt-how-we-got-to-where-we-are-today/

42. Liu, C. (2024, January 17). *Global martech spending will reach $148 billion in 2024*. Forrester. Retrieved December 8, 2024, from https://www.forrester.com/blogs/global-martech-spending-will-reach-148-billion/

43. Santiago, E. (2024, January 15). *The top marketing trends of 2024 & how they've changed since 2023 [Data from 1400+ global marketers]*. HubSpot. Retrieved December 8, 2024, from https://blog.hubspot.com/marketing/marketing-trends

CHAPTER 3

MarTech and Marketing Teams

With a better understanding of the historical context behind us, it is time to explore how MarTech is used in a modern setting. The users of MarTech can be separated into two distinct groups:

- Internal users, consisting of marketing teams and other groups within a company
- External users, consisting of prospective, current, and lapsed customers, and other audiences that are outside the company.

This chapter will primarily focus on internal uses, and within that group, on marketing teams. In Chapter 4, we will discuss other internal users and stakeholders that rely on MarTech to be successful in their work.

Marketing teams seem a readily apparent audience when discussing users of MarTech, though the makeup of marketing teams can vary greatly. Depending on the organization, a marketing team may range from single to triple digits, when considering all of the different aspects that can be involved, including the following and potentially many more:

- Strategy and planning
- Creative and content
- Marketing operations
- Analytics and reporting
- Customer experience

While every organization is structured differently, it is highly likely that MarTech platforms are utilized by most if not all teams within any modern organization. Which platforms are used by which teams can vary drastically, and it is likely that some platforms are used within some teams and not others.

The rest of this chapter will explore several of the groups that make up a marketing team. Depending on the organization, some of these may be collapsed together or further subdivided, but the intent is to illustrate the breadth of use cases of MarTech. For each group, we will look at how they use MarTech platforms and the most commonly used platforms.

MARKETING LEADERSHIP

Marketing leadership, including CMOs, VPs, and senior directors, plays a critical role in steering the overall marketing strategy and ensuring alignment with organizational goals. These leaders rely on MarTech platforms to gain high-level insights into campaign performance, measure ROI, and make data-driven decisions that impact budget allocation, team priorities, and long-term strategy.

They also ultimately make budgetary decisions related to how their teams invest in MarTech platforms and their long-term usage, which makes their understanding of MarTech critical.

How They Use MarTech

Marketing leadership's focus is not on day-to-day operational tasks but on ensuring that marketing efforts contribute to broader business objectives. This can include the following:

- Reviewing dashboards and reports to evaluate campaign effectiveness and ROI
- Analyzing customer insights to identify new market opportunities or refine existing strategies
- Allocating marketing budgets across channels and campaigns based on performance data
- Setting benchmarks and KPIs for marketing teams to track and achieve
- Evaluating the effectiveness of the existing MarTech stack to ensure scalability and value

Types of MarTech Platforms Used

Marketing leaders primarily utilize analytics and reporting platforms that provide high-level, aggregated data on marketing performance across channels. Tools such as Google Analytics, Tableau, and marketing attribution platforms enable them to visualize and interpret trends quickly. While some reporting platforms, such as Microsoft Power BI, are not specifically considered MarTech platforms, their usage can be made specific to the marketing function.

They also rely on customer data platforms (CDPs) to understand customer behavior and purchasing patterns, helping them make informed decisions on strategy and targeting. Budgeting and resource allocation tools, often integrated with project management systems, are key for tracking spend and optimizing resource distribution.

Another critical component for marketing leadership is multi-channel attribution platforms, which help determine which channels and campaigns contribute most significantly to business outcomes and ROI. These insights are often paired with predictive analytics tools that use historical data to forecast future trends, enabling leaders to plan proactively.

Additional Considerations

To improve their experience with MarTech, marketing leaders should ensure that platforms are integrated seamlessly to provide a unified view of all marketing activities. Disparate data sources or siloed platforms can lead to inefficiency and incomplete insights. Implementing tools that provide real-time updates and automated reporting can help leaders stay informed without relying on manual data pulls.

Another consideration is the customization of dashboards to reflect specific KPIs and metrics that align with the organization's goals. Marketing leaders often need tailored views that provide strategic insights rather than operational details. Regular training and collaboration with data teams can also enhance their ability to interpret analytics effectively and make more impactful decisions. Ensuring their involvement in MarTech evaluation processes ensures that the tools selected meet the scalability, reliability, and strategic alignment needs of the organization.

Finally, while marketing leaders do not use all the products in a MarTech stack on a regular basis, they are ultimately held responsible for investments made in those platforms. Therefore, it is important for marketing teams to ensure their leadership has a full understanding of their contributing value.

CUSTOMER ACQUISITION

The primary function of a customer acquisition team is to drive new leads and convert them into paying customers. They play a pivotal role in expanding the customer base and ensuring a steady flow of prospects into the sales funnel. By leveraging MarTech platforms, this team tracks lead behavior, optimizes touchpoints, and creates strategies to guide potential customers from awareness to conversion. Their goal is to maximize efficiency in lead generation and to nurture processes while aligning their efforts with broader organizational revenue goals.

How They Use MarTech

Customer acquisition teams utilize platforms that reach prospective and new customers and assist in the measurement of acquisition strategies and tactics, such as the following:

- Identifying and capturing leads through landing pages, forms, and paid advertising campaigns
- Scoring and prioritizing leads based on behavior and engagement using lead scoring models
- Running email nurturing campaigns to guide leads through the sales funnel
- Analyzing campaign performance to optimize ad spend and conversion rates
- Integrating customer relationship management (CRM) data with sales teams to ensure seamless lead handoff and follow-up

Platforms Utilized

Customer acquisition teams rely heavily on CRM platforms such as Salesforce or HubSpot to manage lead data and track interactions across the buyer journey. These platforms help to ensure that leads are organized, scored, and distributed to the appropriate teams for follow-up. Marketing automation tools such as Marketo, Pardot, or ActiveCampaign are crucial for automating nurturing campaigns, such as sending targeted emails or SMS messages based on a lead's behavior and stage in the funnel.

Additionally, customer acquisition teams utilize advertising platforms such as Google Ads, Facebook Ads Manager, and LinkedIn Campaign Manager to execute and track paid campaigns that bring in leads. Analytics tools play a key role in assessing performance, providing insights into which channels, campaigns, or keywords deliver the best results. Finally, integration tools or middleware such as Zapier or native connectors ensure data flows seamlessly between platforms, enabling the team to maintain a unified view of all lead activities.

Additional Considerations

To improve their effectiveness with MarTech, customer acquisition teams should find ways to create the most seamless integration possible between CRM and marketing automation platforms to ensure that leads are nurtured and handed off without data loss or miscommunication. Automated workflows that trigger personalized follow-ups based on lead behavior can help streamline the nurturing process and prevent valuable leads from slipping through the cracks.

Additionally, the team should leverage advanced analytics tools to better understand which channels and campaigns yield the highest ROI. Building a robust attribution model can help pinpoint the touchpoints that are most influential in converting leads. Regular collaboration with the sales team is essential to align expectations, improve lead quality, and gather feedback on what tactics are working. Continuous training and experimentation with MarTech tools can further help the team optimize their approach and adapt to changing consumer behavior.

DIGITAL MARKETING CAMPAIGNS

Digital marketing teams are responsible for creating and executing campaigns that span multiple online channels, including social media, email, Web sites, and paid advertising. Their goal is to engage audiences, build brand awareness, and drive measurable results, such as traffic, conversions, and sales.

How They Use MarTech

These teams rely heavily on MarTech platforms to manage campaigns, track performance, and optimize their strategies in real time in the following ways:

- Creating and managing digital campaigns across channels such as email, social media, and digital ads
- Monitoring Web site traffic and user engagement through analytics tools
- Implementing A/B testing to optimize ad creatives, landing pages, and call-to-action (CTA) buttons

- Tracking customer journeys across channels to improve attribution and understand touchpoint effectiveness
- Using programmatic advertising tools to automate and optimize ad placements

Platforms Utilized

Digital marketers depend on a broad range of MarTech tools to manage campaigns and achieve their goals. Multi-channel campaign management platforms, such as HubSpot or Adobe Campaign, allow teams to coordinate campaigns across various channels from a single interface. Social media management tools such as Hootsuite and Sprout Social are essential for scheduling posts, monitoring engagement, and analyzing performance on platforms such as Instagram, Facebook, and LinkedIn.

Web site and traffic analytics platforms, including Google Analytics and Hotjar, help digital marketers track user behavior and measure campaign success. Programmatic advertising platforms such as The Trade Desk and Google Ads automate ad placements and optimize performance based on AI-driven insights. Additionally, A/B testing tools such as Optimizely or Google Optimize empower digital marketers to continuously refine and improve their digital assets based on data-driven experiments.

Additional Considerations

To enhance their effectiveness, digital marketers should ensure their MarTech platforms are well-integrated to provide a unified view of campaign performance. Disjointed tools can lead to inefficiency and fragmented data, making it harder to understand the complete customer journey. Investing in platforms with advanced automation and AI capabilities can also streamline workflows and optimize campaigns at scale, freeing up time for creative and strategic work.

It's also essential for digital marketers to stay current on emerging trends and tools, as the digital marketing landscape evolves rapidly. For example, incorporating AI tools for content generation or adopting tools for video-first marketing could provide a competitive edge. Finally, marketers should regularly evaluate their attribution models to ensure they accurately capture the impact of each channel and touchpoint, enabling better decision-making and resource allocation.

CONTENT MARKETING

Content marketing teams are tasked with creating, managing, and distributing valuable content that attracts and engages target audiences while supporting broader business objectives. This includes a wide variety of assets, such as blog posts, videos, whitepapers, infographics, and case studies.

How They Use MarTech

Content marketers rely on MarTech platforms to streamline the creation and management of these assets, ensure they are effectively distributed across channels, and measure their performance to understand their impact on audience engagement and conversion rates in the following ways:

- Creating, editing, and publishing blog posts, videos, and other forms of marketing content
- Managing and organizing digital assets, such as images and videos, for easy access and reuse
- Using analytics to measure the performance of content, such as traffic, engagement, and lead generation
- Coordinating content calendars and workflows to ensure timely production and publication
- Personalizing content delivery based on audience segments and customer journeys

Platforms Utilized

Content marketers rely on a variety of tools to execute their strategies effectively. Content management systems (CMSs), such as WordPress or Adobe Experience Manager, are central to managing and publishing digital content. These platforms allow marketers to create, schedule, and update content seamlessly, ensuring consistency across Web properties. Digital asset management (DAM) platforms, such as Bynder or Widen, are essential for organizing and storing digital files, ensuring that teams can access the latest versions of assets and maintain brand consistency.

Analytics tools such as Google Analytics, Semrush, and HubSpot provide insights into how content performs, helping marketers track metrics such as page views, time on page, and conversion rates. Additionally, tools such as CoSchedule or Trello are often used to coordinate content calendars and streamline production workflows. For personalization, platforms such as Optimizely or dynamic content tools enable the delivery of tailored content based on audience preferences and behavior.

Additional Considerations

To improve their effectiveness, content marketers should focus on integrating their MarTech tools to create a unified workflow from content creation to performance measurement. For example, connecting their CMS with analytics and personalization tools can provide real-time feedback on what resonates with audiences and allow for on-the-fly optimizations. Having seamless integration with DAM platforms can also enhance productivity by eliminating time spent searching for assets.

Content teams should also prioritize automation where possible, such as automating social media distribution or using AI tools such as ChatGPT for initial drafts or idea generation. Additionally, regular audits of existing content can ensure that evergreen assets are updated and reused effectively, maximizing ROI. Finally, providing training on analytics tools can help marketers go beyond surface metrics, equipping them to make more strategic decisions about their content strategy.

WEB SITE MARKETING

Web site marketing teams are responsible for ensuring that an organization's Web site serves as a highly effective tool for attracting, engaging, and converting visitors. These teams focus on creating user-friendly, visually appealing, and conversion-optimized experiences that align with business objectives. They leverage MarTech platforms such as analytics tools, heatmaps, and A/B testing solutions.

How They Use MarTech

Web site marketers gain insights into user behavior and optimize key elements such as navigation, content layout, and callstoaction to drive results in the following ways:

- Monitoring Web site traffic and user behavior using analytics tools such as Google Analytics
- Running A/B tests to determine the best-performing page designs, headlines, or CTAs

- Utilizing heatmaps to analyze how users interact with specific elements of a Webpage
- Optimizing Web site speed and mobile responsiveness to enhance user experiences
- Personalizing Web site content based on visitor behavior, demographics, or referral source

Platforms Utilized

Web site marketing teams rely heavily on analytics platforms such as Google Analytics and Adobe Analytics to understand user behavior, traffic sources, and site performance. These tools provide insights into critical metrics such as bounce rates, time on site, and conversion rates, which inform optimization strategies. Heatmap tools such as Hotjar or Crazy Egg are used to visualize user interactions, highlighting areas where visitors click, scroll, or drop off, enabling targeted adjustments to improve usability.

Additionally, A/B testing platforms such as Optimizely or Google Optimize allow teams to experiment with different versions of Web pages or elements to identify the most effective designs and messaging. Content personalization platforms, such as Dynamic Yield or Evergage, enable Web site teams to deliver tailored experiences by dynamically adapting content based on user preferences and behaviors. SEO tools such as SEMrush or Ahrefs are also critical for ensuring that the Web site attracts organic traffic and ranks well in search engine results.

Additional Considerations

To improve their effectiveness, Web site marketing teams should ensure their tools are seamlessly integrated to provide a comprehensive view of performance and user behavior. For example, integrating analytics tools with personalization platforms can help teams understand how different audience segments interact with the site and tailor content accordingly. Establishing clear KPIs, such as conversion rates or user engagement metrics, can also help focus optimization efforts on the areas with the greatest potential impact.

Teams should also prioritize mobile-first design, as a significant portion of Web traffic now comes from mobile devices. Regular Web site audits, including technical performance checks, SEO updates, and content reviews, can help maintain optimal performance and ensure the site stays aligned with business objectives. Finally, fostering collaboration with content and digital marketing teams can ensure that Web site updates and campaigns are cohesive, delivering a seamless experience across all touchpoints.

SOCIAL MEDIA MARKETING

Social media marketing teams are responsible for managing a brand's presence on platforms such as Instagram, Facebook, LinkedIn, Twitter, and TikTok. They use social media to build brand awareness, foster engagement, and drive traffic or conversions. This involves creating and scheduling posts, responding to comments and messages, and analyzing campaign performance.

How They Use MarTech

By leveraging MarTech platforms, social media marketers can efficiently manage multiple accounts, track real-time engagement metrics, and optimize strategies to achieve their goals in the following ways:

- Scheduling and automating posts across platforms using social media management tools
- Monitoring engagement metrics such as likes, shares, and comments to assess content performance
- Tracking brand mentions and sentiment through social listening tools
- Analyzing campaign performance to determine ROI and inform future strategies
- Creating and managing paid social media campaigns to drive conversions or leads

Platforms Utilized

Social media marketers rely on tools such as Hootsuite, Buffer, and Sprout Social to schedule posts, track engagement, and manage multiple accounts from a single dashboard. These platforms simplify workflow and ensure consistent content distribution. Social listening tools such as Brandwatch and Mention allow marketers to monitor brand mentions, track trends, and analyze customer sentiment, providing valuable insights into audience perceptions.

For campaign performance analysis, tools such as Facebook Ads Manager, LinkedIn Campaign Manager, and TikTok Ads provide detailed metrics on ad performance, audience reach, and conversions. Advanced analytics platforms such as Sprinklr or Socialbakers enable social media marketers to gain deeper insights into audience behavior, helping them tailor content strategies. Many teams also use Canva or Adobe Creative Suite to create visually appealing, on-brand social media graphics quickly.

Additional Considerations

To maximize their effectiveness, social media marketers should prioritize integrating their social media tools with analytics platforms and CRM systems. This integration allows a more complete understanding of how social media efforts contribute to broader marketing goals and customer journeys. Additionally, marketers should take advantage of automation tools to streamline repetitive tasks such as scheduling posts, allowing more time for creative and strategic initiatives.

Social media marketers should also stay current with platform-specific trends and algorithm updates to ensure their strategies remain effective. Regular experimentation with new formats, such as Reels or Stories, can help brands stay relevant and engage audiences in innovative ways. Finally, fostering collaboration with content teams ensures that messaging and visuals align with overall brand guidelines and marketing campaigns, creating a cohesive customer experience across all channels.

DIGITAL ADVERTISING

Digital advertising teams are responsible for creating and managing paid campaigns across platforms such as Google Ads, social media networks, and programmatic ad exchanges. Their primary goal is to maximize ROI by reaching the right audience at the right time with compelling ad creatives. These teams rely on MarTech and ad tech platforms to execute campaigns, optimize performance, and track results in real time.

How They Use MarTech

By leveraging programmatic tools and data-driven targeting, digital advertising teams can ensure efficient spending and high campaign impact in the following ways:

- Running search engine campaigns using Google Ads to target high-intent audiences
- Creating and optimizing display and video ads on programmatic platforms such as The Trade Desk

- Leveraging audience data to refine targeting and reach specific demographics
- Using A/B testing to identify high-performing ad creatives and calls to action
- Monitoring ad performance in real time to adjust bids, budgets, or targeting strategies for better results

Platforms Utilized

Digital advertisers heavily rely on ad tech platforms such as Google Ads, Meta Ads Manager, and LinkedIn Campaign Manager for running campaigns and managing ad placements. Programmatic advertising platforms such as The Trade Desk or DV360 enable advanced targeting and real-time bidding, ensuring that ads reach the most relevant audience. These tools often integrate with data management platforms (DMPs) to leverage third-party data for enhanced audience segmentation.

Analytics tools, such as Google Analytics and platform-specific reporting dashboards, are crucial for measuring campaign success. These platforms help advertisers track conversions, click-through rates, and overall ROI. Additionally, creative optimization tools such as Canva or Adobe Creative Cloud are often used to design compelling visuals that resonate with target audiences. Integration with CRM or CDP platforms can also provide first-party data insights, allowing for more precise targeting and retargeting efforts.

Additional Considerations

To enhance their efficiency, digital advertising teams should ensure their ad tech tools are integrated with analytics and CRM platforms. This allows seamless tracking of the customer journey, from ad impression to conversion, providing insights that can improve future campaigns. Leveraging advanced attribution models, such as multi-touch or data-driven attribution, can also help identify the true impact of ads across various channels.

Digital advertisers should focus on automation features within programmatic platforms, such as AI-driven bid optimization and predictive analytics, to improve campaign performance. Regular training on evolving ad formats and platform features ensures the team stays competitive, especially as privacy regulations and cookie deprecation reshape targeting methods. Finally, collaborating with creative teams can result in more impactful ad designs that align with audience preferences and brand identity.

EMAIL MARKETING

Email marketing teams focus on using email as a strategic channel for nurturing leads, engaging with existing customers, and driving conversions. By crafting personalized and targeted campaigns, they aim to strengthen customer relationships and encourage repeat business.

How They Use MarTech

Email marketers depend on MarTech platforms, particularly marketing automation tools, to streamline the creation, segmentation, and tracking of email campaigns. These platforms enable the delivery of timely and relevant messages that resonate with audiences and drive measurable results by doing the following:

- Designing and sending promotional email campaigns to announce new products, offers, or events
- Segmenting email lists based on customer demographics, behavior, or purchase history to improve targeting
- Automating drip campaigns to nurture leads and guide them through the sales funnel
- Tracking email performance metrics, such as open rates, click-through rates, and conversions
- Running A/B tests to optimize subject lines, email content, or calls-to-action for better engagement

Platforms Utilized

Email marketers rely heavily on marketing automation platforms such as Mailchimp, HubSpot, Marketo, or Klaviyo. These tools provide an all-in-one solution for designing email templates, segmenting audiences, and automating workflows. Advanced features such as dynamic content and behavioral triggers allow marketers to send personalized emails that respond to specific user actions, such as abandoned carts or completed purchases.

Analytics platforms integrated with email tools offer deep insights into campaign performance. Marketers can track key metrics, such as open rates, click-through rates, and overall engagement, to assess the effectiveness of their strategies. Additionally, email verification tools such as NeverBounce or

ZeroBounce ensure that contact lists remain clean and reduce the chances of sending emails to invalid or inactive email addresses, maintaining the sender's reputation. Integration with CRM platforms allows email marketers to use enriched customer data for more precise targeting and segmentation.

Additional Considerations

To improve the effectiveness of email marketing, teams should prioritize personalization and relevance. Leveraging customer data from CRM or CDP platforms enables email marketers to tailor messages to individual preferences and behaviors. Testing and optimizing subject lines, layouts, and CTAs through A/B testing ensures that campaigns continuously improve over time.

Another critical focus area is maintaining a healthy email sender reputation, which involves monitoring bounce rates, avoiding spam filters, and ensuring compliance with privacy regulations such as GDPR and CAN-SPAM. Teams should also explore advanced automation techniques, such as predictive sending, which uses AI to determine the optimal time to send emails for maximum engagement. Finally, collaboration with content teams ensures that email copy and visuals align with brand messaging and broader marketing campaigns, creating a cohesive customer experience.

EVENTS MARKETING

Events marketing teams are responsible for planning, promoting, and executing events, such as webinars, trade shows, conferences, and product launches, to engage customers and prospects. These teams aim to create memorable experiences that drive brand awareness, generate leads, and foster customer loyalty.

How They Use MarTech

To manage the complex logistics of events, teams rely on MarTech tools, including event management software and CRM platforms, to coordinate promotion, streamline registration processes, and track attendee engagement before, during, and after the event in the following ways:

- Promoting events through email campaigns, social media, and digital advertising to drive registrations
- Managing event registration and ticketing through event management platforms such as Eventbrite or Cvent

- Tracking attendee engagement during events, including session participation and interactions with speakers or booths
- Collecting post-event feedback via surveys to evaluate the success of the event and gather insights for future improvements
- Syncing attendee data with CRM platforms to nurture leads and maintain engagement after the event

Platforms Utilized

Event management software such as Cvent, Bizzabo, or Eventbrite is central to the operations of events marketing teams. These platforms handle everything from creating registration forms and tracking ticket sales to managing event schedules and communications. They also offer features such as mobile event apps and attendee check-in tools to streamline the on-site experience.

CRM tools such as Salesforce or HubSpot play a vital role in tracking and nurturing attendee engagement. By integrating event data with CRM platforms, marketers can gain a deeper understanding of attendee behavior and tailor post-event follow-ups, such as personalized emails or targeted offers. For virtual events, platforms such as Zoom, Hopin, or ON24 enable live streaming, interactive sessions, and audience engagement features, such as polls and Q&As. Additionally, analytics tools embedded in these platforms provide valuable insights into registration rates, attendance levels, and participant feedback.

Additional Considerations

To improve the effectiveness of their MarTech tools, events teams should focus on integration between event management software, CRM platforms, and marketing automation tools. This ensures a seamless flow of data, enabling personalized communication and nurturing campaigns for attendees and leads generated during the event. Automated workflows can also save time by syncing data in real time, such as updating contact details or tagging attendees based on their level of engagement.

Teams should also prioritize creating engaging event experiences, whether virtual or in-person, by leveraging interactive tools such as live polls, gamification, and breakout sessions. Post-event surveys are essential for collecting attendee feedback and identifying areas for improvement. Finally, events teams should continuously analyze their performance metrics, such as attendee conversion rates or session engagement levels, to refine their strategies and deliver more impactful events in the future.

SEARCH MARKETING

Search marketing teams focus on increasing a brand's visibility in search engine results, driving traffic, and converting visitors into customers. This involves two key areas: search engine optimization (SEO), which enhances organic rankings, and paid search advertising, which leverages pay-per-click (PPC) campaigns to appear at the top of search results for targeted keywords.

How They Use MarTech

Search marketers rely on a combination of SEO platforms and paid search management solutions to track performance, identify opportunities, and continuously refine their strategies to maximize campaign effectiveness in the following ways:

- Conducting keyword research to identify high-value terms that align with customer search behavior
- Optimizing Web site content, meta tags, and site architecture to improve organic search rankings
- Managing paid search campaigns on platforms such as Google Ads and Bing Ads to attract high-intent users
- Tracking and analyzing key metrics, such as click-through rates, cost-per-click, and conversion rates, to optimize campaigns
- Performing competitive analysis to identify gaps and opportunities in both organic and paid search strategies

Platforms Utilized

Search marketing teams rely on SEO platforms such as SEMrush, Ahrefs, and Moz to conduct keyword research, analyze backlinks, and monitor organic search rankings. These tools provide comprehensive insights into search performance, helping marketers identify opportunities to improve content and optimize technical aspects of their Web sites. Additionally, tools such as Google Search Console allow teams to track and troubleshoot site performance directly within Google's ecosystem.

For paid search efforts, platforms such as Google Ads and Microsoft Ads Manager are essential for creating, managing, and optimizing PPC campaigns. These tools provide granular control over ad targeting, bidding, and budget allocation. Advanced solutions such as Optmyzr or Marin Software enable search marketers to automate routine tasks and optimize campaigns at scale. Analytics platforms, such as Google Analytics, play a critical role in tying search performance to broader business outcomes, allowing teams to assess ROI and track customer journeys from search queries to conversions.

Additional Considerations

Search marketers should prioritize integrating their tools to gain a holistic view of performance across organic and paid channels. For instance, combining SEO platform data with paid search analytics can help teams identify opportunities to target specific keywords through both strategies. Investing in tools with AI and machine learning capabilities, such as automated bidding or predictive analytics, can further enhance efficiency and campaign performance.

Teams should also remain proactive in adapting to algorithm updates and changes in search behavior, such as the increasing importance of voice search and mobile-first indexing. Conducting regular audits of Web site content and technical SEO can ensure that sites remain optimized for current best practices.

Finally, search marketers should collaborate with content teams to create high-quality, relevant content that aligns with target keywords and customer intent, as this strengthens both organic and paid search performance.

E-COMMERCE

E-commerce teams are responsible for managing online shopping experiences and driving revenue through digital storefronts. These teams work to create seamless, personalized, and engaging interactions for customers, from product discovery to checkout.

How They Use MarTech

By integrating MarTech with e-commerce platforms, these teams can deliver tailored shopping experiences, efficiently manage promotions, and gain actionable insights into customer behavior. Their ultimate goal is to optimize the online buying journey and maximize sales, and they can do so in the following ways:

- Delivering personalized product recommendations based on customer browsing and purchase history
- Managing and promoting discounts, seasonal campaigns, and loyalty rewards programs
- Analyzing customer behavior, such as cart abandonment rates and site interactions, to improve the shopping experience
- Running email campaigns to re-engage customers with abandoned carts or to promote personalized offers
- Using A/B testing to optimize product pages, checkout flows, and promotional banners for higher conversions

Platforms Utilized

E-commerce teams depend on a combination of e-commerce platforms and specialized MarTech tools to achieve their goals. Platforms such as Shopify, Magento, or WooCommerce serve as the backbone of online stores, managing product listings, transactions, and overall storefront operations. These platforms are often integrated with CDPs or personalization tools such as Dynamic Yield or Salesforce Commerce Cloud, which enable tailored recommendations and dynamic content for individual shoppers.

Analytics tools such as Google Analytics, Hotjar, and Mixpanel are essential for tracking customer behavior, including traffic sources, click paths, and conversion rates. Marketing automation tools such as Klaviyo and Mailchimp allow teams to run email campaigns targeting specific customer segments, such as those who abandoned their carts or have shown interest in particular products. Additionally, A/B testing tools such as Optimizely and Google Optimize help fine-tune the shopping experience by testing variations in product pages, checkout processes, or promotional strategies.

Additional Considerations

To maximize the effectiveness of their MarTech stack, e-commerce teams should ensure tight integration between their e-commerce platforms and analytics, personalization, and marketing tools. For example, connecting customer behavior data with email automation platforms can enable highly personalized campaigns that re-engage shoppers with timely and relevant offers.

Investing in tools that provide real-time analytics is also critical for making rapid adjustments during promotions or high-traffic periods.

Another key consideration is mobile optimization, as a significant percentage of online shopping occurs on mobile devices. Teams should prioritize creating fast, responsive, and intuitive mobile shopping experiences to capture this growing audience.

ANALYTICS TEAMS

Analytics teams play a crucial role in marketing by collecting, analyzing, and interpreting data to provide actionable insights that drive decision-making. These teams focus on understanding marketing performance, customer behavior, and overall business trends, ensuring that strategies are data-driven and effective.

How They Use MarTech

By leveraging data aggregation and visualization tools, analytics teams help organizations identify opportunities, optimize campaigns, and measure ROI across channels, which they do in the following ways

- Aggregating data from multiple marketing platforms to create comprehensive performance dashboards

- Identifying trends and patterns in customer behavior to inform targeting and segmentation strategies

- Performing multi-channel attribution analysis to determine the effectiveness of campaigns across touchpoints

- Generating forecasts and predictive insights to guide future marketing strategies

- Analyzing key performance indicators (KPIs) to evaluate the success of campaigns and identify areas for improvement

Platforms Utilized

Analytics teams rely on a suite of tools to manage and analyze data effectively. Data aggregation platforms such as Google Analytics 360, Adobe Analytics, and Tableau allow teams to collect data from multiple sources and visualize it in a user-friendly format. These tools provide insights into traffic, engagement, and conversion rates, helping marketers understand the effectiveness of their campaigns.

For deeper analysis, business intelligence platforms such as Power BI or Looker offer advanced capabilities such as data modeling, predictive analytics, and machine learning. Multi-channel attribution tools such as Ruler Analytics and Improvado help analytics teams understand the contribution of different marketing channels to overall success. Additionally, CDPs such as Segment and Tealium enable the integration and unification of data from various touchpoints, creating a holistic view of customer interactions.

Additional Considerations

To improve their effectiveness, analytics teams should focus on ensuring that data flows seamlessly between platforms, enabling a unified view of marketing performance. Integrating analytics tools with CDPs, CRMs, and campaign management platforms allows more accurate reporting and better insights. Automating data aggregation processes can save time and reduce errors, allowing teams to focus on higher-level analysis.

Additionally, teams should prioritize clear communication of insights through well-designed dashboards and reports tailored to different stakeholders, from executives to operational teams. Regular training on the latest analytics tools and techniques ensures that teams stay ahead of industry trends and can leverage emerging capabilities such as AI-driven predictive analytics. Encouraging collaboration with marketing teams ensures that insights are actionable, aligning analysis with business objectives and driving more impactful decision-making.

CRM TEAM

CRM teams are tasked with ensuring that customer data is accurate, accessible, and actionable across the organization. They focus on maintaining customer data integrity and creating a unified view of customer interactions, enabling seamless collaboration between sales, marketing, and support teams.

How They Use MarTech

By leveraging CRM platforms and related MarTech tools, these teams play a vital role in fostering strong customer relationships and improving the customer experience in the following ways:

- Centralizing customer data from multiple touchpoints, such as email, social media, and purchase history
- Managing lead and contact records to ensure data accuracy and prevent duplication
- Enabling sales and marketing teams to access real-time customer insights for more targeted interactions
- Automating customer communication workflows, such as follow-ups and reminders
- Generating reports on customer trends, engagement, and retention for strategic planning

Platforms Utilized

CRM teams primarily rely on robust CRM platforms such as Salesforce, HubSpot, or Microsoft Dynamics to manage and organize customer data. These platforms serve as the central hub for all customer-related information, enabling teams to track interactions, segment audiences, and coordinate efforts across departments. Many CRM systems also include built-in automation features for managing workflows, sending reminders, and triggering communications based on customer actions.

Integration tools such as Zapier or MuleSoft are often used to connect CRM platforms with other MarTech solutions, such as marketing automation tools or customer service platforms. Data enrichment tools such as Clearbit or ZoomInfo help CRM teams enhance customer records by adding missing or updated information. Analytics tools integrated with CRMs provide insights into customer behavior and trends, enabling data-driven decision-making. Additionally, collaboration tools, such as Slack or Microsoft Teams, may be linked to CRMs to streamline communication across teams.

Additional Considerations

To maximize the effectiveness of their CRM systems, teams should prioritize data quality by implementing regular data hygiene practices, such as deduplication and validation. Integrating the CRM with other key platforms, such as marketing automation and analytics tools, ensures a seamless flow of data and a more complete view of customer interactions. Investing in training for team members across departments can also improve CRM adoption and ensure consistent usage.

Another critical focus area is personalization. By leveraging CRM data, teams can enable more tailored interactions across all touchpoints, enhancing the customer experience and driving loyalty. Regular audits of CRM workflows and integrations can help identify inefficiency and areas for improvement, ensuring that the platform evolves alongside the organization's needs. Collaboration between CRM, sales, and marketing teams is essential to align strategies and ensure that customer relationship efforts are cohesive and effective.

MARKETING OPERATIONS

Marketing operations teams are the backbone of marketing organizations, ensuring that processes, tools, and systems work efficiently to support overall marketing efforts. They manage the implementation, integration, and optimization of MarTech platforms while overseeing data quality, workflows, and performance reporting. By enabling seamless collaboration across teams and ensuring operational excellence, marketing operations professionals help marketers execute campaigns effectively and at scale.

How They Use MarTech

Marketing operations teams use MarTech in the following ways:

- Managing and integrating MarTech platforms to ensure a unified and efficient ecosystem
- Creating workflows and processes that streamline campaign execution across teams
- Ensuring data integrity by implementing regular audits and data governance policies
- Building and maintaining performance dashboards for real-time tracking of marketing KPIs
- Training marketing teams on how to use MarTech tools effectively

Platforms Utilized

Marketing operations teams rely on a diverse array of MarTech tools to manage workflows and maintain operational efficiency. Project management platforms such as Asana, Trello, or Workfront help coordinate tasks and campaigns across departments, ensuring timelines and deliverables are met. Integration tools such as Zapier, MuleSoft, or native API connectors play a critical role in connecting disparate systems, such as CRM, analytics, and marketing automation platforms, to create a unified marketing ecosystem.

DMPs, such as Segment or Tealium, enable marketing operations teams to collect, clean, and unify data from multiple sources. Analytics and reporting tools such as Tableau, Looker, or Power BI provide dashboards and insights to track campaign performance and marketing efficiency. Additionally, marketing automation platforms such as HubSpot or Marketo are managed and optimized by operations teams to ensure workflows, email campaigns, and lead management are running smoothly.

Additional Considerations

To enhance the effectiveness of their efforts, marketing operations teams should focus on seamless integration across all tools in the MarTech stack. Ensuring that platforms communicate effectively reduces inefficiency and provides marketers with a single source of truth for customer data and campaign performance. Regular audits of the MarTech stack can help identify redundant or underutilized tools, optimizing both costs and productivity.

Investing in training for both the operations team and end-users across marketing ensures that everyone understands how to use the tools effectively, reducing friction and errors. Additionally, marketing operations should prioritize scalability by building flexible processes and workflows that can adapt as the organization grows.

Collaboration with other teams, such as IT and data analytics, ensures that the technical infrastructure supports marketing goals and enhances overall operational excellence. Finally, staying updated on emerging MarTech trends and innovations helps marketing operations teams maintain a competitive edge and future-proof their stack.

CONCLUSION

While the composition of marketing teams varies from organization to organization, the use cases for MarTech remain similar, though varying in complexity depending on the size and specifics of a company.

The next chapter will look at how non-marketing teams interact with and depend on MarTech to be successful in their work.

CHAPTER 4

How Other Teams in an Organization Utilize MarTech

While marketing teams are the primary users of MarTech platforms, they are rarely the only ones who interact with them in an organization. In addition to the marketing teams reviewed in Chapter 3, this chapter will explore some of the other teams that might utilize these software tools and platforms on a regular basis, and how they use MarTech to do their work.

NON-MARKETING EXECUTIVES

As explored in Chapter 3, chief marketing officers (CMOs) are regular users of MarTech, yet they are not the only people in the C-suite who benefit from these platforms.

The executive team, including the CEO, COO, and CFO, plays a critical role in shaping organizational strategy and ensuring alignment across all business functions, including marketing. While they may not directly interact with MarTech platforms on a regular basis (in some cases, ever), executives often rely on the insights generated by these tools to make informed decisions about resource allocation, growth opportunities, and market positioning. The executive team collaborates with marketing leaders to understand key performance metrics and uses this data to guide broader business strategies.

How They Use MarTech

Similar to CMOs, these executives are unlikely to be regular daily users of individual MarTech platforms, yet they benefit from the outputs on a continuing basis. This includes the following:

- Reviewing high-level performance dashboards created by data from—or within— MarTech platforms to assess the ROI of marketing initiatives
- Using customer insights derived from MarTech platforms to identify new market opportunities or gaps
- Aligning marketing spend with company-wide financial goals based on data-driven recommendations
- Evaluating the scalability and value of MarTech investments as part of strategic planning
- Incorporating marketing analytics into board-level presentations to communicate business performance

Platforms Utilized

While executives typically do not use MarTech platforms directly, they rely heavily on the outputs of these tools. Reporting and dashboard platforms such as Tableau, Power BI, or Looker often serve as visualization layers for data aggregated from MarTech systems such as CRMs, customer data platforms (CDPs), and analytics tools. These dashboards provide executives with an overview of key metrics, such as marketing ROI, customer acquisition costs (CAC), and customer lifetime value (CLV), allowing them to make high-level decisions based on clear, actionable data.

Marketing automation platforms and journey orchestration tools indirectly impact executive decision-making by providing insights into customer adoption and engagement, marketing efficiency, and customer retention.

For instance, executives might rely on reports generated from tools such as Salesforce or Adobe Analytics to understand customer trends and behavior at a strategic level. These insights often inform broader organizational strategies, including product development, geographic expansion, or operational improvements.

Additional Considerations

While in many companies, marketing is getting more attention in positive ways, as shown in Figure 4.1. CFO and CEOs are still considered the most skeptical of marketing's value (Gartner, 2024)[1] and there are both continuing challenges and opportunities here for marketing teams that increasingly rely on MarTech.

To maximize the value that executives derive from MarTech, it is essential for marketing teams to present data in a concise and actionable format tailored to executive priorities. This may involve creating custom dashboards that highlight metrics aligned with business goals, such as revenue growth or market share. Regular communication between marketing leadership and the executive team ensures that MarTech insights are integrated into strategic discussions.

Non-marketing executives should also consider the scalability and future potential of MarTech investments, ensuring that these platforms align with long-term business objectives. Collaborating with marketing teams to understand the impact of these tools on customer engagement and market positioning can provide a clearer picture of their strategic value.

Finally, fostering cross-departmental collaboration helps ensure that the insights generated by MarTech platforms are used effectively across the organization, reinforcing alignment between marketing and broader business goals.

SALES TEAMS

Often working in close proximity with marketing teams, sales teams are focused on converting marketing qualified leads (MQLs) into sales qualified leads (SQLs), then into customers, and building strong relationships with existing customers and clients. While their primary tools are often sales-focused platforms, they rely heavily on insights and data from MarTech to guide their strategies. CRM systems are often used by both marketing and sales, as well aslead scoring tools, and analytics provide valuable information that helps sales professionals prioritize prospects, personalize outreach, and understand customer needs. The partnership between sales and marketing teams, facilitated by MarTech, ensures a seamless flow of leads and consistent messaging across the buyer journey.

How They Use MarTech

Since sales teams are focused on leads that are further down the funnel than a marketing team, their primary focus is on how to convert MQLs into paying customers, including the following:

- Accessing lead scoring data from marketing automation platforms to prioritize outreach efforts
- Using CRM tools to track customer interactions and identify opportunities for follow-ups
- Personalizing outreach messages with insights derived from CDPs
- Reviewing marketing campaign engagement data to tailor sales pitches based on individual interests
- Collaborating with marketing teams to refine lead handoff processes and ensure a smooth transition

Platforms Utilized

Sales teams are deeply reliant on CRM systems such as Salesforce, HubSpot, or Microsoft Dynamics, which serve as the central hub for tracking customer interactions and managing relationships. These platforms provide visibility into lead activity, including engagement with marketing campaigns, browsing behavior, and purchase history. Lead scoring and qualification tools integrated with CRMs help sales professionals prioritize their outreach, focusing on prospects most likely to convert.

In addition to CRM tools, sales teams often leverage marketing automation platforms such as Marketo or Pardot for real-time updates on lead activity, such as when a lead downloads a whitepaper or clicks on an email link. Integration with analytics platforms such as Tableau or Power BI enables sales teams to track trends and align their efforts with broader marketing and business goals. These insights help sales teams personalize their approach, improving the likelihood of closing deals.

Additional Considerations

To enhance the effectiveness of their collaboration with marketing, sales teams should ensure that MarTech platforms are seamlessly integrated, enabling a continuous flow of information between departments and teams. For instance, automated notifications about lead activity or scoring changes can help sales teams act quickly on high-priority opportunities. Clear processes for lead handoff and shared KPIs can further strengthen the alignment between sales and marketing.

Sales teams should also invest time in understanding the data and insights provided by MarTech platforms. Regular training on CRM usage and analytics tools ensures that sales professionals can fully leverage the information at their disposal. Finally, fostering open communication between sales and marketing teams allows feedback on lead quality and campaign effectiveness, creating a feedback loop that improves the overall customer journey and strengthens the partnership between departments.

DATA TEAMS

Data teams play a critical role in an organization by ensuring data quality, consistency, and usability across departments. They work to integrate MarTech data with other organizational data sources to provide a unified view of performance and trends. Data teams leverage their expertise to uncover actionable insights, enabling predictive and prescriptive analysis that supports better decision-making across the business. While not always direct users of MarTech platforms, they partner closely with marketing teams to ensure data from these tools is accurate, well structured, and aligned with broader analytics efforts.

How They Use MarTech

Data teams don't so much *use* MarTech as they support the *best usage of* these MarTech platforms by ensuring that the best possible data infrastructure, cleansing, and upkeep is maintained. This includes the following:

- Integrating MarTech data, such as campaign performance metrics or customer insights, into the organization's data warehouse
- Validating and cleaning data from MarTech platforms to ensure accuracy and consistency
- Building dashboards that combine MarTech data with other business KPIs for cross-functional reporting
- Applying predictive models to MarTech data to forecast customer behavior or campaign performance

- Conducting exploratory data analysis to uncover trends and opportunities in customer engagement or marketing spend

Platforms Utilized

Data teams may not directly manage MarTech platforms but they often work to support data ingestion, or interact with data exported from systems such as CRMs, marketing automation platforms, and analytics tools. Platforms such as Salesforce, HubSpot, Google Analytics, and Adobe Analytics provide rich data sets that data teams integrate into centralized repositories using tools such as Snowflake, BigQuery, and AWS.

Data transformation and integration platforms such as Tableau Prep, Alteryx, and Apache Airflow help data teams clean, structure, and merge data from multiple platform sources for broader organizational use. Advanced analytics and machine learning platforms such as Python and R, or cloud-based tools such as Google Vertex AI enable data teams to perform predictive and prescriptive analyses on MarTech data. These insights are often visualized using business intelligence platforms such as Tableau, Microsoft Power BI, or Google Looker to support decision-making.

Additional Considerations

To maximize their effectiveness, data teams need close collaboration with marketing teams to ensure that the data captured by MarTech tools is complete and well structured. Establishing clear data governance policies, including standards for data collection, cleaning, and validation, helps ensure accuracy and consistency across systems. Regular audits of MarTech platform integrations can also identify gaps or inefficiency in data flows that may need to be addressed.

Data teams should prioritize automation to streamline processes for ingesting and transforming MarTech data, enabling real-time or near-real-time reporting. Investing in training for both data and marketing teams can improve communication and ensure alignment around shared goals. Finally, data teams should work to present insights in ways that are accessible and actionable for stakeholders, tailoring dashboards or reports to different audiences, from marketing leaders to the executive team. This partnership ensures that MarTech-related data delivers maximum value to the organization.

Technology Teams

Similar to the way that data teams enable data to be transmitted across business applications and sources, technology teams are responsible for the technical infrastructure that supports an organization's MarTech stack. They ensure the successful implementation, integration, and maintenance of MarTech platforms, making certain these tools align with the organization's overall IT architecture and comply with security and privacy regulations. Technology teams play a pivotal role in enabling seamless data flow across platforms, addressing technical issues, and ensuring that MarTech tools perform reliably and scale effectively as business needs evolve.

How They Use MarTech

Technology teams use MarTech in the following ways:

- Integrating MarTech platforms with existing IT systems, such as CRMs, CDPs, or data warehouses
- Managing API connections and data pipelines to ensure seamless data flow between tools
- Ensuring platforms comply with privacy regulations such as GDPR or CCPA
- Performing regular security audits to protect sensitive customer and organizational data
- Scaling MarTech infrastructure to accommodate increased data volumes and business growth

Platforms Utilized

While technology teams are not end users of MarTech platforms, they may work closely with tools such as Salesforce, HubSpot, or Adobe Experience Manager to implement and maintain them. Integration platforms such as MuleSoft, Zapier, or Workato are commonly used to connect disparate systems and ensure data flows smoothly between marketing tools and other business systems, such as ERP or finance platforms.

Technology teams also leverage infrastructure and cloud-based tools such as AWS, Microsoft Azure, or Google Cloud to host and scale MarTech applications. For security and compliance, tools such as Okta (for identity management) and Varonis (for data protection) are utilized to safeguard customer data and ensure regulatory adherence. Monitoring and logging tools, such as Splunk and Datadog, are employed to track system performance, detect anomalies, and resolve technical issues proactively.

Additional Considerations

With the goal of ensuring the consistent and effective use of MarTech platforms, technology teams should focus on establishing robust integration and data governance frameworks. Creating clear documentation and standardizing APIs and data pipelines help minimize disruptions and ensure systems remain interoperable as new tools are added to the stack. Regular testing and audits of MarTech infrastructure can identify vulnerabilities or inefficiency that may need addressing.

Collaboration with marketing and data teams is critical for aligning MarTech capabilities with business needs. By participating in platform selection processes and understanding the goals of end users, technology teams can better support tool implementation and optimization. Investing in ongoing training and staying updated on the latest marketing innovations can also help technology teams future-proof their infrastructure and maintain a competitive edge. Finally, ensuring scalability and flexibility in the MarTech architecture allows organizations to adapt quickly to changing market conditions and growth opportunities.

CUSTOMER SERVICE TEAMS

Most often working with existing users of a product or service, customer service teams play a critical role in ensuring customer satisfaction by addressing inquiries, resolving issues, and building long-term loyalty. These teams increasingly rely on insights from MarTech platforms to personalize support interactions, understand customer history, and provide consistent service across channels. While they may not directly manage marketing campaigns, their access to data from CRM, CDPs, and automation tools enables them to offer efficient and tailored solutions, enhancing the overall customer experience.

How They Use MarTech

Being focused on customer needs, including questions at critical points in the usage of a product or service, customer service teams primarily use MarTech platforms to better understand customer interactions, sentiment, and their purchase or product usage history, such as the following:

- Accessing customer history and preferences through CRM platforms to provide personalized support
- Using chatbots and automated ticketing systems to handle common customer inquiries and route complex issues to agents
- Leveraging feedback tools, such as post-interaction surveys, to measure customer satisfaction and improve service
- Identifying trends in customer pain points through sentiment analysis and customer interaction data
- Collaborating with marketing teams to provide insights into recurring issues that could inform campaigns or product development

Platforms Utilized

Customer service teams depend heavily on CRM platforms such as Salesforce, Zendesk, or HubSpot to access a centralized repository of customer information. These platforms allow service agents to view detailed interaction histories, preferences, and past purchases, enabling faster and more personalized resolutions. Automation tools such as Zendesk Support or Freshdesk streamline ticket management and integrate with live chat and chatbot solutions such as Intercom or Drift to handle basic inquiries, freeing agents to focus on more complex tasks.

Sentiment analysis tools and voice-of-the-customer platforms such as Medallia or Qualtrics provide valuable insights into customer satisfaction, helping teams identify and address common issues. CDPs and marketing analytics tools indirectly support customer service by offering richer customer profiles and insights into behavioral data. Collaboration tools such as Slack or Microsoft Teams, integrated with customer support platforms, facilitate quick communication between service teams and other departments, such as marketing or sales.

Additional Considerations

To maximize the effectiveness of their interactions with MarTech platforms, customer service teams should prioritize integration with marketing and sales tools. A well-integrated system allows customer service representatives to access a complete view of the customer journey, including campaign interactions, past inquiries, and purchase history, enabling more contextual and effective support.

Regular training on CRM and automation tools can help service agents fully utilize available features, improving response times and overall customer satisfaction. Customer service teams should also provide feedback to marketing teams regarding common issues or questions, helping refine messaging and product offerings. Finally, investing in real-time sentiment analysis and feedback loops ensures that customer service efforts are continuously improving, fostering stronger relationships and contributing to broader organizational goals.

CREATIVE TEAMS

As any marketer knows, creative teams are essential to marketing success, crafting visually appealing and compelling content that aligns with brand messaging and resonates with target audiences. These teams work across a wide range of media, producing everything from graphics and videos to copy and interactive designs. By collaborating closely with marketing teams, creative professionals ensure that campaigns stand out and drive engagement. They rely on MarTech platforms to streamline workflows, manage assets, and facilitate efficient collaboration.

Additionally, creative teams are increasingly relying on insights gathered from marketing analytics to understand how different creative approaches perform, to continuously improve their work. Increasingly this also involves utilizing AI-based tools that are able to suggest optimizations and improvements in real time based on customer data and campaign performance.

How They Use MarTech

Creative teams make the original source material for the campaigns that marketers distribute throughout their MarTech infrastructure. This includes the following:

- Designing graphics, videos, and animations for digital ads, social media, Web sites, and email campaigns, which increasingly utilize generative AI to automatically resize for different form factors
- Managing digital assets such as logos, images, and videos through a centralized digital asset management (DAM) platform

- Collaborating with marketing teams to ensure creative aligns with campaign goals and brand guidelines
- Using project management tools to track the progress of creative deliverables and meet deadlines
- Iterating on creative assets based on performance data from A/B testing and analytics

Platforms Utilized

Creative teams depend on tools such as Adobe Creative Cloud (Photoshop, Illustrator, Premiere Pro) or Canva for designing and producing visual assets. For video editing and animation, platforms like Final Cut Pro or After Effects are commonly used. These tools allow the creation of professional-quality content tailored to the needs of various marketing channels.

DAM systems, such as Bynderand Widen, play a critical role in storing, organizing, and sharing creative assets across teams. These platforms ensure that team members can easily access the latest versions of files and maintain brand consistency. Collaboration and project management tools, such as Trello, Asana, and Monday.com, help streamline workflows, track progress, and facilitate communication between creative and marketing teams. Additionally, analytics platforms integrated with MarTech tools provide insights into the performance of creative assets, enabling teams to iterate and optimize their designs based on data.

Additional Considerations

Creative teams can improve their efficiency by ensuring seamless integration between their design tools, DAM systems, and project management platforms. This connectivity reduces bottlenecks, enables faster approvals, and ensures that all stakeholders have access to the latest assets and updates. Establishing clear workflows and timelines can also enhance collaboration and keep projects on track, particularly in fast-paced marketing environments.

Another key area for improvement is leveraging performance data to refine creative strategies. By integrating analytics tools with design workflows, creative teams can gain insights into which types of visuals, messaging, or formats perform best. Regularly updating skills and staying current with design trends and emerging technologies, such as AI-assisted design tools, can also help creative teams deliver innovative and impactful content. Encouraging collaboration and communication between creative, marketing, and operations teams ensures that creative efforts align closely with overall campaign goals and brand strategy.

CUSTOMER EXPERIENCE

While Customer experience (CX) teams were more likely to report to the CMO in the past, since 2019, when 22% of CX teams reported to the CMO (Forrester, 2019)[2], as recently as 2024 (Forrester, 2023)[3], a study found that only 8% of CX leads said they reported to marketing. One potential outcome of this is a sharper focus and differentiation between the practices of marketing and CX, with the latter focused on creating seamless, enjoyable, and consistent interactions between a brand and its customers across all touchpoints. They aim to ensure that customers feel valued and satisfied at every stage of their journey, from awareness to post-purchase support. Often reporting within the marketing function, CX teams work closely with other departments to align strategies and leverage MarTech tools to gather insights, personalize experiences, and address pain points effectively.

How They Use MarTech

CX professionals are keenly focused on the quality of service and satisfaction that customers are receiving, as well as the processes and steps it takes to complete key transactions, including the following:

- Mapping and analyzing the customer journey to identify friction points and opportunities for improvement
- Collecting customer feedback through surveys, reviews, and sentiment analysis tools to inform strategy
- Implementing personalization strategies to deliver tailored content, offers, and recommendations
- Using chatbots and live chat tools to provide instant customer support and enhance engagement
- Tracking customer satisfaction metrics, such as NPS (Net Promoter Score) and CSAT (Customer Satisfaction Score), to gauge performance

Platforms Utilized

CX teams rely on CDPs such as Segment or Salesforce Customer 360 to gather and unify data from multiple touchpoints, creating a comprehensive view of customer behavior and preferences. These platforms enable CX teams to personalize interactions and provide consistent experiences across channels. Feedback and sentiment analysis tools such as Qualtrics, Medallia, and Sprinklr allow teams to collect and analyze customer opinions, offering insights into satisfaction and areas for improvement.

Additionally, journey orchestration tools such as Adobe Journey Optimizer and Thunderhead help map and manage customer journeys in real time, enabling CX teams to identify and address pain points. Live chat and chatbot solutions, such as Zendesk Chat or Intercom, enhance customer support by providing immediate responses to inquiries. Analytics platforms integrated with MarTech tools help CX teams measure the effectiveness of their strategies, tracking key metrics such as NPS, retention rates, and lifetime value.

Additional Considerations

To optimize their impact, CX teams should ensure tight integration between MarTech tools such as CDPs, analytics platforms, and journey orchestration tools. This integration enables a seamless flow of data, helping teams gain a holistic view of customer interactions and deliver more consistent and personalized experiences. CX teams should also collaborate closely with marketing, sales, and customer support teams to align goals and ensure a unified approach to customer engagement.

Investing in training on emerging technologies, such as AI-driven personalization or voice-of-the-customer analytics, can help CX teams stay ahead of customer expectations. Regularly updating and refining customer journey maps based on real-time data ensures that strategies remain relevant and effective. Finally, fostering a culture of feedback and continuous improvement can help CX teams adapt to evolving customer needs and maintain high levels of satisfaction and loyalty.

CONCLUSION

Despite marketing teams being the primary users of MarTech within an organization, they are by no means the only ones who both support and benefit from these platforms. The insights and information derived from these tools enable many other teams to gain a better understanding of customer and company performance.

In the next chapter, we will explore the evolution of CX and how MarTech has evolved to support the way that consumers want to interact with brands.

NOTES

1. Gartner. (2024, September 18). *Gartner survey finds only 52% of senior marketing leaders can prove marketing's value and receive credit for its contribution to business outcomes.* Gartner. https://www.gartner.com/en/newsroom/press-releases/2024-09-18-gartner-survey-finds-only-52-of-senior-marketing-leaders-can-prove-marketings-value-and-receive-credit-for-its-contribution-to-business-outcomes

2. Forrester. (2020). *The state of CX teams.* Forrester. https://www.forrester.com/blogs/the-state-of-cx-teams/

3. Forrester. (2023). *The state of customer experience teams, 2023.* Forrester. https://www.forrester.com/report/the-state-of-customer-experience-teams-2023/RES180035

CHAPTER 5

MarTech and the Evolving Customer Relationship

After exploring the relationship between MarTech platforms and internal teams in Chapters 3 and 4, it is now time to look at the recipients of much of the output of MarTech: prospective, current, and lapsed customers. To do this, it is important to understand that, much like the evolution of marketing (and MarTech) over time, there has been an evolution of customer expectations, behavior, and the overall customer experience.

This chapter will explore this evolution, and while the components discussed follow a mostly chronological order, there have been several evolutions in each component. Thus, instead of a timeline, this chapter will present these factors more as a series of elements that continue to inform both consumers' behavior as well as how marketers rely on technology to meet these expectations.

GREATER ADOPTION OF DIGITAL CHANNELS

Until the popular adoption of the Internet and all of the interactive channels that followed from it, the customer experience could be thought of as having two primary components. The first was primarily based on marketing by using a broadcast, or one-to-many approach, from media such as print advertising, radio, television, direct mail, or other channels that did not lend themselves to receiving immediate feedback, and which would often provide challenges to determining a direct return on investment (ROI).

The second was on a one-to-one personal communication basis, whether a frontline employee in a retail store, a hotel concierge, a doctor's receptionist, or many other instances of a single person interacting with a customer. This extended to customer service phone lines and even written correspondence. Yet, as personal as the interactions were (and still are), the ability to scale these interactions has proven difficult without sacrificing quality, attention to detail, and other factors.

The adoption of digital channels marked a pivotal moment in consumer behavior, as the Internet became a trusted space for activities ranging from product research to online purchases. While it was initially met with skepticism due to concerns over security and unfamiliarity, consumers steadily grew more comfortable engaging with digital platforms. Today, nearly every buying journey involves digital touchpoints, whether it's reading product reviews, comparing prices, or purchasing items directly online. In fact, as of the Fall of 2024, a report from 1WorldSync found that 72% of consumers researched a product online before buying, regardless of whether they ended up buying that product online or at a brick-and-mortar retail store (1WorldSync, 2024)[1]. This means that digital channels are part of the customer acquisition process even if the sale doesn't originate or end online.

What This Means for Brands

For brands, this shift underscores the critical need for a strong digital presence. Customers expect easy-to-navigate Web sites, detailed product information, and smooth e-commerce experiences. They also expect this regardless of whether they plan to purchase through that same Web site or through a different channel altogether, whether that is through a retail store, a third-party marketplace such as Amazon, a brand, or social media. Brands must also invest in digital advertising to capture attention and create seamless transitions from discovery to purchase.

A notable example is Amazon, which capitalized early on consumer trust in digital channels, first opening its virtual doors to the public in 1995[2]. By offering extensive product selections, user-friendly interfaces, and secure payment systems, Amazon revolutionized e-commerce and set the benchmark for online shopping experiences.

How This Affects MarTech

Over the decades that followed some early adopters, brands have embraced e-commerce platforms, integrated digital advertising strategies, and enhanced their online security measures. Many have also incorporated customer reviews and user-generated content to build credibility and foster trust among digital consumers.

SOCIAL MEDIA MINDSET

Social media has transformed from a casual communication tool to a vital part of daily life, with billions of users engaging in activities such as commenting, sharing, and purchasing products. As of October 2024, out of the 5.52 billion worldwide Internet users, 5.22 billion were also social media users (Statista, 2024)[3].

In addition to sharing content with one another, consumers now rely heavily on social media for product recommendations and reviews, often valuing the opinions of influencers over traditional brand messaging.

This has also extended from simply getting product recommendations to consumers becoming increasingly more comfortable purchasing products directly through social media channels, with the social commerce global market estimated at $1.3 trillion in 2023 and projected to reach $8.5 trillion in 2030 (ResearchandMarkets, 2024)[4].

What This Means for Brands

Much like the Internet more generally, social media is a channel that continues to grow and expand in its utility and in its utilization by consumers. Brands must recognize the power of social media as both a communication channel and as a sales channel, and additionally leverage the power of word of mouth and positive reviews from other people on these platforms. This involves fostering authentic connections with audiences, leveraging influencers to build credibility, and investing in social commerce capabilities to drive purchases directly through platforms such as Instagram and TikTok.

Glossier, a beauty brand launched in October 2014, built its empire largely through social media by engaging directly with customers, collaborating with influencers, and encouraging user-generated content.

Their approach has fostered a loyal community and driven significant sales through social platforms, translating into a billion-dollar business within a decade after its launch (Fischer, 2023)[5].

How This Affects MarTech

To capture the growing social media audience, brands have shifted focus to create visually appealing, shareable content and formed partnerships with influencers.

Additionally, to support the growing desire of many consumers for social commerce, many have integrated direct-purchase options on platforms such as Facebook, TikTok, and Instagram, making social media a key driver of sales.

MOBILE-FIRST BEHAVIOR

The release of the iPhone and subsequent mobile innovation ushered in a mobile-first era, where consumers rely on their smartphones for everything from research to purchasing. With an estimated 7.12 billion smartphones in use as of the end of 2024 (Gill, 2025)[6] these devices are now central to everyday life, functioning as research tools, social hubs, and shopping devices, with the average consumer in the United States checking their phone 205 times per day as of 2024, a 42% increase from the previous year[7].

This mobile behavior is not simply relegated to checking social media feeds from their friends either. According to Pew Research, 32% of US adults make a purchase using their smartphone at least once a week, with some demographics (ages 30–49) as high as 49% using this mobile behavior[8].

What This Means for Brands

With the ubiquity of smartphones in daily life in many countries, brands need to prioritize mobile optimization in all aspects of their marketing strategy. This includes ensuring that Web sites are mobile friendly, developing apps where necessary, and utilizing mobile ad formats to capture attention effectively.

This includes enabling better e-commerce experiences on mobile devices, and in many cases, creating mobile apps that offer both new and existing functionalities to users that extend engagement, customer acquisition, and purchase opportunities to mobile users.

Starbucks has capitalized on mobile behavior through its app, offering convenient features such as mobile ordering, payments, and loyalty rewards. This strategy has increased customer engagement and streamlined the purchasing process.

By building a mobile app that appealed to its customers, it was able to see a 48% rise in customer engagement and a 67% increase in loyalty program signups[9]. This also contributed to nearly one-fifth of transactions at US-based Starbucks locations being mobile orders[10].

How This Affects MarTech

Mobile-first strategies have led brands to invest heavily in responsive Web design, app development, and mobile payment systems. Many also prioritize mobile analytics to understand consumer behavior and optimize campaigns accordingly.

INCREASED IMPORTANCE OF THE OMNICHANNEL CUSTOMER EXPERIENCE

Consumers now expect seamless, consistent experiences across the multiple channels they use throughout their buying journey. Whether interacting with a brand via mobile, desktop, in-store, or social media, customers demand a unified and cohesive experience.

In addition to preferences about *where* they receive their experiences with brands, consumers are also increasingly focused on the value of the experiences they receive in their interactions with a brand in general, regardless of the channel.

Research has shown that four-fifths of customers have indicated that they switched brands due to receiving poor customer experience, with 43% saying they were at least somewhat likely to switch brands after only a single poor customer service interaction (Qualtrics, 2021)[11]. The same percentage (43%) indicated that they would be willing to pay more for greater convenience, and 65% of US consumers have indicated that a positive experience with a brand is more influential in their purchase decisions than great advertising (Puthiyamadam and Reyes, 2018)[12].

What This Means for Brands

Companies have invested in measuring customer experience through measurements such as net promoter score (NPS) and customer satisfaction (CSAT) and invested billions of dollars in digital transformations, many of which have been designed to improve the internal delivery of products and services as well as improving external, customer-facing customer service, product, and service delivery.

Delivering an *omnichannel* experience requires brands to integrate their data, processes, and platforms. MarTech plays a crucial role in ensuring that campaigns, messaging, and customer data are consistent across all touchpoints.

Nike has excelled in creating an omnichannel experience by integrating its Web site, app, and physical stores. Customers can shop online, use the app to locate products instore, or access exclusive app-only features, creating a seamless journey.

How This Affects MarTech

To adapt to these changing customer priorities, brands have adopted customer journey orchestration tools and integrated their systems to provide unified experiences. Investments in cross-channel analytics and customer data platforms (CDPs) to better understand customer behavior and deliver more personalized customer experiences have also become critical for success.

CONCERN ABOUT CONSUMER DATA PRIVACY

Consumers have grown increasingly concerned about data privacy following high-profile breaches and misuse of personal information. As of 2019, they viewed their personal data as less secure than it had been five years previously, and while this has slightly decreased in the years since then, in a 2023 study, 81% were concerned about how companies use the data they collect about them (Pew Research, 2023)[13].

Consumers now demand transparency, control over their data, and assurances of secure practices, which has led in part to government regulations such as the General Data Protection Regulation (GDPR) in the European Union (EU), or the California Consumer Privacy Act (CCPA) in the state of California in the United States, which went into effect on January 1, 2023[14]. While these regulations have their limitations, they have both highlighted the need for companies to implement best practices, and have helped organizations to take tangible steps toward safeguarding consumer's private information. These regulations have also inspired other regulations and legislation around the world.

What This Means for Brands

Brands must prioritize privacy compliance and ethical data practices. Building trust through transparency and respecting consumer privacy is not only a regulatory requirement but also a competitive differentiator.

Despite setbacks that have plagued most companies at one time or another, companies such as Apple have made privacy a core part of their brand positioning, introducing features such as App Tracking Transparency and promoting user control over data. This approach has resonated with privacy-conscious consumers.

How This Affects MarTech

Many brands have implemented stricter data security measures, adopted privacy-first marketing strategies, and ensured compliance with regulations such as GDPR and CCPA. Clear communication about data usage has also become a key priority.

EXPECTATION OF PERSONALIZATION

Despite privacy concerns, consumers increasingly expect personalized experiences tailored to their preferences and behavior. They reward brands that deliver relevance and seamless engagement across channels. In fact, according to a McKinsey & Company report, 71% of consumers have an expectation of personalization when receiving communications and products or services from brands, and 76% are frustrated when they don't receive personalized experiences[15].

What This Means for Brands

Brands need to leverage data and MarTech to deliver personalization at scale. This requires understanding customer preferences, segmenting audiences, and implementing AI-driven tools for tailored content and recommendations.

Loyalty programs, such as those used by many hotel brands or airlines, are built to provide tailored offers and experiences to customers who are more likely to share repeat experiences and more detailed information with companies that, in turn, reward those customers with special deals, discounts, and access to products and services that are above and beyond what other customers might receive.

How This Affects MarTech

Greater personalization requires that a company has a better understanding of its customers, including their past behaviors and their propensity to take specific actions. This requires greater amounts of data collection directly from consumers, as well as some third-party augmentation of their own data.

Brands have adopted AI and machine learning technologies to analyze customer data and predict preferences. Many have also focused on dynamic content creation and real-time personalization on channels such as their Web sites and mobile apps to engage audiences more effectively.

This personalization has extended beyond simple substitutions of a first name in an email greeting, to providing content, offers, and actions that are tailored to a customer's behavior and indicated preferences. While much personalization still falls in the realm of broader audience segments, which might all receive similarly personalized content, the goal is an "audience of one" approach[16], where each individual receives content that is specifically tailored to them, their recent purchase behaviors, online habits, in-store visitation preferences, and more.

DESIRE FOR SELF-SERVE OPTIONS

Consumers increasingly prefer self-serve options, whether instore or on digital channels, such as FAQs, chatbots, and online account management, to solve problems and answer questions without relying on customer service representatives. Research has found that 81% of customers attempt to take care of issues themselves before reaching out to a live representative at a company, regardless of the industry (Ponomareff et al., 2017)[17].

Even in brick-and-mortar environments, a study by PYMNTS showed that 84% of consumers in the United States enjoy self-service kiosks, with 66% preferring them to staffed checkout counters[18].

What This Means for Brands

Brands must provide intuitive, user-friendly self-service tools to meet customer expectations. These solutions not only enhance customer satisfaction but also reduce the burden on support teams.

Platforms such as Zendesk enable companies to create robust self-service portals, including knowledge bases and AI-powered chatbots, empowering customers to find answers independently.

How This Affects MarTech

Brands have integrated AI-driven chatbots, expanded online help centers such as knowledge bases, and designed user-friendly apps and Web sites to cater to the self-serve mindset. Continuous improvements based on user feedback have further enhanced these tools.

Because of the diverse customer sets that many brands cater to, this also necessitates a need for better customer data and personalization features in their communication channels.

CONCLUSION

The factors explored in this chapter comprise many of the facets that have gone into the evolution of the customer experience and consumer expectations over the past several years.

While consumer preferences will continue to evolve with changes in technology and access to information, it is important for brands to build a MarTech infrastructure that allows them to adapt to the needs of prospective and current customers.

Beyond solely catering to customers, there are several compelling reasons to make critical investments in MarTech infrastructure from a business perspective. We will explore them in the next chapter.

NOTES

1. 1WorldSync. (2024). *Product content benchmark 2024*. 1WorldSync. https://resources.1worldsync.com/product-content-benchmark-2024/

2. Milliot, J. (2015, July 16). *20 years of Amazon.com bookselling*. Publishers Weekly. https://www.publishersweekly.com/pw/by-topic/industry-news/bookselling/article/67986-20-years-of-amazon-com-bookselling.html

3. Statista. (2024). *Worldwide digital population 2024*. Statista. https://www.statista.com/statistics/617136/digital-population-worldwide/#:~:text=Worldwide%20digital%20population%202024&text=Of%20this%20total%2C%205.22%20billion,population%2C%20were%20social%20media%20users.

4. Research and Markets. (2023). *Social commerce: Global strategic business report*. Research and Markets. https://www.researchandmarkets.com/reports/5140143/social-commerce-global-strategic-business-report

5. Schwartz, L. (2023, September 25). *How Glossier made effortlessness a billion-dollar brand*. The New Yorker. https://www.newyorker.com/magazine/2023/09/25/how-glossier-made-effortlessness-a-billion-dollar-brand

6. Gill, Sunil. (2025, January 1). *How many people own smartphones in the world? (2024-2029)*. Prioridata. https://prioridata.com/data/smartphone-stats/

7. NBC Palm Springs. (2024, December 18). *Americans check their phones 205 times a day in 2024, report reveals*. NBC Palm Springs. https://www.nbcpalmsprings.com/2024/12/18/americans-check-their-phones-205-times-a-day-in-2024-report-reveals

8. Pew Research Center. (2022, November 21). *For shopping, phones are common and influencers have become a factor, especially for young adults*. Pew Research Center. https://www.pewresearch.org/short-reads/2022/11/21/for-shopping-phones-are-common-and-influencers-have-become-a-factor-especially-for-young-adults/

9. Inviqa. (n.d.). *Starbucks case study*. Inviqa. https://inviqa.com/case-studies/starbucks

10. Hure, Denis. (2024, September 21.). *Starbucks Rewards: A model for customer loyalty success*. Reward the World. https://rewardtheworld.net/starbucks-rewards-a-model-for-customer-loyalty-success/

11. Qualtrics. (2021, December 14). *Qualtrics-Servicenow customer service research*. Qualtrics. https://www.qualtrics.com/blog/qualtrics-servicenow-customer-service-research/

12. Puthiyamadam, Tom and Joeé Reyes. (2018). *Experience is everything: Here's how to get it right*. PwC. https://www.pwc.com/us/en/advisory-services/publications/consumer-intelligence-series/pwc-consumer-intelligence-series-customer-experience.pdf

13. Pew Research Center. (2023, October 18). *How Americans view data privacy*. Pew Research Center. https://www.pewresearch.org/internet/2023/10/18/how-americans-view-data-privacy/

14. California Office of the Attorney General. (n.d.). *California Consumer Privacy Act (CCPA)*. California Department of Justice. https://oag.ca.gov/privacy/ccpa

15. McKinsey & Company. (2021, November 17). *The value of getting personalization right—or wrong—is multiplying.* McKinsey & Company. https://www.mckinsey.com/capabilities/growth-marketing-and-sales/our-insights/the-value-of-getting-personalization-right-or-wrong-is-multiplying

16. Turner, J. & Moxley, Ch. (2021, October 22). *An Audience of One: Drive Superior Results By Making the Radical Shift from Mass Marketing to One-to-One Marketing.* McGraw Hill.

17. Dixon, M., Ponomareff, L., Turner, S., & DeLisi, R. (2017, January). *Kick-ass customer service. Harvard Business Review.* https://hbr.org/2017/01/kick-ass-customer-service

18. PYMNTS. (2024, January). Unattended: The Payments Technology Shifting the Future of Commerce. PYMNTS. https://www.pymnts.com/tracker_posts/unattended-the-payments-technology-shifting-the-future-of-commerce/

CHAPTER 6

THE BUSINESS CASE FOR MARTECH INVESTMENTS

In Chapter 5, the customer needs and expectations of MarTech were explored. In this chapter, the benefits that MarTech can provide for the needs of the business will be the focus.

There are, of course, several facets of a business that MarTech can address and provide solutions for, and some of the most prevalent perceived benefits related to customer needs and demands include the ability to analyze customer data to deliver on personalization, solving for multichannel consistency, and automating tasks to create greater efficiencies and speed within marketing teams[1].

EVALUATING WHERE MARTECH CAN MAKE THE BIGGEST IMPACT

Along with a good understanding of the potential MarTech tools that you need, along with an understanding of the best ways to implement them within your organization, you also need to evaluate these platforms and methods against your business needs. This includes current gaps in your ability to reach your audiences, as well as planning for future growth, consumer expectations, and competitive pressure.

Defining what your business needs in terms of MarTech can be looked at in five ways:

- Ensuring marketing and communication effectiveness
- Closing the customer experience gap

- Increasing competitive advantage
- Creating greater internal efficiency and collaboration
- Enabling data-driven decision-making

Together, these areas highlight how MarTech can transform not only marketing teams but entire organizations. Throughout this chapter, we will explore each of these areas in detail, providing relatable hypothetical examples to illustrate the potential of MarTech in action. These scenarios will demonstrate how businesses of all types can leverage MarTech to drive growth, innovation, and success in a rapidly changing marketplace.

Ensuring Marketing and Communication Effectiveness

Effective marketing is critical to any business. By enhancing the reach, consistency, and relevance of campaigns across multiple channels, MarTech ensures that businesses can deliver the right message to the right audience at the right time. From enabling precise audience segmentation to providing real-time insights into campaign performance, MarTech platforms have the potential to empower marketing teams to be more strategic, efficient, and impactful in their efforts. This includes some of the following benefits.

Improved Audience Segmentation

MarTech platforms allow businesses to go beyond broad demographic targeting and segment their audiences based on behavior, preferences, and past interactions. This level of granularity ensures that campaigns resonate with individual customers, increasing engagement and conversions.

Real-Time Performance Tracking

Tools such as analytics dashboards and campaign tracking software enable marketers to monitor campaign performance in real time. Teams can identify what's working and what's not, making data-driven adjustments on the fly to maximize impact and ROI.

Personalization Tools

Personalization is no longer a luxury; it's a necessity, with 65% of customers expecting companies to adapt to their changing needs and preferences (Salesforce, 2023)[2]. MarTech tools enable businesses to deliver tailored content, offers, and experiences that strengthen customer connections and build brand loyalty. Personalized marketing can turn one-time buyers into repeat customers and advocates.

Hypothetical Examples

Some of the ways that MarTech can be used to support effective communication and marketing are as follows:

- A *fashion retailer* uses a marketing automation platform to analyze browsing behavior on its Web site. Based on customer preferences, the retailer sends personalized email promotions featuring products similar to those a customer recently viewed. This approach not only increases open and click-through rates but also drives higher sales.
- A *software company* tests two landing page designs for a new product demo using an A/B testing tool. By measuring which design generates more demo sign-ups, the company optimizes its landing page strategy, ultimately boosting lead conversions and improving campaign efficiency.

Challenges and Solutions

While incorporating MarTech platforms can provide significant advantages to an organization, it's not without challenges. Tool overload is a common issue, with many businesses implementing more platforms than they can effectively manage. This can lead to inefficiency and underutilized capabilities. To mitigate this, organizations should conduct regular audits to identify redundant or underperforming tools and streamline their stack.

Another challenge is skill gaps within teams. Without proper training, marketers may struggle to maximize the potential of sophisticated MarTech platforms. Investing in ongoing training programs and leveraging user-friendly tools can bridge this gap and ensure that teams feel confident using their technology. Additionally, organizations should prioritize platforms that integrate seamlessly with existing tools to avoid data silos and enhance overall efficiency.

By addressing these challenges, businesses can fully leverage MarTech platforms to enhance their marketing and communication efforts, ensuring campaigns are not only impactful but also aligned with their broader organizational goals.

Closing the Customer Experience Gap Between Customer Expectations and Product/Service Delivery

As noted in the exploration of customer preferences and expectations found in Chapter 5, there is a growing need for brands to provide exceptional customer experiences, in addition to stellar products and services. Many businesses struggle to meet these demands, however. This gap between customer expectations and actual service delivery can lead to frustration, dissatisfaction, and lost revenue.

Effective adoption of MarTech tools can provide the tools to help an organization bridge this gap by enabling it to deliver tailored, consistent, and more frictionless customer experiences. From personalized communications to real-time journey orchestration, the right MarTech empowers organizations to meet—and often exceed—customer expectations, including in the following business applications.

Personalization at Scale

Implementing the right platforms makes it possible to deliver highly personalized experiences to millions of customers. Whether it's personalized product recommendations, targeted email content based on recent purchase behavior, or dynamic Web site features, businesses can use data to cater to individual needs and preferences, fostering stronger customer relationships.

Unified Customer Profiles

By centralizing data from multiple touchpoints, tools such as customer data platforms (CDPs) create comprehensive, unified profiles of each customer. This ensures that omnichannel interactions across email, social media, Web sites, and customer support channels are consistent and relevant, eliminating silos and improving the overall experience.

Journey Mapping and Customer Journey Orchestration

These tools enable businesses to visualize and manage the customer journey from start to finish. The process of journey mapping identifies pain points and areas for potential improvements from an internal or outward-facing perspective.

Customer journey orchestration (CJO) platforms automate actions to address these issues, ensuring smooth transitions between channels such as mobile devices and in-store experiences, reducing friction for the customer. After all, according to Salesforce, 79% of customers expect consistent interactions, whether that is between departments or channels[3].

Hypothetical Examples

A few potential applications of platforms that can help close the customer experience gap include the following:

- A *subscription box service* uses a CDP to analyze purchasing habits and preferences. Based on this data, the service sends personalized email recommendations suggesting additional products that align with customers' tastes. For instance, a customer who frequently orders skincare products might receive an offer for a new line of organic face creams, increasing both engagement and sales.
- A luxury *hotel brand* employs journey orchestration tools to enhance the post-stay experience. After a guest checks out, the system automatically sends a personalized follow-up email with a discount code for their next visit, along with tailored recommendations for nearby attractions. This not only encourages repeat bookings but also strengthens the brand's relationship with the customer.

Future Implications

As technology evolves along with customer expectations, the potential for technology platforms to enhance customer experiences will only grow. AI tools, including predictive analytics, automation, and generative AI, are already being used to anticipate customer needs and provide proactive solutions. For example, AI-driven chatbots can provide real-time assistance, while predictive analytics can forecast a customer's next likely purchase or need, enabling businesses to act before the customer even expresses interest.

Additionally, the integration of augmented reality (AR) and virtual reality (VR) into customer experiences has the potential to further reduce the gap between expectations and delivery. With current technology, it is possible for a furniture retailer to offer customers an AR tool to visualize how a sofa would look in their living room before making a purchase. These innovations will enable businesses to provide increasingly immersive and satisfying experiences.

By leveraging these advanced tools, businesses can ensure they remain at the forefront of customer experience innovation, meeting and exceeding expectations in a rapidly changing market while delivering exceptional service at scale.

Increasing Competitive Advantage

Of course, you are not just looking at your MarTech stack in terms of relative growth to your internal metrics. You also need to ensure that you are keeping up with both customer expectations and competitive demands.

Staying competitive requires agility, insight, and innovation. Strategic implementation of MarTech has the potential to provide businesses with the tools they need to outpace competitors by streamlining workflows, harnessing data to identify opportunities, and positioning brands strategically. By enabling faster time-to-market, leveraging advanced analytics, and delivering actionable insights, the right MarTech empowers organizations to anticipate market shifts and respond effectively.

Faster Time-to-Market

Automation and more agile workflows powered by project management and collaboration tools can reduce the time it takes to launch campaigns, develop content, or pivot strategies. Businesses can respond to changing market conditions and customer needs faster, giving them a decisive edge over slower competitors.

Advanced Analytics to Act on Market Trends

Platforms equipped with predictive and advanced analytics tools help businesses identify trends and patterns before they become mainstream. This foresight allows companies to take proactive actions, whether that's adjusting product offerings or launching timely campaigns.

Enhanced Brand Positioning Through Data-Driven Insights

With a wealth of customer and market data, businesses can refine their messaging, target the right audience segments, and position their brand as a leader in the industry. Platforms that enable precise targeting and personalized experiences that differentiate a brand from its competitors.

Hypothetical Examples

Consider these potential use cases:

- A *food delivery startup* uses predictive analytics to analyze order patterns during peak hours, such as lunch breaks or weekend evenings. By anticipating demand surges, the startup adjusts its marketing campaigns, sending targeted discounts or push notifications to potential customers at just the right time, outpacing competitors in attracting orders.

- An *e-commerce brand* leverages social media listening tools to monitor conversations about trending products on social media. For example, upon noticing increased chatter about a specific fitness gadget, the brand quickly launches a targeted ad campaign, positioning itself as the go-to retailer for the product while competitors lag behind.

Practical Tips

To maximize competitive advantage, businesses should embrace continuous experimentation and innovation with the tools within their MarTech infrastructures. For example, regularly running A/B tests on campaigns or experimenting with new marketing channels can provide insights into what resonates most with target audiences. Organizations should also invest in tools that provide access to real-time data, enabling them to act quickly when market conditions shift.

Another key strategy is staying ahead of industry trends by integrating emerging technologies, such as AI-powered personalization or voice search optimization, into their technology stack. Businesses should also foster a culture of collaboration across teams, ensuring that insights from marketing, sales, and customer service are shared to create a unified, competitive strategy.

By leveraging the agility, insights, and tools offered by the best MarTech platforms, businesses can maintain a competitive edge in dynamic markets, securing their position as industry leaders while leaving slower-moving competitors behind.

Creating Greater Internal Efficiency and Collaboration

MarTech does more than optimize external-facing activities; it also has the potential to evolve and improve how internal teams work together. By improving collaboration, streamlining workflows, and automating repetitive tasks, an effective MarTech infrastructure enables teams to focus on high-value strategic and creative work while automating much of what might have previously taken up precious time with negligible returns. With shared platforms and integration capabilities, organizations can ensure alignment across departments, enhance productivity, and allocate resources more effectively.

Streamlined Workflows with Project Management and Automation Tools

MarTech platforms such as project management tools and automation software simplify complex workflows, reducing bottlenecks and ensuring the timely execution of campaigns. Teams can monitor progress, manage dependencies, and ensure all stakeholders remain on the same page.

Better Cross-Department Communication

Shared platforms, such as CRMs or digital asset management (DAM) systems, act as central hubs for collaboration, enabling seamless communication between departments such as marketing, sales, and customer service. This improves coordination, reduces redundancy, and ensures consistent customer interactions.

Reduction of Repetitive Tasks

By automating time-consuming processes, such as email distribution or social media scheduling, MarTech allows teams to redirect their energy toward strategy and creativity, leading to more innovative and impactful campaigns.

Hypothetical Examples

Consider the following potential applications of MarTech to create greater efficiency:

- A *marketing team at a healthcare company* uses a project management platform integrated with a DAM system to streamline campaign production. By centralizing asset storage and tracking tasks, the team reduces campaign launch times by 30%, ensuring they stay ahead of regulatory deadlines.
- A *global enterprise* adopts a shared CRM to align sales, marketing, and customer support teams. With all customer interactions stored in one system, the organization eliminates duplicate outreach to the same customers, improving efficiency and enhancing the customer experience.

Common Pitfalls

Despite its benefits, implementing software tools to improve efficiency and collaboration is not without challenges. One common issue is poor adoption rates among team members, often caused by a lack of training or unclear workflows. To address this, organizations should invest in comprehensive onboarding programs and ongoing support to ensure employees are comfortable using new tools.

Resistance to change is another potential barrier, particularly when teams are accustomed to existing processes. This can be mitigated by involving key stakeholders early in the selection and implementation process, ensuring that the chosen tools address real pain points and are seen as valuable additions rather than burdens.

Lastly, organizations should guard against over-reliance on technology by maintaining a balance between automation and human input. While MarTech can automate many tasks, creativity, strategic thinking, and interpersonal collaboration remain critical to marketing success. By addressing these challenges proactively, businesses can unlock the full potential of MarTech to create more efficient and collaborative teams.

Enabling More Data-Driven Decisions

In a world where competition is fierce and customer expectations are constantly evolving, data-driven decision-making has become a cornerstone of successful business strategies. MarTech empowers organizations to move beyond intuition and guesswork by providing access to real-time data, actionable insights, and comprehensive analytics. By leveraging MarTech tools, businesses can make informed decisions that improve campaign effectiveness, enhance customer experiences, and drive measurable outcomes.

Access to Real-Time Data

MarTech platforms deliver real-time insights, allowing businesses to monitor performance and make adjustments on the fly. This agility ensures campaigns stay relevant and effective, even as market conditions change.

Insights into Customer Behavior and Preferences

Advanced analytics tools provide deep insights into customer behavior, including what drives engagement, conversions, and loyalty. By understanding these patterns, businesses can refine their strategies to better meet customer needs.

Multi-Touch Attribution (MTA) and Media Mix Modelling (MMM)

Tools intended to help organizations determine their return on ad spend (ROAS) such as multi-touch attribution or media mix models help organizations understand which channels and touchpoints contribute most to campaign success.

This enables the smarter allocation of resources such as advertising budgets and ensures every dollar spent delivers maximum impact.

Hypothetical Examples

Consider the following examples of how these tools can be applied:

- A *B2B SaaS company* uses multi-channel attribution tools to identify that 60% of demo requests originate from email campaigns, while 30% are driven by paid search ads. Armed with this data, the company reallocates its budget to emphasize these high-performing channels, increasing demo requests by 20%.
- A *nonprofit organization* monitors real-time engagement data during a live fundraising event. By analyzing which social media posts and email appeals generate the most donations, the team adjusts its messaging mid-event, resulting in a 15% boost in contributions.

Future Trends

The future of data-driven decision-making will be shaped by advancements in AI and machine learning. These technologies will improve the accuracy of predictions, helping businesses forecast customer behavior and campaign performance with unprecedented precision. For instance, AI-driven tools could analyze past data to recommend the optimal time and channel for launching a campaign, maximizing reach and ROI.

Predictive analytics will also enable businesses to anticipate customer needs and act proactively. For example, a retail company could predict when a customer is likely to reorder a product and send a timely reminder, increasing both satisfaction and revenue. As these tools become more accessible and user-friendly, even small organizations will be able to harness the power of data-driven insights.

By integrating AI and machine learning into their MarTech stack, businesses can unlock new opportunities for growth and innovation, ensuring they stay ahead in an increasingly data-centric marketplace. Embracing these tools now will not only enhance decision-making today but also position organizations to thrive in the future.

INTRODUCING A MARTECH MATURITY MODEL

In Chapter 25, this model will be discussed in more depth, which will enable all of the terms within it to be fully explored and defined. Until then, it is important to introduce the model to show how the business case and goals can translate into a long-term measurement of an organization's successful implementation of MarTech.

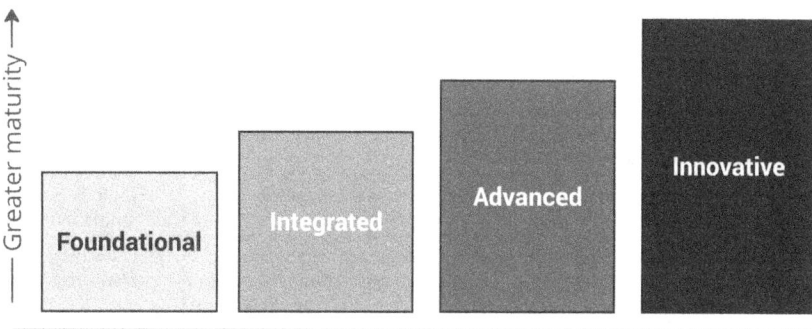

FIGURE 6.2 MarTech maturity model

Each stage in the maturity model provides a description of what a typical organization fitting that criteria is experiencing, as well as how MarTech plays a role in their organization.

TABLE 6.1 A MarTech maturity model

Technology Ecosystem	Primary Challenges	Main Opportunities
1. Foundational		
Minimal toolset (often single-channel)	Data inconsistencies and duplicates	Establish a stable platform
Siloed or spreadsheet-based data	High manual workload	Automate simple tasks
Limited automation and integrations	Poor visibility into performance	Basic integrations (e.g., CRM and email)

(Continued)

Technology Ecosystem	Primary Challenges	Main Opportunities
2. Integrated		
Core platforms connected (CRM, marketing automation, analytics)	Growing complexity of integrations	Streamline cross-platform data flow
Basic centralized data layer	Potential skill gaps in data/automation	Deeper data-driven insights
Introductory personalization and workflow automations	Governance policies may lag behind adoption of new tools	Set the stage for more robust automation and analytics
3. Advanced		
Sophisticated data infrastructure (CDP, data warehouse)	Technical overhead (data engineering, ML expertise)	AI-driven personalization and lead scoring
Highly automated marketing, sales, and service workflows	Complex compliance requirements	Holistic, cross-department collaboration
Cross-channel personalization and targeting	Risk of vendor or architectural lock-in	Scalable, enterprise-grade campaigns
4. Innovative		
Modular, API-first architecture (microservices, containerization)	Ongoing innovation pressure	Omnichannel, context-aware engagement
Real-time orchestration of customer journeys	High technical complexity and resource demands	Continuous competitive differentiation
Built-in experimentation frameworks (feature flags, A/B testing)	Maintaining advanced data pipelines and real-time analytics	Rapid adaptation to emerging tools (AR, IoT, etc.)

A helpful exercise at this point would be to review the high-level descriptions of the maturity stages (see Table 6.1) and identify where your organization fits within each area. Keep this framework in mind as you read the chapters that follow, and refer to it as you explore some of the areas of MarTech that may be less familiar.

CONCLUSION

Understanding how to develop an effective MarTech stack and having the processes to evaluate it in an ongoing manner according to the needs of the business will ensure it will continue to serve your organization effectively now and for years to come.

The next chapter will explore the growing importance of AI tools and their incorporation into the MarTech stack.

NOTES

1. Harvard Business Review Analytics Services survey, (June 2024) https://hbr.org/resources/pdfs/comm/sas/CRE5404_HBR_PS_SAS_Sept24%20(1).pdf

2. Salesforce. (2024). *State of the AI Connected Customer: Insights from 16,585 consumers and business buyers on expectations and trust in the age of AI and agents*. Salesforce. https://www.salesforce.com/content/dam/web/en_us/www/documents/research/State-of-the-Connected-Customer.pdf

3. Salesforce. (n.d.). *Customer expectations: How to meet and exceed them*. Salesforce. https://www.salesforce.com/resources/articles/customer-expectations/?sfdc-redirect=369

CHAPTER 7

METHODS OF MARTECH IMPLEMENTATION

Having explored the categories of MarTech platforms at a high level, it is also important for organizations to make some key decisions about how those tools are implemented and integrated into a company's existing technology infrastructure, as well as how individual pieces within the martech stack interact and are integrated with each other.

In addition to the categories of MarTech platforms, there are also three methods that can be employed in order to implement them and integrate them with your other tools and platforms. Quite often in an enterprise, more than one of these approaches is used, though it is a best practice to define a preferred approach, which can help guide strategic decisions, as well as implementation and even hiring strategies.

ALL-IN-ONE, BEST-OF-BREED, AND COMPOSABLE

Now, we're going to review each of the three approaches:

- All-in-one solutions that have multiple functions and are often a combination of several software applications that utilize a single interface and logins or permissions structures.

- Best-of-breed integrations that take the top-performing and most relevant platforms in a number of different areas and combine them to create a customized MarTech stack.

- Composable approaches that, similar to best-of-breed integration utilize a number of different platforms, but differentiate themselves by their focus on an ease of integration.

All-in-one

All-in-one MarTech solutions—sometimes referred to as monolithic solutions—are platforms that offer multiple online marketing functions in an integrated package. They are designed for businesses that want to simplify their marketing processes and avoid using several different systems and the integrations, authentication issues, and other headaches that doing so can cause.

Some of the common functions that all-in-one MarTech solutions provide are marketing automation, CRM, Web site content management, search engine optimization (SEO), social media marketing, and e-commerce capabilities.

These all-in-one marketing platforms aim to help businesses plan, execute, and measure their marketing campaigns across various channels and reach their target audience more effectively. By using all-in-one MarTech solutions, businesses can save time and money, improve productivity and collaboration, and optimize their marketing performance.

Benefits

All-in-one solutions allow a single interface and login in order to perform a wide variety of tasks. It can also take longer to integrate new data sources and gain insights into marketing campaigns, provided the tools are already part of the solution.

Even siloed teams at large companies can benefit from the standardization that these all-in-one platforms offer, and it can often reduce the learning curve for team members to switch between elements of the platform.

Drawbacks

While an all-in-one solution can have a broad array of features that appeal to most of what a marketing team needs, if they don't have the functionality you want "out of the box," it can be costly and in some cases impossible to add them until they make it into the product's roadmap.

Additionally, in some cases, integrations that are not native to the platform can be costly to develop, though there are several all-in-one platforms that have more open application programming interfaces (APIs) that allow relatively easy integrations.

Finally, all-in-one platforms often come to exist because a platform company acquires complimentary products and slowly integrates them one by one. Because of this, the integration of these platforms can sometimes feel clunky and barely any better than if you integrated a separate platform yourselves. Therefore, ensure before you invest in an all-in-one platform that the core functionality you need is easily accessible with a streamlined interface.

Best-of-Breed

Unlike all-in-one solutions where a large system may perform most or all of an enterprise's marketing needs, the best-of-breed approach to MarTech infrastructure is beneficial because it allows marketers to select the most suitable tools for their specific needs and goals. Unlike an all-in-one solution, which may have limitations or gaps in functionality or its ability to integrate with other systems, a best-of-breed approach enables marketers to customize and optimize their technology stack by choosing the best and most appropriate products from different vendors. This way, marketers can leverage the strengths and features of each tool in their MarTech stack, while avoiding the drawbacks and risks of vendor lock-in.

The best-of-breed approach also fosters innovation and agility, as marketers can easily adopt new technologies or switch to better alternatives as the market evolves. Using this approach, marketers can create a more flexible, scalable, and effective MarTech infrastructure that supports their business objectives and delivers superior results.

Benefits

The best-of-breed approach provides several benefits:

- It allows a brand to pick the very best platforms that suit its needs and goals.
- Additionally, there can often be industry considerations (e.g., regulated industries such as healthcare and financial services can have specific needs), or audience considerations (e.g., B2C or B2B can often have different needs) that may drive a brand to need one solution over the other.

Drawbacks

Cost and time to integrate the very best platforms can be a factor with this approach. This is why brands often take a hybrid approach of relying primarily on an all-in-one approach for most of their marketing work, then integrating specific best-in-breed technologies when their needs are specialized, or when results on a particular channel or tactic are mission critical.

Additionally, the best-in-breed approach can have a heavy cost in terms of onboarding and the learning curve for team members. Because they will need to learn how several different systems work, it can often be more difficult and time-consuming than learning a single all-in-one platform.

Composable

A composable approach to software integration is based on the idea of building applications from reusable and modular components that can be easily connected and configured. This differs from traditional software integration approaches that rely on custom code, complex middleware, or rigid architectures to integrate different systems and data sources.

Benefits

A composable approach offers several benefits over traditional approaches:

- *Faster time to market*: Composable components can be quickly assembled and deployed to meet changing business needs, without requiring extensive coding or testing
- *Greater flexibility and scalability*: Composable components can be easily reconfigured, replaced, or added to support new functionality, data sources, or user interfaces
- *Lower cost and risk*: Composable components reduce the dependency on specialized skills, vendor lock-in, and technical debt that often result from traditional integration approaches

Drawbacks

Similar to best-of-breed approaches, the composable approach requires software engineers in order to integrate and incorporate functionality, though composable can often be simpler, quicker, and easier, than integrating an entirely new platform.

While utilizing more all-in-one or monolithic platforms can cause organizations to accumulate "technical debt" or a growing amount of workarounds and software patches or one-off solutions that make it easy to perform upgrades or even to tell the source of some issues, composable approaches have their own equivalent of this.

When an organization has too many composable platforms that are integrated with one another, a phenomenon known as "integration tax" can occur each time a change needs to main. This refers to the cost of time and resources to ensure all of the integrated composable systems continue to function in expected ways once a change has occurred or a new composable platform is added to the infrastructure.

OPEN SOURCE AND PROPRIETARY

Organizations must decide between proprietary software investment, open source adoption, or a hybrid approach when creating their MarTech infrastructure. Different approaches to MarTech infrastructure development offer unique benefits and drawbacks that influence the scalability, flexibility, cost-effectiveness, and long-term viability of the technology stack.

Proprietary software consists of applications that companies develop and distribute through strict licensing terms. Users purchase a license to operate proprietary software without receiving access to the underlying source code. Software vendors maintain responsibility for software updates and provide customer support as well as security patches. The MarTech industry features prominent software solutions such as Adobe Experience Cloud (including Adobe Experience Manager), Salesforce Marketing Cloud, and Oracle Eloqua.

Open source software allows users to access source code publicly to enable modifications and distribution according to their requirements. While open source solutions depend on community support they often receive commercial support packages from vendors. Marketing solutions such as Mautic for marketing automation, Pimcore for DXP and PIM functionality, and Matomo for Web analytics serve as examples of open source software in the marketing field.

Proprietary Software

Proprietary platforms provide multiple benefits particularly for businesses that need ready-made solutions and extensive vendor support. Key benefits include the following:

- *Out-of-the-box functionality*: Proprietary systems deliver powerful features straight from the box that need little modification for deployment
- *Vendor support and SLAs*: Organizations gain advantages from formal support agreements because they provide guaranteed service levels alongside bug fixes and security updates
- *Easier compliance management*: Vendors usually provide built-in features for compliance with regulations such as the GDPR and CCPA, which helps reduce the workload for internal legal and IT departments
- *Integrated ecosystems*: Vendors who hold proprietary software frequently offer integrated suites that allow seamless data transfer between platforms (Adobe Experience Cloud and Salesforce Marketing Cloud serve as examples)

Drawbacks

Organizations need to evaluate the limitations of proprietary systems despite their advantages. Some of these are as follows:

- *High costs*: Large enterprises with complex use cases face significant costs from licensing fees. The costs of proprietary systems rise as data volume grows, along with increased usage and additional features.
- *Vendor lock-in*: Proprietary systems require deep integration, which creates both technical and financial barriers for vendor switching or data migration later on.
- *Limited customization*: Proprietary platforms offer configuration options yet control access to their source code, which prevents organizations from fully tailoring the software to their distinct business requirements.
- *Dependency on vendor roadmap*: The timetable for feature development and innovation follows the vendor's schedule, which might not match an organization's priorities.

Open Source Software

Organizations seeking customizable solutions or wanting to eliminate vendor dependency find open source platforms advantageous because they provide both flexibility and potential cost savings. Some of the benefits include the following:

- *Cost-effectiveness*: Open source software eliminates the licensing expenses organizations must pay for hosting, customization, and support services
- *Customization and flexibility*: Organizations that access source code can modify software according to their unique workflows and industry needs
- *Avoid vendor lock-in*: With open source solutions, organizations achieve independence from a single vendor enabling greater command over the platform's future development and lifespan
- *Active communities and rapid innovation*: Open source platforms gain advantages from active developer communities that deliver ongoing updates alongside new features and integrations

Drawbacks

Open source solutions may appear advantageous but organizations must carefully assess possible drawbacks when adopting them, due to some potential drawback, including the following:

- *Increased internal responsibility*: Organizations need to manage updates and security patches while maintaining platform stability unless they partner with a commercial open source vendor.
- *Skill and resource requirements*: To successfully implement open source software, organizations need specialized technical knowledge for development and maintenance activities.
- *Support variability*: Communities can provide assistance but often fall short in response speed and service guarantees compared to formal SLAs. Some projects may experience slower issue resolution.
- *Integration complexity*: Integrating open source solutions into your MarTech stack often demands custom development to create the necessary system connectors that are not available as pre-built options.

The choice between proprietary and open source MarTech platforms hinges on an organization's business priorities together with its budget constraints and internal expertise, as well as its long-term strategic objectives. Organizations frequently blend proprietary tools' stability and support features with open source solutions' flexibility and cost-effectiveness to form a hybrid approach. A thorough assessment of both proprietary and open source options remains essential for creating a reliable and enduring MarTech infrastructure.

Hybrid Scenarios

Of course, in many cases, a company may choose open source in some areas and proprietary in others. Most organizations should evaluate both open source and proprietary software options as they often find value in a combination approach. Combining both platform types into a single MarTech stack through a hybrid approach results in cost-effective flexibility and open source innovation while ensuring proprietary system stability and enterprise-level support.

Why Organizations Choose Hybrid

Best-of-Breed Functionality

Marketers find a hybrid model advantageous because it lets them pick top tools from the proprietary and open source worlds. A company may choose Adobe Experience Manager as its proprietary digital experience platform (DXP) to handle complex Web experiences and select Apache Unomi as its open source customer data platform (CDP) to achieve flexible customer data management.

Flexibility with Critical Support Where Needed

Open source solutions provide customization options for strategic differentiation, including personalized customer experiences and specialized workflows, while proprietary platforms deliver secure and supported solutions for essential tasks such as CRM and analytics.

Cost Optimization

Organizations save licensing expenses by using open source platforms for non-essential functions while keeping enterprise-level performance in mission-critical tasks through proprietary systems.

Considerations for Hybrid Models

Integration Strategy

The successful functioning of a hybrid stack depends on a clear integration strategy that enables smooth data exchange between open source and proprietary systems. Middleware solutions together with APIs and Integration Platform as a Service (iPaaS) systems create essential functions for this area.

Governance and Security

The governance and security aspects become more complex when managing a hybrid stack. Security and compliance protocols for handling data and granting user access need to be implemented throughout open source and proprietary systems.

Vendor and Community Management

Organizations need to maintain equilibrium between proprietary vendor agreements such as contracts and SLAs and open source community participation through contributions and updates. This requires proactive engagement on both fronts.

CONCLUSION

As demonstrated in this chapter, there are a number of considerations to make when planning and designing a MarTech infrastructure, and these choices can have long-term benefits or recurring challenges.

In the next chapter, we will explore the growing role that AI has on the MarTech stack today and in the future.

CHAPTER 8

AI AND MARTECH

The visibility of AI related to marketing has grown since AI gained prominence with the introduction of OpenAI's ChatGPT on November 30, 2022, which was the most successful software launch of its time, attracting over one million users within 5 days[1] and reaching 100 million monthly active users in January 2023, just two months after launch[2].

As of late 2024, 90% of respondents of a study of business to consumer (B2C) marketers by Invoca said they would be increasing their investments in AI platforms and tools, with 91% saying they would have a 2025 budget dedicated to AI tools, increasing by 5% from the previous year[3].

This chapter will review at a high level the areas where AI is making an impact on MarTech, but in the rest of the book, AI will not be discussed as something separate from MarTech, but instead something that is integrated into the marketing process as well as the MarTech infrastructure.

AI's application to MarTech is far-reaching and as diverse as the different applications that AI can take, including machine learning approaches, predictive analytics, and the more recently popular generative AI. The rest of this chapter will explore several of these applications.

GENERATIVE AI

Generative AI refers to AI models capable of creating original content, such as text, images, videos, and audio. These tools use advanced algorithms such as generative adversarial networks (GANs) or large language models (LLMs) to produce content based on patterns and examples from existing data.

In marketing, generative AI is leveraged to create compelling copy, design visuals, develop personalized email campaigns, and even produce video content tailored to specific audiences.

Generative AI has become a game-changer for marketers because it dramatically reduces the time and effort required to produce high-quality, engaging content. Marketers often face tight deadlines and resource constraints, and generative AI tools enable them to scale their content production efforts efficiently without compromising quality. By generating personalized, audience-specific materials, marketers can connect with their target demographic in a more meaningful way, boosting engagement and conversions.

AI excels at this task because it can analyze vast amounts of data to identify trends, preferences, and patterns that resonate with audiences. For example, a generative AI model can create social media posts aligned with a brand's tone or design visuals that match audience preferences. Additionally, these tools can iterate rapidly, producing multiple variations of content that marketers can test to determine the most effective option. This blend of creativity and efficiency allows marketers to stay competitive in a fast-paced environment.

Generative AI also democratizes access to creative capabilities, enabling smaller teams or organizations to compete with larger, resource-rich competitors. For instance, a startup can use AI-generated video ads or Web site copy to create a professional and polished marketing presence without hiring a large creative team. This technology levels the playing field, making high-quality marketing content accessible to all.

Hypothetical Use Cases

Email Campaign Creation

A retail marketer uses a generative AI platform to draft personalized promotional emails for their entire customer base. The AI tailors subject lines and body copy based on individual shopping preferences and purchase history, improving open and click-through rates.

Social Media Content Generation

A food brand leverages generative AI to create multiple versions of Instagram posts for an upcoming product launch. The tool generates captions, hashtags, and visuals that resonate with different audience segments, streamlining the social media strategy.

Examples of Generative AI Platforms

Here are some generative AI platforms:

- *Jasper AI*: A tool for generating high-quality marketing copy, such as blog posts, ad headlines, and social media content.
- *Canva Magic Write*: A generative AI feature integrated into Canva that helps create visuals and written content for marketing campaigns.
- *Runway ML*: A platform for creating AI-generated videos, animations, and visual content for marketing and advertising purposes.

Generative AI empowers marketers to produce content at scale, maintain creative quality, and engage audiences with personalized messaging, making it an indispensable tool in modern marketing strategies.

PREDICTIVE ANALYTICS

Predictive analytics involves using AI to analyze historical and current data to make predictions about future events, behaviors, or trends. This type of AI leverages machine learning models and statistical algorithms to identify patterns in data and provide insights that marketers can use to make informed decisions. Predictive analytics is widely used in marketing to forecast customer behavior, optimize campaigns, predict churn, and enhance decision-making.

Predictive analytics is vital to marketers because it enables them to make proactive, data-driven decisions rather than reactive ones. By understanding what customers are likely to do next—whether it's making a purchase, unsubscribing from an email list, or responding to a specific offer—marketers can design strategies that meet customer needs at the right time. For example, a predictive model can analyze purchase history and browsing behavior to identify customers who are likely to buy a specific product, allowing marketers to send targeted promotions, create an audience segment, or provide unique one-to-one personalized content, offers, or experiences.

AI is integral to predictive analytics because it can process massive amounts of data far faster than humans can. Machine learning algorithms improve over time as they analyze more data, becoming increasingly accurate in their predictions.

These tools can evaluate customer lifetime value (CLV), predict the performance of marketing campaigns, and identify customers at high risk of churn, enabling marketers to allocate resources more effectively.

Another benefit of predictive analytics is its ability to look at past behavior, data, and similar customers' responses to similar stimuli and use those to personalize customer experiences. By anticipating individual preferences and behavior, marketers can tailor content, offers, and messaging to resonate with specific audience segments. This enhances customer satisfaction, strengthens brand loyalty, and increases ROI. Furthermore, predictive analytics can help marketers identify surfacing trends early, giving them a competitive edge by enabling faster adaptation to market changes.

Hypothetical Use Cases

Customer Retention

A subscription-based video streaming service uses predictive analytics to identify customers who are likely to cancel their subscriptions. By analyzing watch history and engagement data, the company sends personalized offers, such as free trial extensions or content recommendations, to retain these customers.

Sales Forecasting

A B2B technology company employs predictive analytics to analyze historical sales data and predict quarterly revenue. The marketing team uses these insights to adjust campaign budgets and focus efforts on the most promising leads.

Examples of Platforms That Incorporate Predictive Analytics

The following platforms incorporate predictive analytics:

- *Salesforce Einstein Analytics*: A tool integrated with Salesforce CRM that uses AI to provide predictive insights into customer behavior and campaign performance
- *Adobe Sensei*: Adobe's AI-powered analytics platform that predicts customer actions and delivers insights to improve marketing efforts
- *Google Analytics 360*: Advanced analytics capabilities that allow marketers to predict trends, forecast traffic, and optimize customer journeys based on data

Using machine learning algorithms, historical behavior, and other data, predictive analytics empowers marketers to stay ahead of customer needs, optimize resources, and drive greater campaign efficiency, making it an indispensable component of any data-driven marketing strategy.

WORKFLOW AUTOMATION

Although marketers are not alone in engaging in plenty of busywork throughout their days, this is an area where AI can help marketing teams tremendously. Workflow automation refers to the use of AI-powered tools to automate repetitive and time-consuming tasks within marketing processes. These tasks can include scheduling social media posts, triggering email sequences based on customer behavior, and even assigning leads to sales teams. By automating workflows, marketers can streamline their operations, reduce errors, and free up time to focus on their core areas of focus: strategic initiatives, creativity, customer engagement, or other areas.

Workflow automation is crucial for marketers because it allows them to achieve efficiency and scale without increasing operational overhead. Marketing teams—at companies large or small—often juggle multiple tasks, channels, and campaigns simultaneously. Automating processes such as lead nurturing, content distribution, and performance reporting ensures that these tasks are executed consistently and on time. This reliability enhances campaign effectiveness and reduces the likelihood of errors.

AI is a critical enabler of workflow automation because it allows these tools to go beyond simple "if-this-then-that" logic. AI-powered platforms can analyze data, predict outcomes, and dynamically adjust workflows in real time. For instance, if a customer's engagement drops, an AI tool can trigger a re-engagement email campaign without human intervention. This adaptability ensures that marketing workflows remain relevant and impactful, even as customer behavior evolves.

Automation also helps marketers maintain agility in fast-paced environments. By reducing the manual workload, teams can quickly adapt to changes in strategy or market conditions. Additionally, automated workflows ensure consistency across campaigns and channels, strengthening brand messaging and improving the customer experience. Workflow automation doesn't just save time—it enhances the overall quality of marketing efforts.

Hypothetical Use Cases

Lead Nurturing

A SaaS company uses workflow automation to send personalized email sequences to prospects based on their stage in the sales funnel. When a lead downloads a whitepaper, the system automatically triggers follow-up emails with related resources, nurturing them toward a demo request.

Social Media Scheduling

A retail brand leverages workflow automation to schedule and post content across multiple social media platforms. The tool automatically adjusts post timing based on audience engagement trends, maximizing visibility and interaction.

Example Software Platforms That Utilize Workflow Automation

The following platforms use workflow automation:

- *HubSpot*: A comprehensive marketing automation platform that enables workflows for lead nurturing, email marketing, and campaign management
- *Zapier*: A tool that connects different apps and automates workflows between them, such as syncing CRM data with email platforms or project management tools
- *ActiveCampaign*: An automation platform specializing in email workflows, customer segmentation, and lead scoring for personalized marketing

Workflow automation empowers marketing teams to operate more efficiently, maintain consistency, and focus on strategic priorities. While this may be the most established of the uses of AI among marketing teams, with a recent focus on AI-enhanced features by many commonly used platforms, there continue to be new enhancements within the MarTech stack to continually improve the benefits that workflow automation can bring to marketing operations.

PERSONALIZATION

Modern consumers expect companies to deliver tailored interactions and experiences, with 71% sharing this sentiment[4]. Some elements of personalization have existed since the earliest days of marketing, such as addressing a letter to an individual by replacing their first name or sending specific offers to individuals in an email or direct mail database.

AI-driven personalization takes this idea further—tailoring marketing messages, content, and offers to individual customers based on their preferences, behaviors, and needs—using data from multiple touchpoints—such as browsing history, purchase behavior, and demographic information—to deliver highly relevant and individualized experiences at scale. From personalized product recommendations to dynamic email content, personalization aims to make customers feel understood and valued by the brand.

Personalization is a critical component of effective marketing because customers now expect tailored interactions from brands. Generic, one-size-fits-all messaging no longer resonates in today's highly competitive marketplace. AI-driven personalization allows marketers to engage customers in meaningful ways, increasing the likelihood of conversions, customer satisfaction, and loyalty. By delivering the right message at the right time, marketers can create a stronger emotional connection with their audience.

AI is indispensable for personalization because it enables real-time data processing and decision-making. Machine learning algorithms analyze customer data to uncover patterns and predict future behavior. For example, an AI system might identify that a customer frequently purchases certain types of products and recommend similar items or complementary products. Personalization also extends to predictive capabilities, such as suggesting when a customer might need to reorder a product or tailoring offers to align with a customer's anticipated needs.

Furthermore, AI-driven personalization can operate at scale, making it feasible for businesses to deliver unique experiences to thousands or even millions of customers. From dynamic Web site content to individualized ads, personalization powered by AI ensures that every touchpoint feels relevant and aligned with the customer's journey. This level of engagement not only improves ROI but also strengthens long-term relationships with customers.

Hypothetical Use Cases

E-commerce Product Recommendations

An online clothing retailer uses AI to analyze browsing history and purchase data, recommending outfits tailored to each customer's style preferences. For instance, a customer who frequently buys casual wear might see personalized suggestions for new t-shirts and sneakers.

Dynamic Email Content

A travel agency sends personalized email offers based on a customer's past vacation preferences. A customer who recently booked a beach getaway receives offers for similar destinations, while another who booked a ski trip sees winter resort deals.

Example Software Platforms That Utilize Personalization

The following platforms utilize personalization:

- *Dynamic Yield*: A personalization platform that tailors Web site content, product recommendations, and messaging based on real-time customer behavior
- *Segment*: A customer data platform (CDP) that consolidates customer data and powers personalized experiences across multiple channels
- *Adobe Target*: A personalization and testing tool that enables marketers to deliver tailored content, offers, and experiences on Web sites, apps, and more

Personalization is essential for modern marketers aiming to meet rising customer expectations. By leveraging AI to deliver highly relevant and timely experiences, businesses can improve engagement, foster loyalty, and drive revenue growth.

SYNTHETIC PERSONAS AND RESEARCH

Performing market research can be prohibitively expensive for many organizations, putting them at risk of planning products and campaigns that may not resonate effectively with their potential audiences. Synthetic personas are AI-generated representations of hypothetical customers, built using aggregated and anonymized data from existing customer profiles or market research. These personas are designed to mimic the behaviors, preferences, and decision-making processes of real consumers, providing marketers with valuable insights into customer needs and expectations. While many of these platforms have yet to be proven to be as effective as real-world research studies, they have the potential to save market research teams considerable time and resources.

The practice of synthetic research refers to the use of AI to simulate scenarios and test strategies in a controlled, virtual environment, allowing marketers to evaluate potential outcomes without deploying resources in the real world.

Synthetic research and personas are becoming increasingly important for marketers seeking to understand and predict customer behavior. Traditional persona development relies on qualitative data and assumptions, which may not fully capture the complexity of modern consumer interactions. AI-generated personas, on the other hand, are built on large-scale, data-driven insights, making them more accurate and dynamic. These personas evolve as new data is introduced, allowing marketers to stay aligned with changing customer behaviors.

AI plays a crucial role in creating synthetic personas by analyzing patterns in demographic, psychographic, and behavioral data. For example, AI can identify common traits among high-value customers, such as purchasing habits, preferred channels, and pain points, and generate personas that represent these groups. Synthetic research uses similar AI techniques, such as predictive modeling and simulations, to test marketing strategies. For instance, AI can simulate how different audience segments might respond to a new product launch or campaign, helping marketers refine their approach before going live.

These tools allow marketers to test strategies, optimize campaigns, and develop products with reduced risk and cost. By using synthetic personas and research, businesses can better allocate resources, tailor messaging, and improve overall marketing effectiveness. This approach also enables faster decision-making, as simulations and persona analysis provide immediate feedback on potential strategies.

Hypothetical Use Cases

Campaign Testing

A fitness equipment brand creates synthetic personas representing different customer segments, such as busy professionals and at-home fitness enthusiasts. The brand uses AI to simulate how each persona would respond to variations in ad creative, helping refine messaging before launching the campaign.

Product Development

A beauty company uses synthetic research to simulate customer reactions to a new skincare product line. The AI tests scenarios where different price points, packaging designs, and ingredients are introduced, allowing the company to optimize its offering based on predicted consumer preferences.

Example Synthetic Research Platforms

Here are some examples of synthetic research platforms:

- *Evidenza*: A platform that leverages AI to create synthetic personas and simulate customer interactions to assist with marketing strategy and plan development
- *Livepanel*: A tool designed to simulate research panels using data provided by customers to create synthetic personas
- *Simudyne*: A simulation platform that allows financial marketers to run large-scale synthetic research scenarios to test strategies and predict outcomes

While they are still in the early stages of maturity, synthetic personas and research have the potential to provide marketers with a powerful way to predict and adapt to customer behavior. By leveraging AI to generate detailed personas and run simulations, businesses can reduce uncertainty, improve decision-making, and develop strategies that resonate with their audiences.

AI-POWERED CUSTOMER SUPPORT

Consumers have become increasingly impatient when interacting with brands or requesting support. AI-powered customer service refers to the use of AI technologies such as chatbots, virtual assistants, and voice recognition tools to assist customers in real time, eliminating hold-time issues and providing the potential to reduce frustration at the time it takes to make a solution.

These tools leverage natural language processing (NLP) and machine learning to understand customer queries, provide answers, and escalate issues when needed. AI-powered customer support enhances the speed, efficiency, and availability of support services while reducing the workload on human customer service teams.

Customer service incorporating AI is critical for marketers because customer experience is a key differentiator in today's competitive market. Customers now expect instant responses to their questions and issues, and delayed or inadequate service can damage brand perception and loyalty. By deploying AI tools, businesses can provide 24/7 assistance, ensuring customers receive timely and accurate support at all times.

AI has the potential to excel in this domain due to its ability to process and respond to customer queries instantly. Chatbots and virtual assistants use NLP to understand natural language inputs, enabling them to handle a wide variety of customer questions. These tools can also access and analyze data from customer profiles to provide personalized responses, such as order updates or product recommendations. For example, an AI chatbot on an e-commerce Web site can instantly provide information about shipping timelines or suggest complementary products based on a customer's purchase history.

While the interactions during service or support generally fall under the realm of a customer service or technical support organization, there are potential benefits for marketers. AI-powered tools also help marketers by collecting valuable customer data during interactions. These insights can reveal common pain points, frequently asked questions, and customer preferences, which marketers can use to improve messaging, products, and campaigns. Furthermore, AI can reduce the burden on human agents, allowing them to focus on complex or high-value inquiries while the AI handles routine tasks.

Hypothetical Use Cases

E-commerce Support

A fashion retailer uses an AI chatbot to assist customers with sizing questions, shipping updates, and product availability. For instance, a customer searching for a jacket can ask the bot about stock levels and delivery options, receiving an instant, accurate response.

Post-Purchase Assistance

An electronics company deploys an AI-powered virtual assistant to help customers troubleshoot common product issues, such as setting up a new device or updating firmware. If the issue is too complex, the AI seamlessly transfers the conversation to a human agent.

Example Platforms

Here are some examples of AI-powered customer support platforms:

- *Zendesk AI*: A customer support platform that integrates AI chatbots and self-service solutions to improve response times and efficiency
- *Intercom*: A conversational AI tool that enables real-time chat, automated workflows, and personalized customer interactions
- *Ada*: A chatbot platform that automates customer support across multiple channels while providing personalized responses based on customer data

AI-powered customer support enhances customer satisfaction and loyalty by providing fast, accurate, and personalized assistance. For marketers, these tools are invaluable for improving the customer experience, collecting actionable insights, and ensuring consistent brand engagement across touchpoints.

CONTENT OPTIMIZATION AND ANALYSIS

A key component of a marketer's role is to monitor and adjust their work to continually improve the results and return on investment (ROI). Content optimization and analysis involve using AI tools to evaluate and improve the performance of marketing content across channels. These tools analyze metrics such as engagement, readability, SEO effectiveness, and sentiment to identify areas for improvement. By leveraging AI, marketers can ensure their content aligns with audience preferences, ranks highly in search engines, and drives meaningful engagement, ultimately maximizing the ROI of their content marketing efforts.

With generative AI tools making it easier to produce content in a variety of formats at the push of a button, *creating* content is not enough; it must be optimized to reach the right audience at the right time and in the right format. Content optimization and analysis are critical for marketers to stay relevant and competitive. AI excels at this by processing vast amounts of data to uncover insights that would be impossible to identify manually. For example, properly trained AI tools can analyze keyword trends, track competitor content, and recommend the ideal structure and tone to improve a blog post's SEO ranking and engagement.

AI-powered tools such as NLP algorithms also help marketers ensure that their content resonates with their audience. These tools analyze sentiment to determine whether the messaging aligns with the intended emotional impact, and they identify opportunities to refine language for clarity and persuasion. Additionally, AI can dynamically test and optimize content elements, such as headlines or visuals, ensuring maximum impact before publication.

By automating repetitive tasks such as keyword research or performance tracking, AI tools free marketers to focus on creative and strategic work. Furthermore, AI's ability to analyze data in real time allows continuous optimization, ensuring that campaigns remain effective even as audience preferences or market conditions change. This adaptability is vital for marketers looking to maximize engagement and conversions.

Hypothetical Use Cases

SEO Optimization

A tech blog uses an AI-powered content analysis tool to evaluate an article's readability, keyword density, and structure. The tool recommends replacing jargon-heavy language with simpler terms and adding internal links to improve the article's search engine ranking.

Social Media Engagement

A fitness brand analyzes the performance of its Instagram posts using an AI tool that evaluates metrics such as likes, comments, and shares. The tool identifies that posts featuring user-generated content perform better, leading the brand to focus on this strategy for future campaigns.

Examples of Platforms That Support Content Optimization

The following platforms support content optimization:

- *Clearscope*: A content optimization platform that helps improve search engine optimization (SEO) by analyzing keywords, suggesting relevant terms, and scoring content for search performance
- *Grammarly Business*: An AI-powered tool for refining grammar, tone, and clarity in written content to ensure it resonates with target audiences
- *BuzzSumo*: A platform for analyzing the performance of content across social media and identifying trending topics and best-performing formats

Content optimization and analysis enable marketers to create impactful, data-driven content that connects with their audience and achieves strategic goals. By leveraging AI, marketers can refine their efforts in real time, ensuring that every piece of content delivers maximum value and relevance.

AUDIENCE TARGETING AND SEGMENTATION

Audience segmentation is nothing new to marketers, who have been dividing a broader audience into smaller, more specific groups based on characteristics such as demographics, behavior, preferences, or purchase history for decades.

AI-powered audience targeting and segmentation tools enhance this process by analyzing vast datasets to uncover patterns and create highly accurate audience segments. These tools enable marketers to deliver personalized, relevant messages to the right people, maximizing the effectiveness of campaigns and improving overall engagement.

The pivotal role that AI plays in this process is to analyze data far beyond what humans can manage manually. Machine learning algorithms can process historical data, real-time behavior, and even external factors such as seasonality to identify nuanced audience segments. For example, an AI tool might identify a segment of customers who frequently purchase during sales events and another group that buys full-priced items regularly. These insights allow marketers to develop targeted strategies for each group, such as exclusive sale notifications for one segment and early access to new arrivals for the other.

Additionally, AI enables real-time segmentation, meaning marketers can adjust campaigns dynamically based on customer behavior. For instance, if a customer abandons their shopping cart, AI can automatically place them in a re-engagement segment and trigger a personalized email to incentivize completion. This level of precision and adaptability helps marketers maximize ROI while creating a more tailored and enjoyable customer experience.

Hypothetical Use Cases

E-commerce Campaigns

A home goods retailer uses AI to segment customers based on past purchases and browsing behavior. One segment receives targeted ads for kitchenware, while another receives personalized offers on furniture, increasing engagement and sales.

B2B Lead Nurturing

A software company uses AI-powered segmentation to divide its leads into small business owners, mid-sized firms, and enterprise clients. Each group receives tailored content and follow-ups, such as whitepapers or demos, aligned with their unique needs and challenges.

Example Platforms

Here are some examples of platforms for audience targeting and segmentation:

- *Segment*: A CDP that collects, unifies, and segments audience data for targeted marketing campaigns
- *BlueVenn*: A platform that provides advanced audience segmentation and predictive analytics to personalize marketing efforts
- *Salesforce Marketing Cloud*: A robust tool that uses AI-driven segmentation to enable personalized messaging across email, social, and other channels

By leveraging AI for audience targeting and segmentation, marketers can maximize the relevance of their messaging, improve campaign performance, and enhance customer satisfaction. AI-powered segmentation ensures that marketing efforts are not only efficient but also tailored to meet the unique needs of diverse customer groups.

ADVERTISING OPTIMIZATION AND MEDIA BUYING

With global advertising spending estimated to be $785.9 billion in 2025, the industry has seen consistent growth over the previous four years, with digital advertising taking up approximately 60.9% of the total[5].

Using AI to enhance the performance of digital advertising campaigns and streamline the ad placement process allows AI-based platforms to analyze performance metrics, bidding strategies, audience engagement, and market trends to determine the most effective ad formats, channels, and budgets. This ensures that marketers achieve maximum ROI by delivering the right ad to the right audience at the right time.

Ad optimization and media buying are critical for marketers because advertising budgets need to be spent efficiently to generate the best results. With the increasing complexity of digital advertising, including multiple platforms, formats, and audience preferences, manual optimization is not sufficient to achieve peak performance. AI provides marketers with the tools to automate and refine their campaigns, saving time and reducing the risk of human error.

AI-powered ad platforms use real-time data to adjust bids dynamically, ensuring that budgets are allocated to the most promising opportunities. For example, AI can detect when a particular audience segment responds better to video ads and automatically increase spending on those formats while reducing spending on less effective ones. Predictive analytics also help marketers forecast which strategies will deliver the highest ROI based on historical performance data.

Additionally, AI can identify patterns in audience engagement, such as the times of day when users are most likely to interact with ads, and adjust delivery accordingly. It also helps with programmatic media buying, where AI-based algorithms and systems automate the purchasing and placement of ads, often within milliseconds. This ensures that marketers can secure optimal placements at competitive prices, staying ahead in fast-paced digital advertising environments.

Hypothetical Use Cases

Dynamic Ad Bidding

A travel company uses an AI-powered programmatic ad platform to bid on ad placements in real time. The platform analyzes engagement data and increases bids for ads targeting vacation planners during peak hours, while reducing bids during off-peak times, optimizing their ad spend.

Creative Optimization

A gaming company leverages AI to test multiple versions of its ad creative, such as different headlines, visuals, and call-to-actions. The AI identifies that ads featuring user-generated gameplay videos perform best and reallocates the campaign budget to emphasize those formats.

Example Platforms

Here are some examples of advertisement optimization platforms:

- *Google Ads*: Offers AI-driven tools for ad bidding, audience targeting, and campaign optimization to maximize performance

- *The Trade Desk*: A programmatic advertising platform that uses AI to optimize ad placements and bidding strategies across multiple channels
- *AdRoll*: A platform that leverages AI to personalize and optimize display, social, and email ads for better performance and ROI

Ad optimization and media buying powered by AI enable marketers to deliver more effective campaigns while making the best use of their advertising budgets. By leveraging real-time insights and automated decision-making, businesses can achieve greater precision, scalability, and impact in their advertising efforts.

BRAND MONITORING AND SENTIMENT ANALYSIS

Brand perception is critical to many aspects of marketing success, as it directly influences customer trust, loyalty, and purchasing decisions—not to mention word-of-mouth recommendations, reviews, and more. AI-powered brand monitoring tools allow marketers to stay informed about how their brand is being discussed online, enabling them to identify opportunities to amplify positive sentiment and address negative feedback before it escalates.

Brand monitoring and sentiment analysis involve the use of AI tools to track mentions of a brand across digital platforms and analyze the tone of customer conversations. These tools collect and evaluate data from social media, online reviews, news outlets, and other channels to gauge public opinion and identify trends in sentiment, such as positive, neutral, or negative feedback. This helps marketers understand how their brand is perceived and respond effectively to customer sentiment.

AI is particularly valuable in sentiment analysis because it can process vast amounts of unstructured data from diverse sources, such as social media posts, customer reviews, and forums, in real time. NLP enables these tools to understand the context, tone, and nuances of customer feedback, even detecting sarcasm or mixed emotions. For example, AI can flag a sudden surge in negative mentions of a brand's product, helping marketers investigate and address the issue proactively.

By providing actionable insights, AI empowers marketers to make data-driven decisions that enhance brand reputation. For instance, identifying the attributes customers love most about a product can inform future campaigns, while addressing common complaints can improve customer satisfaction and reduce churn. These tools also help marketers measure the impact of campaigns on public sentiment, ensuring their efforts align with customer expectations and brand values.

Hypothetical Use Cases

Crisis Management

A beverage company detects a spike in negative sentiment after a product recall announcement. AI tools identify recurring complaints on social media, enabling the company to release a targeted apology and update about the resolution, mitigating further backlash.

Campaign Impact Analysis

A fashion brand launches a sustainability-focused ad campaign and uses sentiment analysis to gauge customer reactions. AI identifies that the campaign is resonating well with younger audiences, prompting the brand to expand the campaign's scope on platforms frequented by that demographic.

Example Platforms

Here are some examples of brand monitoring and sentiment analysis platforms:

- *Brandwatch*: A social listening platform that uses AI to monitor brand mentions and analyze sentiment across digital channels
- *Sprinklr*: A customer experience management platform that includes sentiment analysis for monitoring brand reputation and customer feedback
- *Hootsuite Insights*: An AI-powered social media listening tool that tracks brand mentions and evaluates sentiment trends in real time

Utilizing AI-based tools for brand monitoring and sentiment analysis provides marketers with the insights needed to protect and enhance their brand reputation.

CROSS-CHANNEL ATTRIBUTION

Modern customer journeys are rarely linear, as consumers interact with multiple channels both online and offline before converting. This makes it difficult for marketers to continuously improve their efforts because without proper attribution, it can be challenging to understand which efforts are driving results. AI-powered attribution models provide a clear picture of the customer journey, stopping marketers from over-investing in low-performing channels or underestimating the value of high-impact touchpoints.

Cross-channel attribution involves using AI to analyze and understand the impact of various marketing touchpoints on a customer's journey, from initial awareness to final conversion. It helps marketers determine which channels (e.g., social media, email, paid ads, or direct Web site traffic) contributed most to achieving a specific goal, such as a sale or sign-up. AI-powered attribution models, such as multi-touch attribution, assign value to each interaction, enabling marketers to optimize their strategies and allocate resources more effectively.

AI plays a pivotal role by processing complex data from multiple sources to uncover insights that would be impossible to analyze manually. Machine learning algorithms evaluate customer interactions in real time, identifying patterns and assigning value to each touchpoint. For instance, AI can recognize that a customer who clicked a Facebook ad, visited the company's Web site, and received a retargeting email ultimately converted because of the combined effect of these interactions.

With AI-driven cross-channel attribution, marketers can better allocate budgets, refine messaging, and align campaigns to optimize performance. Additionally, these tools can provide predictive insights, such as forecasting the effectiveness of future campaigns based on historical data. By using AI for attribution, marketers can move beyond gut instincts and make data-driven decisions that improve ROI and drive business growth.

Hypothetical Use Cases

E-commerce Analysis

An online retailer uses AI-driven cross-channel attribution to evaluate the impact of their Black Friday campaigns. The AI identifies that social media ads create initial interest, email campaigns drive repeat visits, and paid search ads lead to final conversions, enabling the retailer to allocate more budget to these high-performing channels.

Event Promotion

A software company promoting a webinar uses attribution tools to track interactions across LinkedIn ads, email invites, and blog posts. The analysis reveals that blog posts are the most effective for generating registrations, prompting the company to increase their blog content production for future events.

Example Platforms

Here are some examples of cross-channel attribution platforms:

- *Google Analytics 360*: Provides advanced attribution modeling to track the performance of marketing campaigns across multiple channels
- *Adobe Analytics*: Offers AI-driven cross-channel attribution and predictive analytics to help marketers understand and optimize customer journeys
- *Ruler Analytics*: Specializes in multi-touch attribution, connecting marketing touchpoints to revenue to provide a clear view of ROI

Cross-channel attribution allows marketers to understand how their efforts across multiple channels work together to drive conversions. By leveraging AI, businesses can optimize their strategies, improve resource allocation, and achieve greater campaign success with precise and actionable insights.

VISUAL RECOGNITION AND IMAGE PROCESSING

Understanding how customers interact with visual content and leveraging those insights can significantly enhance engagement and conversion rates, making visual recognition are increasingly important in marketing as the digital landscape becomes more visually driven, with social media platforms, e-commerce sites, and video streaming services relying heavily on visuals to engage users.

Visual recognition and image processing involve the use of AI to analyze, categorize, and interpret visual content, such as photos, videos, and graphics. These tools use computer vision algorithms to identify objects, detect faces, analyze scenes, and even assess visual trends. In marketing, visual recognition helps brands gain insights into user-generated content (UGC), monitor the effectiveness of visual campaigns, and streamline creative development processes.

AI-powered tools excel at processing large volumes of visual data in real time, identifying patterns, and providing actionable insights. For example, AI can analyze customer photos tagged with a brand on social media to identify trends, such as the most popular product features or settings where the products are used. Similarly, image recognition can help marketers monitor the effectiveness of ad placements or identify how their products are being used in UGC.

AI also supports the creative process by automating repetitive tasks such as tagging and categorizing digital assets, enabling marketers to focus on strategy and innovation. For instance, AI can sort thousands of images in a digital asset management (DAM) system based on visual attributes, making it easier for creative teams to find the assets they need. Visual recognition tools also assist in identifying visual trends, helping marketers stay relevant in fast-moving markets.

Hypothetical Use Cases

UGC Insights

A clothing brand uses AI-powered visual recognition to analyze customer photos shared on Instagram. The tool identifies that customers frequently pair a particular jacket with jeans and boots, prompting the brand to create a campaign showcasing similar looks.

Ad Placement Monitoring

A beverage company leverages AI to scan social media for visual mentions of its logo in influencer content. The tool assesses how prominently the logo is featured, helping the company measure the impact of its sponsorship campaigns.

Example Platforms

Here are some examples of platforms that use visual recognition and image processing:

- *Google Vision AI*: A tool for analyzing images and detecting objects, text, and logos, useful for monitoring brand visibility in visual content
- *Clarifai*: A visual recognition platform that categorizes and analyzes images and videos for trends and insights
- *Bynder*: A DAM platform with AI-powered tagging and categorization features to streamline the management of digital assets

Visual recognition and image processing tools enable marketers to unlock valuable insights from visual data, optimize creative workflows, and monitor brand presence across digital channels. By leveraging AI, businesses can harness the power of visuals to engage customers and drive impactful marketing strategies.

CUSTOMER JOURNEY ORCHESTRATION

Similar to the need to measure customers' behaviors across channels, customer journey orchestration is critical for marketers because customers today interact with brands across numerous channels, often in unpredictable ways. Ensuring a cohesive experience across these touchpoints is essential for building trust, satisfaction, and loyalty. Without effective orchestration, customers may encounter disjointed or irrelevant interactions, leading to frustration and missed opportunities.

Customer journey orchestration involves using AI methods such as automation and predictive analytics to manage and optimize the interactions customers have with a brand across multiple touchpoints. AI-powered tools enable marketers to create seamless, personalized experiences by analyzing customer behavior in real time and dynamically adjusting messaging, offers, and content based on each individual's journey. These tools help ensure that every step of the customer's experience is aligned with their preferences and needs, from initial awareness to post-purchase engagement.

AI excels at orchestrating customer journeys because it can process large volumes of data in real time and predict customer intent based on behavior. For instance, AI can recognize when a customer who abandoned their shopping cart revisits the Web site and dynamically triggers a personalized discount offer to encourage conversion. AI-powered journey orchestration platforms also use predictive analytics to anticipate the next steps in a customer's journey, such as sending an email with related products after a purchase or directing customers to relevant content on social media.

These tools allow marketers to automate and refine the customer experience while maintaining personalization at scale. By providing a consistent and tailored journey, marketers can improve customer satisfaction, increase conversions, and foster long-term relationships. Additionally, journey orchestration helps optimize resource allocation by focusing efforts on high-value touchpoints that drive engagement and revenue.

Hypothetical Use Cases

Retail Personalization

A fashion retailer uses an AI-powered journey orchestration platform to deliver personalized experiences across channels. For example, a customer browsing sneakers on the Web site receives a follow-up email featuring similar styles, and if they don't convert, they later see a targeted Instagram ad with a limited-time discount.

Subscription Re-Engagement

A streaming service uses AI to orchestrate journeys for lapsed subscribers. Based on past viewing habits, the system sends a personalized email featuring new releases in genres the customer enjoyed, followed by a discounted re-subscription offer.

Example Platforms

- *Adobe Journey Optimizer*: An AI-powered platform for managing and personalizing customer journeys across multiple channels in real time
- *Medallia Experience*: A journey orchestration tool that uses AI to deliver contextually relevant experiences across customer touchpoints
- *Pega Customer Decision Hub*: Uses automation and predictive analytics to provide the next best action (NBA) for customers interacting with channels that utilize the platform

Customer journey orchestration helps marketers provide seamless, tailored experiences that build trust and drive conversions. By leveraging AI to analyze and act on customer data in real time, businesses can optimize every stage of the customer journey, ensuring that each interaction feels relevant and valuable.

CONCLUSION

There are a multitude of ways that AI can be used to enhance and improve the customer experience as well as the return on investment from marketing teams. This chapter has explored several of these ways, though more use cases continue to arise.

In addition to considerations about AI, marketers and the companies they support need to make critical decisions about how their MarTech platforms are implemented, hosted, and served. In the next chapter, the considerations that companies need to make before making MarTech investments will be explored.

NOTES

1. Marr, B. (2023, May 19). *A short history of ChatGPT: How we got to where we are today*. Forbes. https://www.forbes.com/sites/bernardmarr/2023/05/19/a-short-history-of-chatgpt-how-we-got-to-where-we-are-today/

2. Reuters. (2023, February 1). *ChatGPT sets record for fastest-growing user base: Analyst note*. Reuters. https://www.reuters.com/technology/chatgpt-sets-record-fastest-growing-user-base-analyst-note-2023-02-01/

3. Invoca. (2024). The State of AI in B2C Digital Marketing Report. Invoca. https://cdn.prod.website-files.com/5d82e225060d003d65ddae98/67606c157ca7ffc0305e5419_The_State_of_AI_in_B2C_Marketing_Report-US-v2-compressed.pdf

4. McKinsey & Company. (2021, November 17). *The value of getting personalization right—or wrong—is multiplying*. McKinsey & Company. https://www.mckinsey.com/capabilities/growth-marketing-and-sales/our-insights/the-value-of-getting-personalization-right-or-wrong-is-multiplying

5. Dentsu. (2024, May 29). *Ad spend growth tracks ahead of the economy*. Dentsu. https://www.dentsu.com/news-releases/ad-spend-growth-tracks-ahead-of-the-economy

PART 2

CUSTOMER DATA

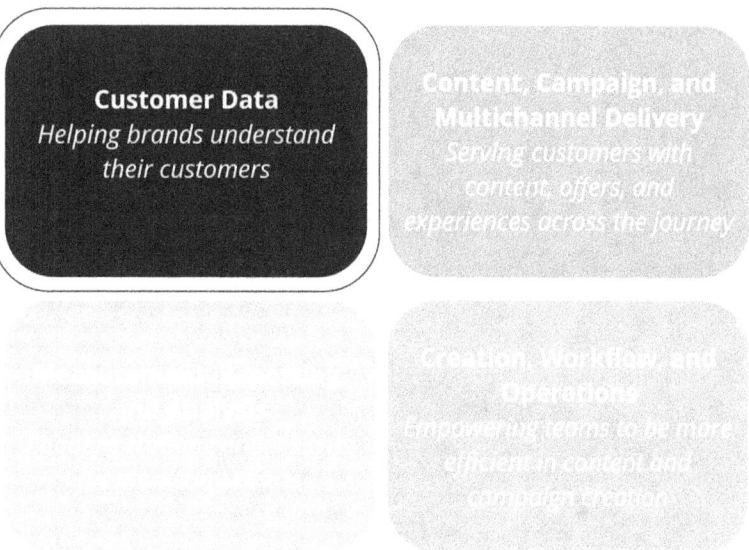

FIGURE P2.1 Customer data: Helping brands understand their customers

This part of the book will discuss the first of the four categories of MarTech platforms introduced in Chapter 5: customer data. Platforms in this category provide the potential to help businesses to better understand their customers and include platforms such as customer relationship management (CRM) and customer data platforms (CDPs).

Additionally, considerations around customer data, from quality and relevance to privacy and compliance, will be discussed.

CHAPTER 9

Key Customer Data Considerations

As most marketers are already aware, customer data is the lifeblood of building effective marketing efforts, from initial customer awareness through retaining loyal customers, and understanding those customers through the entire customer lifecycle. Access to robust customer data enables businesses to understand their audiences, personalize experiences, and measure the success of their campaigns.

However, as the volume, variety, and velocity of data continue to grow, managing customer data effectively has become increasingly complex. This can be due to many factors, such as a variety of marketing and communication channels involved, to different teams within an organization communicating, to the sheer number of software platforms with their own data sources involved in key steps in the process. Thus, having the right MarTech infrastructure is essential to harnessing the full potential of customer data while navigating challenges such as privacy regulations, data integration, and quality assurance. This chapter will explore the critical considerations marketers must keep in mind when evaluating and optimizing their customer data strategies within their MarTech stack.

From understanding the types of customer data and their ownership to addressing privacy concerns and data integration challenges, marketers must build a strong foundation for leveraging data effectively. As consumers demand greater transparency and personalized experiences, businesses must strike a balance between using data ethically and maximizing its strategic value. By exploring the nuances of data ownership, privacy, storage, and access, this chapter is intended to equip marketers with the insights they need to design a robust customer data strategy that supports both their organizational goals and their customers' expectations.

UNDERSTANDING DATA OWNERSHIP

Data ownership is a critical consideration for marketers because it directly impacts how customer data is collected, used, and maintained. Knowing who owns the data and how it's sourced not only helps businesses comply with regulations but also builds trust with customers. In marketing, data ownership refers to the control a business has over the data it collects, processes, and analyzes. Different types of customer data offer varying levels of ownership and strategic advantages. By understanding the nuances of these data types, marketers can make informed decisions about their marketing strategies and MarTech investments.

Types of Customer Data

While there are other ways of understanding and labeling customer data, a particularly useful one is to categorize it based on the owner of the data within the process. This framework will be used here to provide a method of differentiating between the types of customer data.

Zero-Party Data

The first category is information that customers intentionally and proactively share with a business. Zero-party data could include details such as preferences, feedback, or interests that a customer provides through surveys, preference centers, or interactions with the brand. Unlike other data types, zero-party data is entirely permission-based and reflects the customer's willingness to engage, making it one of the most accurate and trustworthy sources for personalization.

For example, a fashion retailer might ask customers to select their preferred styles and colors during onboarding, ensuring that future communications are relevant to their tastes.

First-Party Data

First-party data (sometimes referred to as 1P) is collected directly by a business from its own channels, such as Web site activity, purchase history, customer relationship management (CRM) records, or email interactions. This type of data is highly valuable because it is accurate, privacy-compliant (when collected with consent), and specific to the organization's customers. First-party data provides actionable insights that help marketers build personalized campaigns and improve customer experiences. For instance, an e-commerce company might track browsing behavior to recommend products or offer discounts on items left in a shopping cart.

It should also be noted that, in many cases, both first-party and zero-party data are referred to under the umbrella of first-party data, specifically when referring to a company's focus on a "first-party data strategy," which often refers to the organization's intention to rely on more information collected direct from customers, rather than relying on third-party sources, some of which may be unreliable, and others that may have less than ethical practices of procuring their data. The topic of first-party data strategies will be discussed in more detail later in this chapter.

Second-Party Data

The most common source of second-party data software platforms such as social networks, e.g., Facebook, TikTok, YouTube, or X, and shared with brands that have accounts on those platforms. The terms of service of the platforms dictate the company's ability to use the data, and the functionality of the data collection reporting can also constrain the usage here.

Another form of second-party data is data shared between two organizations with a mutual agreement. This data often comes from trusted partners, such as co-marketing relationships or industry collaborations. For example, a travel agency might share data with a hotel chain to offer complementary services to overlapping customer segments. While second-party data can be highly relevant and valuable, it requires strong partnerships and clear agreements to ensure compliance with data privacy regulations and ethical practices.

Third-Party Data

Third-party data is aggregated and sold by external providers who collect information from a variety of sources, such as Web sites, social media, and apps. It is often used to supplement first-party data by providing broader audience insights, such as demographic or behavioral trends. While third-party data can be useful for market research or targeting new audiences, it comes with challenges, including concerns about accuracy, relevance, and privacy compliance. Recent restrictions on third-party cookies and increasing consumer demand for transparency have made reliance on this type of data less viable in the long term.

While third-party data has faced criticism due to concerns about privacy, accuracy, and regulatory compliance, not all third-party data is inherently bad or unethically acquired. In fact, when sourced legitimately and used ethically, third-party data can provide valuable insights that enhance marketing strategies and drive better decision-making. Reputable third-party data providers adhere to strict privacy standards, ensuring that the data they offer is anonymized, aggregated, and compliant with regulations such as the GDPR and CCPA. This allows businesses to use third-party data responsibly without compromising consumer trust or violating privacy laws.

For example, third-party demographic data purchased from trusted sources such as Nielsen or Experian can help brands understand broader market trends and target new audience segments effectively. Similarly, third-party behavioral data from industry-compliant platforms, such as social media audience analytics from Meta, can provide insights into consumer preferences and online activity that supplement a brand's own data. Another use case is geographic data from third-party providers such as Google Maps, which helps businesses optimize store locations or create hyper-localized marketing campaigns. When used to complement first- and second-party data, ethically sourced third-party data can provide a broader perspective, enabling businesses to expand their reach and refine their strategies without compromising integrity.

To get a deeper understanding of these types of data, refer to Table 8.1.

TABLE 8.1 Data ownership and examples

Data Ownership Type	Examples
Third-Party Data	- Demographic data purchased from external data providers (e.g., age, income, location) - Behavioral data collected from cookies tracking users across unrelated Web sites - Audience segmentation data from programmatic advertising platforms
Second-Party Data	- Social media follower information on a platform such as Facebook, Instagram, or YouTube - Travel data shared between a hotel chain and an airline partner to offer joint promotions - Retail purchase data shared between a grocery store and a product manufacturer for co-marketing campaigns

(*Continued*)

Data Ownership Type	Examples
First-Party Data	• Web site analytics data capturing user behavior, such as pages visited and time spent on the site • Purchase history from an e-commerce platform, including products bought and transaction details • Email engagement data, such as open rates, click-throughs, and unsubscribes
Zero-Party Data	• Customer preferences collected through an onboarding quiz (e.g., favorite styles or colors for a fashion retailer) • Feedback from a customer survey about product satisfaction or future interests • Explicitly shared wish lists or goals, such as fitness objectives provided to a gym or health app

Understanding these types of customer data and their unique characteristics helps marketers make strategic decisions about which data to prioritize, how to collect it ethically, and how to integrate it into their MarTech infrastructure. Prioritizing zero-party and first-party data, in particular, positions businesses for long-term success in a privacy-conscious and consumer-driven market.

Benefits of a First-Party Data Strategy

Despite the ongoing benefits of some third-party data sources, a primarily first-party data strategy has emerged as a key strategic goal for many organizations in an increasingly privacy-conscious and customer-driven world. Unlike third-party data, which often comes from external sources and can be inconsistent, first-party data—or zero-party data—is directly collected by the brand, ensuring higher levels of accuracy and relevance due to its recency and the directness of the collection process. By investing in a robust first-party data strategy, businesses gain numerous advantages that support both compliance and customer centricity.

Thus, there are several benefits to a first-party data strategy, which we will discuss in the next few subsections.

Enhanced Data Accuracy and Control

First-party data is collected directly from a brand's own channels, such as Web sites, mobile apps, or email interactions. Because it originates from direct customer engagement, it is more likely to be reliable and accurate compared to third-party data.

Marketers also retain full control over how the data is collected, managed, and used, enabling them to build strategies based on precise and actionable insights. This level of control reduces the risks of inaccuracy or misinterpretations and allows businesses to confidently drive decisions.

Better Compliance with Privacy Regulations

With increasing global regulations such as the GDPR in the European Union, the CCPA and others in the United States, and others around the world, relying on first-party data is a safer path to compliance. Since this data is collected with direct customer consent, it minimizes privacy risks and ensures adherence to legal requirements. Businesses using first-party data can demonstrate transparency in their data practices, reducing the likelihood of regulatory penalties and fostering consumer trust.

Stronger Customer Trust and Loyalty

When customers understand how their data is being used and actively consent to sharing it, they are more likely to trust the brand. A first-party data strategy fosters this transparency by prioritizing consent-based data collection, and the most successful brands ask for only the information they need to provide more valuable customer experiences, offering transparent explanations of why they are asking for the information.

In turn, customers appreciate brands that value their privacy, which can lead to stronger relationships, improved retention rates, and higher lifetime value. Trust is further enhanced when brands use first-party data to deliver meaningful, personalized experiences.

Greater Potential for Personalized Marketing Efforts

Receiving tailored content is important to consumers, as according to McKinsey & Company, 71% of consumers expect personalization, and 76% of consumers get frustrated when they don't receive personalized content and offers[1]. First-party data provides marketers with the insights they need to create highly tailored campaigns. By analyzing customer behaviors, preferences, and interactions, brands can deliver personalized offers, messages, and experiences that resonate deeply with their audience. For example, a retail brand using first-party data can recommend products based on past purchases or send birthday discounts, driving engagement and conversions.

The Role Third-Party Data Plays in a First-Party Data Strategy

Despite the growing importance of first-party data, third-party data still holds value in specific contexts. Third-party data can provide a broader perspective by offering demographic, geographic, and behavioral insights that complement a brand's first-party data. This is particularly useful when targeting new customer segments or entering new markets where first-party data may be limited.

Additionally, third-party data remains beneficial for programmatic advertising and audience expansion. For example, when launching a new product, a brand might use third-party data to identify lookalike audiences or reach untapped demographics. While reliance on third-party data is decreasing due to privacy concerns and regulatory changes, it will probably continue to play a supplemental role for marketers seeking to broaden their reach or validate trends.

By strategically combining the precision of first-party data with the scope of third-party data where appropriate, marketers can create a balanced approach that maximizes insights and effectiveness while maintaining compliance and trust.

Factors Affecting Customer Data Quality

Customer data quality is critical to the success of any marketing strategy. Poor-quality data can lead to misguided campaigns, wasted resources, and damaged customer relationships. Ensuring data quality requires attention to several factors that influence how well data reflects customer behavior and supports actionable insights. By focusing on recency, hygiene, consistency, enrichment, and source reliability, marketers can maximize the value of their customer data and build more effective campaigns.

Recency and Relevance

No matter the completeness of the information, the value of customer data diminishes over time if it is not regularly updated. Recency ensures that data reflects the most current behavior, preferences, and interactions of customers. For instance, a customer's purchase behavior from three years ago may no longer be relevant to their current interests. Marketers must prioritize collecting and analyzing up-to-date data to ensure that their strategies remain aligned with customer needs. Using automation and real-time data collection tools, brands can maintain the relevance of their insights, driving better personalization and engagement.

Data Hygiene

Similar to recency and relevance, the effects of data over time can have an effect on its efficacy in other ways. Maintaining clean and accurate data is essential for reliable marketing operations. Data hygiene involves regular processes to clean, deduplicate, and validate data. This reduces errors caused by outdated information, duplicate records, or incorrect entries. For example, ensuring that a customer's email address is accurate and unique in the database can prevent sending duplicate or undeliverable emails.

Tools such as data validation software and CRM cleaning features help streamline this process, ensuring high-quality data across systems.

Consistency Across Channels

As omnichannel engagement with customers increasingly becomes the norm, customers interact with brands through various touchpoints, including email, social media, Web sites, and in-store visits.

Data consistency across these channels ensures that customer information aligns seamlessly, providing a unified view of each individual. For instance, a customer's preferences shared during a chatbot interaction should be reflected in their email communications, as well as when they walk into a brick-and-mortar retail store to buy a new product. Achieving consistency requires robust integration between systems, such as connecting a CRM with social media platforms or e-commerce tools.

Data Enrichment

Although companies may take care to plan out their data collection strategies, there are often gaps to be filled in the information they have about their customers. Data enrichment involves adding missing or supplemental information to existing customer profiles to create a more comprehensive view. For example, enhancing a basic customer record with demographic information, purchase history, or preferences enables more precise targeting and personalization.

Enrichment can be achieved through first-party data collection, partnerships with trusted data providers (which might take the form of second-party or third-party data relationships), or the use of AI-powered tools that analyze customer behavior to fill in gaps. Enriched data empowers marketers to create highly tailored campaigns that resonate with their audience.

Data Source Reliability

One often-relied-upon measure of the quality of data is that it is only as good as its source. Thus, data sourced from unreliable or unverified providers can lead to inaccuracy and damage trust, which leads many organizations towards a first-party data approach. Since solely relying on first- and zero-party data is not feasible for most organizations, evaluating and ensuring the credibility of data sources is therefore critical.

For instance, data from a reputable third-party provider with strict compliance policies is more reliable than data obtained from questionable sources. Brands should implement stringent vetting processes for third-party data and use technology to validate the accuracy of data collected from their own channels.

Additionally, the data collection practices even from first-party data sources should be regularly reviewed to ensure that there are no reliability issues such as loss of connectivity, duplication, or other potential issues.

By addressing these factors, marketers can ensure their customer data is accurate, reliable, and actionable. High-quality data not only supports better decision-making but also enhances the effectiveness of marketing efforts, ultimately driving stronger customer relationships and improved business outcomes.

DATA STORAGE AND ACCESS

The way customer data is stored and accessed plays a pivotal role in shaping an organization's marketing strategy and overall effectiveness. In an era of data-driven decision-making, businesses rely on accurate, accessible, and integrated data to deliver seamless customer experiences and optimize campaigns. As the volume and variety of customer data grow, however, managing it effectively has become increasingly complex. From siloed systems to integration challenges, businesses must navigate several hurdles to ensure their data infrastructure supports their marketing goals.

A robust approach to data storage and access enables marketers to gain a unified view of their customers, breaking down barriers between departments and systems. It empowers organizations to connect insights from multiple touchpoints, make real-time decisions, and personalize customer interactions at scale. This section explores the challenges associated with siloed data, the limitations of traditional storage solutions, and the importance of data integration in building a comprehensive and efficient MarTech infrastructure. By addressing these issues, businesses can unlock the full potential of their customer data and drive better outcomes.

Challenges with Siloed Data

Data silos occur when information is stored in isolated systems that cannot communicate with one another. These silos often develop when different teams or departments use separate tools or platforms to manage their specific data needs. For marketers, data silos pose a significant challenge by preventing the integration and accessibility of customer information, which are critical for delivering cohesive and personalized experiences. Addressing silos is essential to unlocking the full potential of MarTech and achieving cross-channel marketing success.

How Data Silos Hinder Customer Insights

Siloed data creates fragmented customer profiles, making it difficult for marketers to gain a unified view of their audience, particularly when those audiences are active across several channels both online and offline. For example, if customer interactions with email campaigns are stored in one system, while purchase history resides in another, marketers cannot connect the dots to understand the full customer journey. This lack of visibility hinders cross-channel marketing efforts, resulting in inconsistent messaging, missed opportunities for personalization, and a poor customer experience.

Data silos also prevent real-time decision-making. In today's fast-paced marketing environment, teams need to act quickly based on customer behavior and market trends. Siloed systems delay data access and processing, making it difficult to adapt campaigns or seize time-sensitive opportunities. Moreover, silos can lead to redundant or conflicting strategies when teams operate independently, wasting resources and diluting the overall impact of marketing efforts.

Examples of Customer Data Silos

CRM systems: CRMs often contain detailed records of interactions, such as sales calls, emails, and support tickets. If this data isn't integrated with marketing systems, however, campaigns may fail to account for recent customer interactions, leading to irrelevant or repetitive messaging.

Social media platforms: Social networks generate valuable engagement data, such as likes, shares, and comments. When this data remains isolated, marketers lose insights into how social interactions influence the broader customer journey or connect with other channels.

E-commerce platforms: These platforms often hold key transactional data, such as purchase history and cart abandonment rates. Without integration with other tools, marketers may struggle to use this data for personalized email campaigns or retargeting ads, limiting their ability to re-engage customers effectively.

By addressing data silos, businesses can achieve a more integrated approach to data storage and access. This not only improves operational efficiency but also enables the creation of seamless, personalized customer experiences across all touchpoints. Breaking down silos is a fundamental step in building a robust MarTech infrastructure that drives meaningful insights and results.

CASE STUDY: VOLTAGE MOTORS AND THE CUSTOMER DATA DILEMMA

VoltAge Motors, a growing electric vehicle startup, has been making waves with its innovative approach to sustainable transportation. Behind its sleek designs and cutting-edge technology lies a pressing issue, however: a fragmented and underdeveloped customer data strategy. Despite a growing customer base, VoltAge has struggled to leverage data effectively, hindering its ability to personalize experiences, drive customer retention, and unlock new revenue opportunities.

The Absence of a First-Party Data Strategy

VoltAge's challenges began with the lack of a clear first-party data strategy. While the company collected some customer data through its Web site, app, and dealerships, it failed to define a cohesive approach to gathering, managing, and utilizing this information. As a result, data was inconsistent and often incomplete, making it difficult to create accurate customer profiles. For example, VoltAge could not track customers' charging habits or preferences for subscription services such as advanced driving assistance or premium infotainment packages. This lack of insight prevented the marketing team from crafting personalized campaigns, leaving customers feeling disconnected and underserved.

Data Siloes Between Marketing and Customer Retention Teams

Adding to the problem, VoltAge's marketing and customer retention teams operated in silos, each relying on separate tools and systems. The marketing team used data from Web site analytics and email campaigns to target potential customers, while the retention team focused on service histories and subscription renewals in their CRM. This disconnect created a significant blind spot: neither team could see a complete picture of the customer journey.

For instance, a customer who purchased a VoltAge vehicle but did not subscribe to the premium charging network was considered a "closed lead" by the marketing team. Meanwhile, the retention team had no visibility into that customer's behavior before their purchase, thereby missing critical opportunities to up-sell or cross-sell services. This lack of coordination not only impacted revenue but also frustrated customers who received generic, irrelevant offers instead of tailored recommendations that matched their needs.

Taking a Strategic Turn

Recognizing the inefficiency caused by their fragmented data strategy, VoltAge Motors decided to address the root of the problem. The company's leadership initiated a comprehensive analysis of its customer data systems and processes, focusing on four critical areas:

- *Customer data platform (CDP)*: VoltAge began evaluating CDP solutions that could unify data from multiple sources, providing a single view of each customer. By consolidating data from marketing campaigns, CRM systems, and app usage, VoltAge aimed to build accurate and actionable customer profiles.

- *CRM system enhancements*: The company also looked at upgrading its CRM system to improve collaboration between marketing and customer retention teams. This would allow both teams to access the same customer insights, enabling consistent messaging and coordinated efforts to up-sell services.

- *First-party data strategy*: VoltAge explored new ways to collect first-party data directly from customers, such as through app-based surveys, preference centers, and enhanced Web site interactions. These efforts would help build trust with customers while ensuring compliance with privacy regulations.
- *Cross-team data sharing*: Lastly, VoltAge implemented processes and tools to encourage better collaboration between teams. Regular data-sharing meetings, shared dashboards, and integrated workflows were introduced to ensure that all departments worked from the same dataset, enabling a more cohesive customer strategy.

Looking Ahead

With these changes underway, VoltAge Motors is taking its first steps toward a stronger customer data strategy. The company hopes to reduce data silos, improve personalization, and unlock new revenue streams by leveraging a more unified approach to customer insights. As we revisit VoltAge's journey later in this book, we'll explore how these efforts evolve and the impact they have on the company's growth and customer satisfaction.

CONCLUSION

Customer data plays a key role in the success of a company's marketing efforts, and thus it is the foundation of successful customer relationships.

The next chapter will look at some of the most important MarTech platforms that collect customer data for use by marketing teams and others within an organization.

NOTE

1. McKinsey & Company. (2021, November 17). *The value of getting personalization right—or wrong—is multiplying*. McKinsey & Company. https://www.mckinsey.com/capabilities/growth-marketing-and-sales/our-insights/the-value-of-getting-personalization-right-or-wrong-is-multiplying

CHAPTER 10

CRM, CDP, AND DMP

The contemporary marketer is exposed to a wide population of customers with interactions and points of contact across a multitude of platforms, such as social media, email, e-commerce, and mobile apps. Every encounter—such as click, purchase, and feedback—generates valuable data that describes customer behavior and habits. While this information is valuable, because of the growing quantity and diversity of information and the places where it is collected and stored, it becomes increasingly challenging for marketers to work with it to maximum effectiveness.

To succeed, businesses must unify this fragmented data into a singular, cohesive view of the customer journey. Without integration among the variety of touchpoints, customer interactions remain siloed, preventing marketers from understanding how individual marketing channels influence purchasing decisions. For instance, an analysis of a social media campaign may not be related to purchase actions in an e-commerce store, which makes it difficult for marketers to evaluate the effectiveness of their campaign.

CHALLENGES OF MANAGING CUSTOMER DATA

While there can be many potential areas that pose challenges, fragmentation is one of the biggest obstacles in managing customer data. Siloed systems, often used by different teams or departments, store information in disconnected databases that fail to communicate with each other. This disconnect limits marketers' ability to analyze data comprehensively and leads to duplication, inconsistency, and missed insights.

Poor or inconsistent data also creates barriers to greater personalization and customer engagement. For instance, if a customer receives an email that is not related to their past purchases or interests, they may not engage with the brand again, perceiving the message as spam or the brand as unresponsive to their needs. Such lost opportunities show the need for solid data management systems that support a unified and consistent marketing approach.

PLATFORMS THAT CAN ADDRESS THESE CHALLENGES

Customer data platforms (CDPs), customer relationship management (CRM) systems, and data management platforms (DMPs) address these challenges by providing tools that centralize, organize, and analyze customer data, each in distinctive ways, as will be explored in this chapter. These platforms are intended to break down customer data silos, integrate different data sources, and offer useful information that can help in developing strategies for personalized and effective marketing.

For instance, a CDP can combine data from different touchpoints—for example, information about visits to the Web site and interactions with email—into a single customer profile. CRMs help the teams to manage and develop customer relationships in real time, improving harmony and ensuring that the contact is relevant and timely. DMPs, however, work by providing useful information about the audience for digital advertising campaigns based on anonymized and aggregated data. These platforms enable marketers to work more productively, make decisions based on data, and increase their efforts to meet the needs of the modern customer journey.

The three main types of platforms are CDPs, CRMs, and DMPs, and each has a different role.

CDPs are systems that are used to store and manage first-party data that is collected from customers across different channels. This means that CDPs are able to collect data from a customer's Web site, mobile app, email campaigns, and many other sources and use this data to build a complete picture of each customer's interactions with the company.

CDPs are recommended for businesses that need to deliver personalized experiences and are data driven. Some of the major features of CDPs are data integration, which enables the collection of data from different sources; identity resolution to link customer data collected from different devices and channels; and real-time activation of insights for use in marketing campaigns.

For instance, a retail company can use a CDP to segment its customers based on their purchasing history and send an email campaign to each group of customers recommending specific products. Other top scenarios in which this is applied are segmentation for marketing purposes and ensuring consistency of the omnichannel approach. CDPs are essential for the modern approach to marketing as a way of connecting with customers.

CRM systems are applications that improve and manage relationships with customers. The primary difference between CDP and CRM is that while CDPs focus on managing data to support marketing activism, CRMs are more practical in that they help the organization to keep a record of customer contact information, communication history, and transactions, and in doing so enable the teams to give the customer the right experience at the different touchpoints of the customer life cycle.

Some of the key functions of CRMs are contact management, sales pipeline management, and service management. For example, a SaaS company can use a CRM to drive lead generation by setting up a campaign of follow-up emails and tracking the movement of the leads in the sales funnel. CRMs also aid in the tracking of customer service interactions to ensure that support personnel have the necessary information to resolve issues properly. In this way, CRMs help to ensure that the customer is not passed from person to person and that they receive the care that they deserve. CRMs are important tools for enhancing cooperation between departments to ensure that the customer receives consistent and personalized treatment at each touchpoint.

DMPs are used for managing and analyzing anonymous user data, mainly for advertising. This is because CDPs and CRMs deal with identifiable customer information and DMPs work with anonymous data to identify trends of the audience and develop strategies for ad targeting. These platforms are most useful for business that engage in programmatic advertising and are able to target their audience based on demographic, behavioral, and contextual data.

Some of the major features of DMPs are audience segmentation, the integration of third-party data, and the optimization of ad campaigns. For example, an e-commerce company can use a DMP to identify its most loyal customers and advertise to lookalike audiences on social media platforms. Other applications of DMPs include using lookalike audiences for programmatic buying and analyzing the market to identify emerging trends. Even though the use of DMPs has reduced due to issues with privacy and the growth of first-party data strategies, they are still useful for businesses that focus on audience extension and digital advertising optimization.

The aforementioned platforms (CDPs, CRMs, and DMPs) serve unique purposes in the MarTech stack. They help marketers to collect and arrange data, manage relationships, and reach out to audiences in the most appropriate manner, all while enhancing the overall efficiency of their operations.

How The Platforms Work Together

CDPs, CRMs, and DMPs are all different in their focus and primary usage, but when they are integrated as part of a company's technology infrastructure, they create an ecosystem that enables businesses to engage with customers in a more effective way, as illustrated in Figure 9.1.

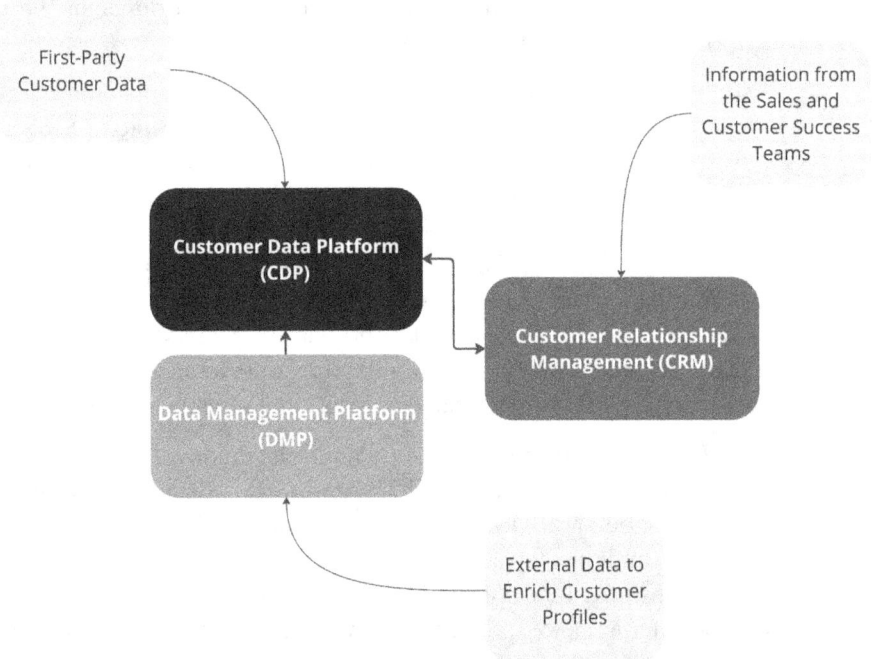

FIGURE 9.1 How CDPs, CRMs, and DMPs work together

This information is passed from CDPs to CRMs and DMPs to ensure the continuity of data. For example, a CDP can collect first-party data from different sources, such as Web site visits or mobile application usage, and send the data to a CRM system that can help with the creation of a complete customer profile, giving sales and customer support teams better information to provide customers more personalized responses. Likewise, DMPs perform the functions of targeting and segmentation using aggregated and anonymized audience data that can drive top-of-the-funnel leads, while CDPs are designed to use identifiable first-party data to personalize the customer experience once individuals are identified. These two purposes guarantee that both targeted advertising and individual-level engagement are being enhanced.

CRMs are operational tools for managing direct customer interactions and using CDP-derived insights to make communication more personalized and to develop the relationship with the customer. For example, a CRM can use the purchase history and behavior data from a CDP to send an automated follow-up or a special offer that is customized to each customer's preferences. This synergy between platforms creates a loop where insights are properly transferred from one system to another, making it easier to come up with better and more effective marketing strategies.

Combined Use Case

As an example, imagine a retail company that wants to enhance its marketing effectiveness through the integration of these systems. The company uses a CDP to monitor customer needs, such as products bought and categories of interest, from their Web site and mobile application. These insights are sent to the CRM system that handles the loyalty program contact information to ensure that every contact, whether through email or physical store contact, is personalized.

At the same time, the company uses a DMP to improve the quality of online advertising by creating lookalike audiences from anonymized data. The DMP works on general audience information and converts this information into specific first-party data that can be used by the CDP. Therefore, these platforms enable the retail company to provide the customer with consistent and personalized experiences while also optimizing advertising costs.

Bringing these platforms together to work in harmony helps businesses improve their performance, enhance their customer experience, and enhance their outcome. The rest of this chapter will explore each platform in more depth.

CUSTOMER RELATIONSHIP MANAGEMENT (CRM)

CRMs are particularly useful for managing the unique and often linear aspects of the customer lifecycle and are widely used in businesses today, with the global market size valued at $91.43 billion in 2023, which is expected to continue to grow[1]. As discussed previously in this chapter, CRMs are systems that act as central repositories for storing customer information, logging all the interactions and communications, and managing the sales, marketing, and customer service processes. This helps the teams to work in a more coordinated manner and ensure that the customer is contacted more intentionally, rather than repeatedly with the same message, or one that is not relevant to that individual or company. In this way, CRMs assist organizations to develop and maintain customer relationships, increase customer loyalty, and, consequently, boost sales.

CRMs and Their Usage by Marketers

CRM is an idea that has been in practice for several decades, and the first systems were developed to help companies keep records of their customers, with Siebel Customer Relationship Management being recognized as the first CRM product, launched in 1993[2]. The initial systems were not very sophisticated and were simply a way of storing contact details, often mimicking the concept of a paper rolodex, which continued to sit on many office desks.

The use of cloud-based CRMs began to emerge in the early 2000s with the advent of companies such as Salesforce, which made these systems available to small and medium enterprises. Over time, CRMs have been integrated more closely with marketing, so that marketers can use customer information to segment, measure the effect of, and target their campaigns.

CRMs are now an essential part of the MarTech landscape, and they form the basis of data-driven marketing and customer-focused approaches to marketing. This article tells the story of the evolution, functions, and best practices for applying CRMs in current marketing.

Key Components of a CRM

CRM systems are software applications that have a robust set of functionalities that help a diverse set of teams within organizations manage many aspects of customer relationships and interactions, as illustrated in Figure 9.2. The following is a detailed review of the key components of a CRM and how they help to improve customer engagement, increase the efficiency of operations, and drive business.

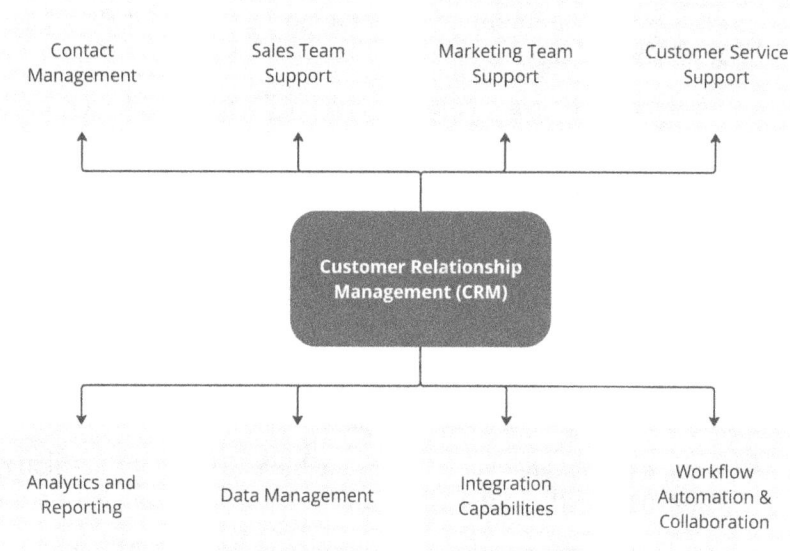

FIGURE 9.2 Components of a CRM

Contact Management

The first and most basic function of any CRM system is contact management, which is a central repository of customer information. This includes the customer's name, address, phone number, email address, and many other details in such a way that everyone in the organization can keep a single record of the customer.

Furthermore, CRMs keep a communication log, including emails, phone calls, and meetings, to make sure that the user has complete information about communication with each customer. In this way, contact management increases the efficiency of work by ensuring that all the teams involved in the organization work in coordination and there is no confusion.

Sales Team Support

CRM systems provide the sales team with automation tools to improve and control the sales process, including the following:

- *Lead management*: CRMs have a record of all the potential customers and where they are in the process.

- *Opportunity management*: The sales personnel can monitor the deals, classify them as important or not important, and follow through with them based on the pipeline.

- *Sales forecasting and pipeline analysis*: The predictive tools help the teams to forecast future sales and identify any constraints in the sales process.

- *Activity management*: CRMs help the sales representatives to perform their daily tasks, such as writing emails, making phone calls, and attending meetings, through reminders and workflows.

Marketing Team Support

Many CRMs of the present day have marketing automation as part of their system to support campaign management and audience targeting:

- *Campaign management*: CRMs enable marketers to develop and manage campaigns within the CRM.

- *Email marketing tools*: CRMs come equipped with email tools for personalized communication and the automation of email campaigns such as welcome or re-engagement campaigns.

- *Customer segmentation capabilities*: Marketers are able to create segments based on customer behavior, preferences, or demographics for targeted marketing.

Customer Service Support

A complete CRM system provides customer service teams with tools to help them manage and solve problems:

- *Case/ticket management*: CRMs provide teams with a platform to log, track, and resolve customer complaints and issues.

- *Knowledge base*: Many systems have a knowledge base that contains FAQs, guides, and other information that can help both customers and support staff.

- *Self-service portals*: These are portals for customers to find solutions to their problems, reducing the load on support teams while increasing customer satisfaction.

Analytics and Reporting

CRMs have analytics and reporting tools that enable organizations to keep track of their performance and make decisions based on the data generated:

- *Customizable dashboards*: Users can select which metrics to display on their dashboard at any given time.
- *Performance metrics and KPIs*: CRMs record sales, marketing, and service performance indicators that describe business outcomes.
- *Data visualization tools*: CRMs use charts, graphs, and other visualizations to identify trends and gain additional insights.

Data Management

It is very important that the data that is to be used in CRM should be accurate and available for use:

- *Data collection and ingestion*: CRMs are able to capture data from various sources including social media, Web sites and email platforms.
- *Data unification and identity resolution*: These systems combine data from different channels to create a single customer profile.
- *Data field customization*: As no two businesses are the same, CRMs are built to enable the customization of data fields to enable a business's unique data collection needs to be addressed.
- *Data cleansing and standardization*: CRMs are used to remove duplicate information, correct poor data, and ensure that all the data is in the right format.

Integration Capabilities

Modern CRMs are open to working with other business systems and applications using the following means:

- *APIs and pre-built connectors*: CRMs can be integrated with third-party applications such as marketing automation tools, e-commerce solutions, and other business software using APIs or pre-built connectors.
- *Data sharing across the organization*: CRMs ensure that every department in the organization has access to the same customer information.

Workflow Automation and Collaboration

Automation is a key feature of CRMs. It helps businesses set up workflows that minimize manual work and increase productivity:

- *Automated processes*: This includes tasks such as following up with customers via email or phone, assigning leads to staff, and reminding them of sales events.

- *Internal messaging*: CRMs have internal messaging, which allows teams to communicate within the platform and get quick responses.

- *File sharing*: CRMs support sharing documents and other assets in a secure manner, and reduce the need for third-party tools.

These components work together to provide a comprehensive system for managing customer relationships across marketing, sales, and customer service functions. The specific features and emphasis may vary between different platforms, but these core elements form the foundation of most CRMs.

HOW TO EVALUATE A CRM

Many businesses rely on their CRM system to manage customer interactions because it supports sales and customer service teams to streamline operations and enhance customer engagement. The selection of an appropriate CRM system is essential to maintain efficient workflow processes and achieve both data precision and growth scalability. This section outlines five fundamental evaluation criteria and provides detailed considerations for each one.

Contact and Relationship Management

As the name suggests, the fundamental purpose of a CRM system is to deliver a centralized database designed for effective management of customer interactions alongside contact details and historical data. Teams can achieve efficient tracking and access to customer information with a robust platform.

Let's have a look at the evaluation criteria:

- *Unified customer profiles*: Does the CRM technology have the ability to store data about customer interactions and present all emails, phone calls, meetings, and transactions in one cohesive view? Keep in mind that this customer profile looks different than what will be explored in the next section with CDPs.

- *Data enrichment capabilities*: Is the platform designed to automatically incorporate social media profiles, purchase history data, and add third-party integrations to contact records?

- *Custom fields and tagging*: Teams have the capability to add custom fields to contact records that match their specific industry or business requirements.

With strong contact management features, a CRM delivers accurate customer records that teams can access and act upon.

Sales and Pipeline Management

A CRM needs to equip sales teams with lead-tracking capabilities, deal management solutions, and sales forecasting tools to enhance work efficiency and accelerate deal closure.

Let's have a look at the evaluation criteria:

- *Lead and opportunity tracking*: The CRM must enable sales teams to monitor prospects throughout the entire sales process starting from initial contact until they convert.

- *Sales forecasting and reporting*: The system should be able to forecast revenue while monitoring sales trends to generate comprehensive reports that assist in decision-making processes.

- *Automation of follow-ups and tasks*: The CRM system supports workflow automation for nurturing leads, sending follow-up reminders, and handling task assignments.

Sales management tools within a CRM system enable businesses to boost conversion rates while concurrently streamlining deal tracking and enhancing sales forecasting accuracy.

Integration and Customization

A CRM needs to operate as part of other business systems while being customizable for an organization's unique operational processes.

Let's have a look at the evaluation criteria:

- *Pre-built integrations*: Check to see if the CRM system has pre-built connections with email marketing tools alongside ERP systems, e-commerce platforms, and customer support software.

- *API and custom development*: Through the platform's API, businesses have the ability to modify existing integrations and develop new features.
- *User customization*: The CRM allows teams to adjust dashboards and workflows and edit fields to suit their business operations.

Strong integration and customization options in a CRM platform eliminate data silos while enabling businesses to personalize their platform experience.

Automation and Workflow Efficiency

A contemporary CRM platform requires automation tools to minimize manual data entry while improving operational efficiency and providing timely customer responses.

Let's have a look at the evaluation criteria:

- *Automated lead scoring*: The CRM system uses automated lead scoring to prioritize leads through engagement metrics, behavioral data, and AI predictions.
- *Marketing and sales automation*: Does the system support automated email dispatches together with lead assignments and workflow activations according to established rules?
- *Task and reminder automation*: The CRM should allow automatic follow-up tasks while syncing calendars and providing overdue task notifications.

A CRM system that features strong automation abilities decreases workload while enhancing response times and maintaining consistent customer engagement.

Analytics, Reporting, and Performance Tracking

The usefulness of a CRM system depends on its ability to deliver insights that enhance decision-making processes. Real-time analytics and reporting enable organizations to monitor customer interactions and assess their sales performance while optimizing marketing campaigns.

Let's have a look at the evaluation criteria:

- *Customizable reports and dashboards*: Do users have the capability to develop custom reports within the system to monitor key performance indicators (KPIs) such as sales trends and lead conversion rates, as well as customer lifetime value?

- *AI-powered insights*: Can the CRM system deliver predictive analytics capabilities alongside automated trend detection and customer behavior analysis?

- *Team performance monitoring*: Does the CRM enable managers to monitor the activity of sales representatives as well as call logs and customer service performance?

A CRM that features robust analytics capabilities enables businesses to base decisions on data, optimize sales strategies, and discover improvement opportunities.

Organizations need to select a CRM system that supports their operational processes and scales with company expansion while delivering valuable data insights for lasting success in customer interaction management.

CUSTOMER DATA PLATFORM (CDP)

Coming to prominence as digital channels proliferated, CDPs are software that has been developed to hold all customer data in one place to ensure that each customer's information is complete and accurate. CDPs are different from CRMs or DMPs as they focus on first-party data such as Web site interactions, app usage, and email engagement to create a real-time single customer profile. This means that CDPs take the dispersed data and turn it into actionable insights, which marketers can then use to deliver personalized experiences, improve targeting, and measure the effectiveness of their campaigns across channels. This makes CDPs a must-have for modern marketing strategies in a world where data is becoming more important.

Brief History and Usage by Marketers

The term "customer data platform" is attributed to a blog post by David Raab in 2013, where he explained some of the current challenges with platforms such as CRMs in solving companies' customer data needs[3]. This idea of a CDP was born from the struggles that marketers had with managing and using customer data with more and more touchpoints in the digital world, perhaps exacerbated by the rapid growth of social media and mobile apps during this period. CRMs and DMPs, the early MarTech platforms, had some level of data management but had some restrictions, such as the centralization of data and a dependence on third-party data. This made marketers need a solution that could easily integrate first-party data and use it in campaigns because of the growing need for personalization and real-time information.

CDPs have gained their place as an integral part of a MarTech infrastructure because they provide a user-friendly interface for marketers that does not involve a person who specializes in information technology. This is because CDPs provide access to data to non-technical personnel and enable them to use it without the help of developers. Today, CDPs are used by marketers in various sectors, including retail, healthcare, and financial services, where customer-oriented strategies are vital for an organization's success. This article explains how CDPs help with the current issues in marketing, what CDPs can do, and how CDPs are changing in the privacy-first world.

Key Features of a CDP

CDPs provide a set of functions for collecting, unifying, analyzing, and activating customer data and thus are an essential part of modern marketing strategies, as shown in Figure 9.3. These platforms serve as a link between raw data and actionable insights, which enable marketers to develop and deliver personalized and impactful customer experiences. While the term "customer data platform" has been applied to software applications that range in the depth and breadth of their overall functionality, the following is a review of the key features CDPs are likely to have.

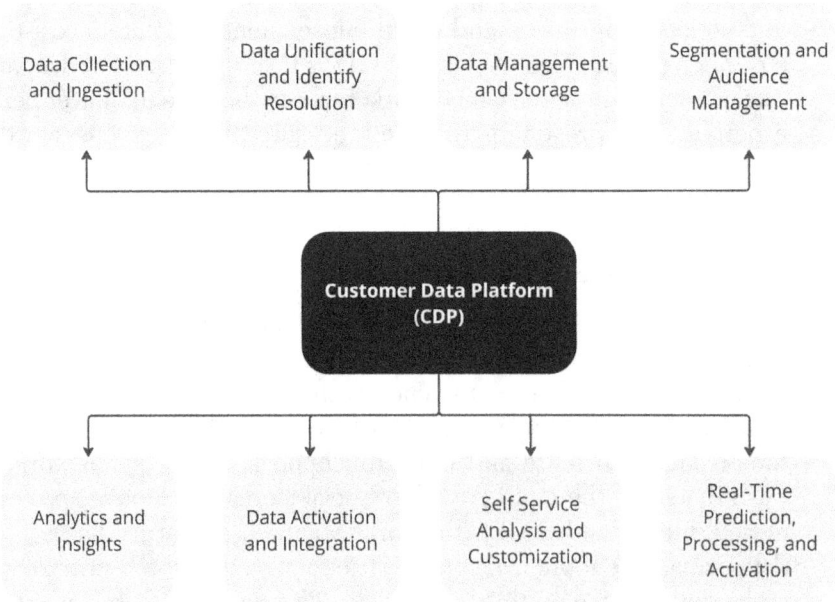

FIGURE 9.3 Components of a CDP

Data Collection and Ingestion

CDPs can collect information from a large number of sources to ensure that there is complete information about the customer:

- *Multiple data sources*: CDPs can ingest data from online channels such as Web sites and social media, mobile apps, and offline channels such as points of sale and call centers.
- *Support for various data types*: These platforms work with structured data (for example, purchase records), unstructured data (for example, social media posts), and semi-structured data (for example, JSON logs).
- *Real-time data collection*: The collection and processing of data in real time ensures that customer profiles are updated in real time and marketing campaigns can be initiated promptly.

Data Unification and Identity Resolution

Another key feature of CDPs is that they are able to combine different data elements into a single customer profile:

- *Unified customer profiles*: Combining data from several sources, CDPs develop complete profiles that contain information on customer behavior and preferences.
- *Identity resolution*: These platforms use sophisticated matching techniques such as using email, cookies, and mobile ID to link different touchpoints of a customer.
- *Cross-device matching*: CDPs ensure continuity by linking customer information across devices, enabling the tracking of customer journeys.

Data Management and Storage

CDPs are customer data management platforms that ensure data accuracy and availability:

- *Centralized database or access to data warehouse storage*: Customer information can be stored in one place, avoiding silos that can hinder marketing efforts. Note that many CDPs now provide support for third-party data warehouses, which, while technically storing customer data outside of the platform, still accomplish the goal of centralized storage of the information.

- *Data cleansing and standardization*: CDPs clean, normalize, and standardize data to ensure that it is consistent and reliable.
- *Persistent storage*: Customer profiles are stored over time, which allows the marketer to see historical behavior and long-term trends.

Segmentation and Audience Management

CDPs give marketers the tools they need for audience targeting and management:

- *Advanced segmentation*: Marketers can create specific audience segments based on behavior, demographics, or preferences.
- *Real-time segmentation*: CDPs update audience segments in real time, which means that businesses can respond to customer activities in real time.
- *Dynamic updates*: Customer attributes and segments are regularly revised based on new data to ensure that the right people are targeted.
- *Automated segmentation*: Machine learning algorithms can also identify patterns in the data and create relevant audience segments.

Analytics and Insights

CDPs come equipped with robust analytics capabilities that help the user gain a deeper understanding of customer behavior and patterns:

- *Customer analytics*: Analysis of customer behavior, preferences, and trends.
- *Predictive analytics*: CDPs use AI to predict customer behavior, such as the chance of the customer leaving or the probability of them making a purchase.
- *Journey mapping*: Marketers can map out the customer journey and identify touchpoints and moments of opportunity.

Data Activation and Integration

CDPs are designed to push customer data into various marketing channels and business systems:

- *Data sharing*: Information about customers can be easily shared with other tools and technologies, such as email marketing tools, ad tech, and CRM.

- *Pre-built connectors*: Many CDPs come with pre-configured integrations with popular platforms, which reduces the time and effort required for integration.
- *Application programming interface (API) access*: CDPs also provide APIs that allow businesses to build their own integrations for custom needs.

Self-Service Analysis and Customization

CDPs are developed to be easy to use for marketers and non-technical users to ensure that everyone can take advantage of customer data:

- *User-friendly interface*: The designs of the user interface are very intuitive, and the teams can easily navigate through the data and build segments.
- *Self-service tools*: Marketers can create and analyze audience segments without involving IT, using self-service tools.
- *Customizable dashboards*: Users can customize their dashboards to display the metrics and insights that are most important to their role.

Real-Time Processing and Activation

The real-time capabilities of CDPs enable them to send responses to customer activities and events in real time:

- *Predictive modeling*: CDPs use AI to create a score that describes the likelihood of customers to convert or churn.
- *Real-time processing*: While some CDPs claim "real-time processing," they only deliver near-real-time processing, though more of these platforms are improving their speed and throughput. Whether this is in literal real time or simply very quick, this feature means that data is processed as it comes in, so the profiles and segments are always current.
- *Real-time activation and orchestration*: CDP can send a personalized message or offer based on a customer's behavior in the moment; for example, sending a discount to a customer who has left their cart abandoned.
- *Anomaly detection*: AI-based tools identify unusual patterns in the data, for example, a sudden decrease in engagement, and allow timely action.

These features allow CDPs to help businesses collect customer data in one place, gain useful insights, and activate personalized marketing strategies effectively.

HOW TO EVALUATE A CDP

The evaluation of a CDP requires an understanding of its ability to centralize customer information and enhance marketing personalization while considering business needs and technical capabilities.

Choosing a suitable CDP is a key decision point for organizations because it enables them to unify customer data and achieve better personalization while boosting marketing results. Businesses need to evaluate different platforms according to their organizational requirements while considering scalability and integration features. Here are five evaluation criteria to assist with this important decision.

Data Collection and Integration

CDPs need to ingest data from various sources efficiently while also offering seamless integration capabilities with existing marketing, sales, and analytics systems. Businesses without strong integration capabilities may experience data fragmentation and obtain inconsistent customer insights.

Let's have a look at the evaluation criteria:

- *Multi-source data collection*: The CDP needs to be able to gather data from multiple channels, including Web sites, mobile apps, CRM systems, email platforms, call centers, and offline sources.
- *Real-time data processing*: Does the platform process customer data instantly with no delays when ingesting and updating data, or does it experience processing lags?
- *Pre-built and custom integrations*: The CDP includes native connectors for major MarTech platforms such as Salesforce and HubSpot while allowing custom connections through API-based integrations.

A CDP provides comprehensive data integration, which enables marketers and analytics teams to access and unify customer interactions from both online and offline channels.

Identity Resolution and Data Unification

A primary function of CDPs involves merging all customer data into one singular profile. A CDP enables marketers to gain a comprehensive and precise picture of each customer, which enhances segmentation capabilities and enables personalized marketing.

Let's have a look at the evaluation criteria:

- *Identity matching across devices and channels*: Does the CDP system have the ability to identify a customer across various touchpoints, such as email, Web site, mobile app, and in-store transactions?
- *Data cleansing and standardization*: The platform needs to demonstrate that it can eliminate duplicate records while standardizing data formats and correcting customer information inconsistencies.
- *Persistent customer profiles*: Does the system create a single customer profile that stays consistent over time while customers engage through numerous devices and channels?

By attributing each customer interaction to the correct individual, a CDP with advanced identity resolution capabilities stops fragmented customer experiences from occurring.

Audience Segmentation and Activation

Marketers should have the ability, through a CDP, to build real-time dynamic audience segments and activate them across various marketing platforms. Businesses must develop precise customer targeting methods that incorporate behavioral analysis along with demographic data and past engagement patterns.

Let's have a look at the evaluation criteria:

- *Advanced segmentation capabilities*: Marketers need to know whether their platform supports the creation of rule-based, AI-driven, or real-time audience segments for campaigns.
- *Cross-channel activation*: Does the CDP enable marketers to distribute audience segments across advertising platforms, email marketing systems, and personalization engines without requiring manual steps?
- *Automated triggers for personalization*: Does the system enable automatic updates to customer segments and initiate personalized messages in response to customer behavior?

Marketers gain the ability to deliver highly personalized campaigns that adapt to customer behavior instantly through a CDP with powerful segmentation capabilities.

Compliance, Security, and Data Governance

As regulations such as the GDPR and CCPA grow more prevalent alongside international privacy laws, organizations need to make sure their CDP complies with rigorous data security standards.

Let's have a look at the evaluation criteria:

- *Privacy and consent management*: The platform enables users to execute opt-ins and opt-outs while also allowing them to make data deletion requests to follow privacy laws.
- *Data encryption and security measures*: What security measures does the CDP implement to protect customer data both at rest and during transmission to prevent breaches?
- *Role-based access and permissions*: Organizations should be able to determine access permissions for teams and individuals who need to handle different categories of customer information.

When a CDP meets compliance standards and maintains security protocols it defends organizational interests while protecting customer data through legal and ethical usage practices.

Analytics and AI Capabilities

The value of a CDP depends entirely on the insights it delivers. Marketers require access to real-time analytics and machine learning-driven recommendations along with predictive modeling capabilities to fully utilize customer data.

Let's have a look at the evaluation criteria:

- *Real-time data insights*: Does the CDP produce live dashboards, customer journey analysis, and behavioral insights?
- *Predictive analytics and AI features*: Does the platform possess predictive analytics capabilities to forecast customer actions while using AI features to suggest personalized content and automate engagement strategies?
- *Attribution and performance measurement*: The CDP technology tracks customer interactions throughout different channels and evaluates the return on marketing investment.

Marketers gain optimized campaign insights through data-driven decision-making from a CDP with advanced analytics and AI capabilities that analyze customer behavior and provide predictive insights.

When choosing a CDP, organizations need to thoroughly assess the various options' capabilities in data integration as well as identity resolution, segmentation capabilities, compliance adherence, and analytics features. Organizations must assess CDPs through their capacity to consolidate customer data while enabling real-time personalization and delivering actionable insights with strong security and compliance measures. Businesses that make informed choices about their data platforms will achieve more effective data utilization, which drives enhanced customer interactions along with higher conversion rates and lasting customer loyalty.

KEY DIFFERENCES BETWEEN A CDP AND A CRM

Although CDPs and CRM systems have the same goal of enhancing customer experiences, they are designed for different purposes and serve different aspects of a business. These can be broken down into ten categories.

Purpose and Focus

The primary purpose of CDP is to manage customer data throughout the customer lifecycle, with a single view of customer behavior, traits, and preferences. CDPs are a source of actionable insights to drive personalization and engagement. In contrast, CRMs are operational tools for managing and supporting one-to-one customer interactions, including sales calls, support tickets, and email exchanges. CRMs are primarily concerned with specific customer contact points, with the aim of building and concluding a sale.

Data Collection and Scope

CDPs collect and integrate data from both online and offline sources in a holistic manner. This includes Web site interactions and social media activity on the online channel, and in-store purchases and call center interactions as examples of the offline channel. A broader perspective on customer behavior is therefore possible. In comparison, CRMs rely on manually entered data from direct interactions such as sales meetings or support conversations. This narrower scope limits CRMs to tracking relationship-specific details rather than capturing a holistic view of customer behavior.

Data Types

CDPs are capable of handling structured data (for example, purchase history), unstructured data (for instance, social media comments), and semi-structured data (for example, JSON logs). This makes CDPs capable of processing and unifying data from a variety of sources. CRMs, on the other hand, are primarily involved with structured data; for instance, contact information, communication history, and sales pipeline metrics, which are usually associated with specific customer interactions.

Real-Time Capabilities

CDPs excel at real-time data processing to enable real-time customer segmentation and marketing campaign activation. For instance, a CDP can immediately trigger a personalized email when a customer leaves their shopping cart unattended. CRMs, however, are usually based on historical data and may not have real-time capabilities; instead, they focus on the history of interactions and the on-going management of relationships.

Intended Users

CDPs are designed for marketers but are also beneficial to other roles, such as product development and analytics, since they ensure data is available across the organization. CRMs, however, are mainly for the sales and customer service teams to help with managing leads, tracking interactions, and resolving customer issues.

Data Integration

Integration is a key feature of CDPs that are developed to gather data from many sources—for example, an e-commerce platform, loyalty programs, and advertising tools—to build a single customer profile. CRMs can also integrate with other systems but are limited in scope, mainly contain data from direct interactions, and lack the comprehensive integration of a CDP.

Scalability

CDPs are designed to handle large amounts of data and are therefore suitable for business-to-consumer companies with thousands of customers, such as retailers and streaming services. CRMs are normally used for business-to-business companies with a limited number of customers for whom relationship management is crucial.

Analytics and Insights

CDPs provide advanced analytics. This includes predictive modeling, customer journey analysis, and real-time reporting. These features enable the marketer to understand customer behaviors and predict customer actions. CRMs provide basic reporting; this mainly involves sales metrics and customer interaction histories. This is useful for evaluating team performance but is not very useful for marketing strategies.

Personalization

The depth of data managed by CDPs allows highly personalized marketing strategies. For example, a CDP can customize the Web site or advertising that a customer is exposed to based on their real-time browsing behavior. CRMs are also capable of personalization but are limited to interactions that are enclosed within the scope of sales or support, such as sending a personalized email to a lead or an existing customer.

Data Accessibility

CDPs are designed to share data with other systems and teams to ensure that insights are easily reachable by everyone in the organization. CRMs have limited data-sharing capabilities and are mainly used for managing interactions within specific departments such as sales and support.

While CRMs are important for the hands-on management of customer interactions, CDPs provide the data foundation to support personalized and targeted marketing, advanced analytics, and real-time engagement. The two systems can work together to provide the best customer experience and enhance business results.

DATA MANAGEMENT PLATFORM (DMP)

DMPs are important in the MarTech ecosystem, especially in programmatic advertising and audience segmentation. They allow businesses to reach out to more customers and help to identify the right customers to approach at the right time using anonymous audience data. This is changing, however, as consumer privacy concerns and regulatory changes make marketers question the usefulness of DMPs in a world that is shifting to first-party data and transparency.

It is important to understand how DMPs align with CDPs and CRMs to build a good data strategy. While CDPs and CRMs are concerned with customer-specific data and interactions, DMPs are useful for understanding anonymous audience behavior and trends. This has become more challenging in recent years, however, due to their dependence on third-party cookies and aggregated data. This paper aims at explaining the role of DMPs, when they should be used, and how they can be effectively used in the light of modern marketing challenges.

Key Features of a DMP

A DMP is a technology solution for collecting, organizing, and analyzing anonymous audience data to support better advertising and marketing decisions. It differs from CDP or CRM systems, which focus on first-party or identifiable customer data, in that it is focused on third-party data and creating audience segments for campaign targeting.

The primary functions of DMPs include audience segmentation, lookalike audience modeling, and programmatic ad targeting. They collect data from various sources, such as Web sites, apps, and third-party vendors, and use it to create anonymized profiles of users. This allows marketers to serve an ad to a specific group of users without tracking personal identifiers. The main difference between a DMP and a CDP is that a CDP gives individual-level insights while a DMP aggregates data to provide audience insights. CRMs manage one-to-one interactions, while DMPs analyze audience trends across large datasets.

Using a DMP

DMPs are most useful when you need anonymous audience data to drive your marketing efforts. The most common use case for DMPs is programmatic advertising, which allows businesses to target ads to people based on characteristics, behaviors, and interests. For example, an airline might use a DMP to identify frequent travelers and encourage them to book holidays.

DMPs are useful for expanding audiences, and they are most effective in industries such as e-commerce, media, and entertainment. For instance, a streaming service could use user engagement information from the DMP to identify lookalike audiences for a new show and advertise to them. There are some limitations of DMPs, however. For example, third-party cookies are being phased out by major browsers, on which DMPs depend. Also, data accuracy can be a challenge because third-party data may not be as accurate as first-party data.

DMPs and a First-Party Data Strategy

While third-party browser cookies have not been completely deprecated, Apple, Microsoft, and Google have all taken steps to minimize their prominence in the Web browsing experience. This has transformed the function of DMPs in the market. As consumers demand greater control over their data and regulatory frameworks such as the GDPR and the CCPA become more stringent, DMPs need to evolve to remain useful. To this end, they are shifting their focus to aggregated insights and context-based targeting that do not heavily rely on unique identifiers of consumers.

Marketers also find ways to complement the DMPs with the CDPs. While the CDP is good at collecting and activating first-party data, the DMP is good at providing audience insights that can help improve the reach of campaigns. For instance, a retailer might use a CDP to identify its high-value customers and a DMP to identify similar audiences for acquisition campaigns. Thus, businesses can develop strategies that balance personalization and privacy by combining these platforms.

The Future of DMPs

How DMPs will be used in the years to come lies in their ability to adapt to a privacy-first digital ecosystem. Some of the emerging trends include the application of AI in audience modeling and predictive analytics. AI-powered DMPs can analyze non-personal data to identify trends and recommend best practices for targeting without the use of personal data.

Another potential sub-path is the uptake of privacy-centered frameworks and techniques such as differential privacy and federated learning, which enable marketers to derive insights from data without compromising user privacy. DMPs may also begin to focus more on context-based targeting, where ads are shown based on the content a user is consuming rather than their browsing history.

In a post-cookie world, DMPs will probably serve a supporting role to CDPs, helping companies with audience expansion while also ensuring compliance and ethical use of data. Therefore, their ability to transform and align with the evolving needs of the market remains an essential factor in determining their future. Therefore, for business owners contemplating the use of

DMPs, it is vital to consider the role of DMPs within the broader MarTech strategy for long-term success.

CASE STUDY: VOLTAGE MOTORS AND THE INTERSECTION OF CDPs, CRMs, AND DMPs

This case study returns us to VoltAge Motors, a rapidly growing electric vehicle manufacturer, which is achieving rapid growth due to a combination of innovative technology for its elective vehicles as well as its captivating marketing campaigns and strategies.

To manage its growing customer base and streamline marketing efforts, VoltAge relies on three critical platforms: a CDP, a CRM system, and a DMP. While these tools have helped the company scale, the overlap in their functionalities has occasionally led to confusion among teams, resulting in inefficiency and missed opportunities.

How VoltAge Uses a CDP, CRM, and DMP

VoltAge's CDP is central to its marketing operations, aggregating first-party data from its Web site, mobile app, and dealership network. This unified data provides detailed customer profiles, tracking everything from test drive requests to charging station usage. Marketers use the CDP for audience segmentation, enabling them to send personalized offers, such as discounts on home charging installations, based on a customer's previous interactions.

The CRM system is primarily used by VoltAge's sales and customer support teams to manage one-to-one interactions. It tracks customer inquiries, purchase histories, and service requests, ensuring that every customer receives timely and relevant follow-ups. For example, when a customer schedules a vehicle maintenance appointment, the CRM triggers an email reminding them of available subscription upgrades, such as premium navigation services.

The DMP supports VoltAge's advertising campaigns, helping the marketing team identify and target anonymous audiences for programmatic advertising. For instance, the DMP enables VoltAge to create lookalike audiences based on its most loyal customers and launch ad campaigns to attract similar individuals. This approach has proven effective in increasing awareness and driving new leads.

Challenges in Differentiating CDP and CRM Usage

Despite the clear roles of these platforms, VoltAge's teams sometimes struggle to determine when to use the CDP or the CRM. The marketing team occasionally attempts to use the CRM for audience segmentation, a task better suited to the CDP. This overlap results in inconsistent messaging, as segmentation created in the CRM lacks the depth and accuracy of the CDP's unified profiles. Similarly, the sales team sometimes requests data from the CDP that is more easily accessible in the CRM, causing delays and redundancies.

These misalignments stem from unclear guidelines on platform usage and insufficient training. Team members are unsure where certain data resides or which platform is best suited for specific tasks. For example, while the marketing team excels at activating CDP data for campaigns, they struggle to leverage CRM insights for one-to-one customer outreach, leaving gaps in their personalization efforts.

Areas for Improvement

VoltAge recognizes the need to address these challenges to fully leverage its MarTech stack. First, the company is developing clearer workflows and documentation to define the unique roles of each platform. For example, guidelines will clarify that the CDP is the source of truth for audience segmentation, while the CRM is the go-to for managing individual customer interactions.

Second, VoltAge plans to invest in training sessions to ensure all teams understand the capabilities and limitations of each platform. By educating employees on when and how to use the CDP, CRM, and DMP effectively, VoltAge aims to eliminate confusion and improve collaboration.

Lastly, VoltAge is exploring ways to enhance integration between the platforms. For instance, they are considering automating data flows between the CDP and CRM to ensure that sales teams can access enriched customer profiles without manually exporting data. Similarly, integrating insights from the DMP into the CDP would enable the marketing team to combine anonymous audience trends with first-party data for more precise targeting.

Looking Ahead

By addressing these issues, VoltAge Motors hopes to unlock the full potential of its CDP, CRM, and DMP. These improvements will not only streamline operations but also enable the company to deliver more cohesive and personalized customer experiences. As VoltAge continues to refine its data strategy, its ability to differentiate and integrate these platforms will be key to sustaining growth and maintaining a competitive edge.

CONCLUSION

While the three systems discussed in this chapter all manage and allow marketers to manage customer data and use it in their marketing efforts, CRMs, CDPs, and DMPs all approach this task in different ways. There is not a single best platform, and many companies will use all three for the foreseeable future.

In the next chapter, the important considerations of customer data privacy and compliance will be discussed.

NOTES

1. Fortune Business Insights. (n.d.). *Customer relationship management (CRM) market size, share & COVID-19 impact analysis.* Fortune Business Insights. https://www.fortunebusinessinsights.com/customer-relationship-management-crm-market-103418

2. Vitek, Chris (10 January 2017). «How Context Sits at Intersection of CRM, ACD». *No Jitter.*

3. Raab, D. (2013, April 25). *I've discovered a new class of system: Customer data platform (CDP). Customer Experience Matrix.* https://customerexperiencematrix.blogspot.com/2013/04/ive-discovered-new-class-of-system.html

CHAPTER 11

PRIVACY AND CONSENT

According to a 2023 report by the Pew Research Center, 62% of Americans do not believe it is possible to go through daily life without having companies collecting data about them, and 81% believe they have little to no control over the data that those companies collect[1].

This causes several challenges for companies that want to provide more personalized customer experiences, yet want to be responsive to those same customers' desire for greater control of the data that companies collect about their behavior and demographics.

THE TRUST GAP BETWEEN CONSUMERS AND BRANDS

As demonstrated by the Pew study, a growing number of consumers are becoming skeptical of brands' ability to protect their data. Research shows that concerns about data misuse and unauthorized access are among the top reasons customers hesitate to share their personal information. High-profile incidents, such as breaches involving millions of records or revelations of data being sold without consent, have widened the trust gap between consumers and brands. This mistrust can manifest as reduced willingness to share data, increased opt-out rates, or even complete disengagement with the brand.

For marketers, this trust gap presents both a challenge and an opportunity; brands that proactively address privacy concerns by being transparent about their data practices and providing customers with control over their data can stand out in a competitive marketplace. For example, implementing clear privacy policies, offering simple opt-out options, and communicating how data is protected can help rebuild consumer confidence.

THE NEED FOR CUSTOMER DATA REMAINS

Despite legitimate consumer concerns about how companies handle their sensitive information, customer data remains indispensable for delivering the personalized experiences that modern customers expect.

As already discussed, marketers use all types of customer data—third-party, second-party, first-party, and zero-party—to understand audience preferences, predict behavior, and tailor messaging to resonate with individual needs. This value exchange—where customers provide data in return for better experiences and offers—can strengthen relationships and drive loyalty when executed ethically and transparently, yet with the frequency of high-profile data breaches that are widely shared across news media, consumers are rightly cautious. By recognizing the growing need for customer data privacy and addressing the trust gap, marketers can create strategies that not only comply with regulations but also foster stronger, more meaningful relationships with their audiences. Building a foundation of trust ensures that customers feel confident sharing their data, enabling brands to deliver value in a way that benefits both parties.

Data Privacy Regulations

The growing and continued use of data-driven marketing has been accompanied by a wave of regulations designed to protect consumer privacy and ensure ethical data practices. These regulations impact how businesses collect, store, and use customer data, making it essential for marketers to stay informed and compliant. Failure to adhere to these guidelines not only results in legal penalties but also risks eroding consumer trust. Understanding the various types of data regulations—public policy, industry-specific, and self-regulation—provides a framework for marketers to navigate this complex landscape.

These regulations can be categorized in the following manner: public policy regulations, industry-specific regulations, and industry (or company) self-regulation, each of which will be explored in further detail in the paragraphs that follow.

Public Policy Regulations

Public policy regulations, such as the General Data Protection Regulation (GDPR), which became effective on May 25, 2018 in the European Union[2], and the California Consumer Privacy Act (CCPA), which was adopted on

June 28, 2018 in the state of California in the United States[3], have set global standards for data privacy and protection. The GDPR, for instance, requires businesses to obtain explicit consent before collecting personal data, offer customers the right to access or delete their data, and ensure robust security measures. Similarly, the CCPA gives California residents greater control over their personal information, including the right to opt out of data sales and request data deletion.

These regulations have forced businesses to prioritize transparency and customer consent, reshaping how data is collected and used. For example, many Web sites now feature clear cookie consent banners, allowing users to manage their preferences. As more regions implement similar regulations—such as Brazil's Lei Geral de Proteção de Dados Pessoais (LGPD)[4] and Canada's Consumer Privacy Protection Act (CPPA)[5]—marketers must adopt privacy-first strategies that can adapt to varying global requirements.

Industry-Specific Regulations

In addition to general privacy laws, certain industries are subject to sector-specific regulations that govern how sensitive data is handled. In healthcare in the United States, the Health Insurance Portability and Accountability Act (HIPAA)[6] mandates strict safeguards for protected health information (PHI), requiring marketers to ensure that data is securely stored and only shared with explicit consent. Similarly, financial institutions must comply with laws such as the United States Federal Trade Commission's Gramm-Leach-Bliley Act (GLBA), which enforces transparency in data-sharing practices[7].

For marketers operating in these regulated industries, compliance means going beyond general privacy practices to address specialized requirements. For example, a healthcare provider in the United States that is running a digital ad campaign must ensure that all patient data remains anonymized and is not inadvertently exposed during retargeting efforts, per HIPAA regulations. Failure to comply with these regulations can result in severe financial penalties and reputational damage.

Industry Self-Regulation

Beyond legal requirements enforced at the federal, state/provincial, or municipal level, many organizations and industry groups have adopted self-regulation frameworks to promote ethical data practices. For example, the Digital Advertising Alliance (DAA) provides guidelines for responsible data use in online advertising, including clear opt-out mechanisms and transparency in data collection[8].

Self-regulation allows industries to set their own standards for data privacy, sometimes going beyond what is legally required, or in other cases filling gaps where there is no regulation at the federal level. For instance, a retail brand might voluntarily adopt stricter data retention policies or limit the use of third-party cookies to align with customer expectations, neither of which may be specifically outlined in existing data privacy regulations. By adhering to these ethical standards, businesses can demonstrate their commitment to protecting consumer privacy, fostering trust and loyalty among their audience.

Data Collection Approaches

With the tightening of data privacy regulations and consumers' growing skepticism of how their data is being safeguarded, the methods used to collect customer data can be as important as the data itself. Transparent and ethical data collection practices not only ensure compliance with privacy regulations but also foster trust and loyalty among customers.

Transparency in Data Collection Methods

Transparency in how data is collected and used is a foundational principle for earning consumer trust. Customers should clearly understand what data is being collected, how it will be used, and the benefits they will receive in return. As an example, Web sites commonly use a pop-up banner to explain that cookies are used to improve the user experience by personalizing content and enabling faster page loads. Providing such clarity reassures customers that their data is being handled responsibly and for their benefit, and in many cases may assure other parties that their data collection approaches meet the local governing body's requirements.

Transparency also means avoiding deceptive practices, such as burying critical information in lengthy privacy policies. Instead, businesses should present key details in plain language and ensure they are easily accessible. By being upfront about their data practices, marketers can strengthen customer relationships and encourage greater participation in data-sharing programs.

The Role of Opt-In and Consent Mechanisms

Opt-in mechanisms are central to ethical data collection, as they ensure that customers actively agree to share their information. Whether it's subscribing to a newsletter, accepting cookies, or signing up for a loyalty program, opt-ins empower consumers to make informed choices about their data. For example, a retail Web site might use a checkbox during account creation to allow users to opt in to receive promotional emails, clearly outlining the types of messages they will receive.

Consent mechanisms should also be flexible and allow users to opt out at any time. Providing a simple way to withdraw consent demonstrates respect for customer preferences and builds trust. Additionally, offering granular options, such as choosing which types of emails to receive or which cookies to accept, further enhances the customer experience by giving them greater control over their data.

Ethical Implications of Tracking Cookies and Pixels

Introduced in the early days of the Web, tracking tools such as browser cookies that store key user information for later use, and pixels that send user behavior information to tracking servers are powerful tools for gathering data about user behavior, but their use comes with ethical considerations.

Browser cookies can track everything from page visits to shopping cart contents, while pixels can provide insights into how users engage with ads and emails. While these tools are invaluable for optimizing campaigns and improving the customer experience, they must be used responsibly to prevent breaching consumer trust.

For instance, third-party cookies, which track users across multiple Web sites, have come under scrutiny for their perceived invasiveness. Many browsers now block third-party cookies by default, and consumers are increasingly using ad blockers to protect their privacy. Marketers must adapt by focusing on first-party cookies and transparent practices that respect user choices. Similarly, businesses should ensure that tracking pixels are not used to collect sensitive information or create overly intrusive profiles.

Ethical data collection is not just about compliance—it's about aligning with customer expectations and values. By prioritizing transparency, consent, and the ethical use of tracking tools, marketers can create a data collection strategy that supports both business goals and long-term customer trust.

Data Retention

How long a business retains customer data is a critical consideration that affects both compliance with privacy regulations and the trust of customers in a company to safeguard their sensitive information. Thus, customer data retention policies must strike a balance between leveraging historical data insights for marketing strategies and respecting customer privacy. Thoughtful, secure, and compliant data retention practices can help organizations optimize their marketing efforts while reducing the risks associated with holding unnecessary or outdated data.

Determining Appropriate Customer Data Retention Periods

One of the first steps in effective data retention is determining how long data should be stored. This decision should align with business objectives, regulatory requirements, and customer expectations. For instance, financial institutions may be required by law to retain customer data for a certain number of years, while an e-commerce brand may only need to keep transactional data for as long as it is useful for marketing or customer support purposes.

Marketers should regularly evaluate whether the data they retain continues to add value. Data that has not been accessed or used in years is unlikely to contribute to actionable insights, and holding on to it unnecessarily increases storage costs and risks. Establishing clear retention periods ensures that data remains relevant and valuable to the organization while avoiding unnecessary liabilities.

Balancing Historical Data Insights with Privacy Concerns

Historical data can be invaluable for identifying long-term trends, forecasting customer behavior, and refining marketing strategies. Retaining data indefinitely, however, raises significant privacy concerns, especially as consumers become increasingly aware of how their information is used. Customers may view excessive data retention as invasive, eroding trust and damaging the brand's reputation.

To address this, marketers must balance the need for historical insights with the principles of data minimization and privacy. For example, anonymizing or aggregating older data can preserve its analytical value while protecting individual privacy. Additionally, businesses should clearly communicate their retention policies to customers, demonstrating transparency and accountability. By showing a commitment to ethical data practices, organizations can maintain customer trust while leveraging valuable insights.

Strategies for Secure and Compliant Data Deletion

Once data has reached the end of its retention period, it must be securely and permanently deleted to comply with regulations such as the GDPR and

CCPA. Effective data deletion practices reduce the risk of data breaches and demonstrate a business's commitment to protecting customer privacy.

Secure deletion involves more than simply erasing files; it requires processes that ensure data cannot be recovered or accessed by unauthorized parties. For instance, businesses can use advanced deletion software that overwrites data multiple times, ensuring it is permanently destroyed. Organizations should also establish protocols for deleting data across all storage systems, including backups and third-party platforms, to ensure complete compliance.

Maintaining accurate records of data deletion is equally important, as it helps businesses demonstrate compliance during audits or customer inquiries. Regular audits of retention practices and deletion processes can further ensure that organizations remain aligned with regulations and best practices.

Data Minimization

Focusing on the core objectives of data acquisition helps to prevent many risks and is also compliant with the law and ethical principles. It is also important that companies only request the data that is required for the specified purposes and do not accumulate more data than is necessary.

It is essential to have routine examinations of the data usage. For example, a subscription service that asks for birthdates at the start but never uses them in marketing campaigns should erase this field to simplify processes and decrease legal risks. This way, things are kept simple and, thus, there is no need to worry about a huge amount of data as well as the resources that would be required to analyze it.

Cross-Border Data Transfers

Moving data from one country to another presents unique issues that need to be understood from a legal perspective. For instance, the GDPR has strict rules on the transfer of data from the EU to third countries, which presents a problem to global businesses.

Some of these challenges can be addressed by data localization strategies. For instance, a SaaS company may choose to store data of European users in the EU and use Standard Contractual Clauses (SCCs) for the movement of data between different jurisdictions. These approaches are legal and do not jeopardize business operations.

AI and Privacy

AI is turning out to be useful in the area of privacy and compliance management. For example, AI-based tools can help with tasks such as identifying areas where data collection can be minimized, helping to ensure that the retention policies are properly enforced and that there are mechanisms to alert the organization about potential compliance risks. For instance, an AI system could be used to identify data that has been kept for more than the specified period, or to detect cases where there is a change in access controls.

While it can be used to prevent privacy issues, the use of AI in marketing also raises ethical issues of its own. The use of automated processes for the personalization of customer experiences or the collection and analysis of data must consider the business objectives versus the individual rights. This is why it is important to be transparent about how AI works and to ensure that algorithms are not biased or intrusive.

MarTech Platforms for Privacy and Compliance

There are several MarTech platforms that play key roles in helping organizations to meet a company's privacy and compliance needs without compromising on marketing performance. These tools make sure that the customer's information is handled in the right manner, in the right way, and at the right time, in order to meet the regulatory requirements and the expectations of the customer.

Consent Management Platforms (CMPs)

CMPs are used for collecting, managing, and recording customer consent. These platforms assist companies with complying with data protection regulations such as the GDPR and the CCPA, which call for consent prior to collecting or using personal information. These platforms provide a centralized way of managing consent preferences across different channels so that customers' rights are respected.

The features of CMPs include cookie banners that inform the users of the data collection practices, opt-in/opt-out options for specific data use, and a detailed record of all the activities performed during the regulatory examination. These CMPs can be integrated with other MarTech platforms, such as customer data platforms and analytics tools, to ensure that the consent preferences are respected across all the touchpoints of the customer journey. For example, when a user opts out of targeted advertising, the CMP communicates this preference to the DMP or ad tech platforms to exclude the individual from future campaigns.

Customer Data Platforms (CDPs)

For a more complete overview of what a CDP does, please refer to Chapter 9. In addition to other functions, CDPs also support the concept of privacy and compliance by collecting and storing customer data in a single, secure environment. It ensures that marketing professionals have an entire picture of the customer while ensuring that they are complying with the set regulations.

CDPs help a company enforce data minimization by collecting and using only the data that is required for marketing purposes and having integrations with data retention policies. For instance, a CDP can automatically delete old customer profiles after a certain period of time in order to prevent the accumulation of data that is not needed. Furthermore, CDPs assist in the management of customer preferences and permissions, which marketers can use when consent is given through the CMP.

Identity Resolution Platforms

While there are some standalone identity resolution platforms, CDPs often provide this functionality and promote compliance by identifying customers across different channels without compromising on the customer's right to privacy. These platforms work with hashed identifiers, which is a form of anonymization of data points such as emails, cookies, or a mobile device ID. Thus, these platforms are not likely to expose personal identifiable information (PII).

For instance, an identity resolution platform might associate a customer's behavior on a brand's Web site with their physical store activity so that the marketing teams can design custom campaigns without revealing real user identities. This enables the user to have a seamless experience across different touchpoints while ensuring that the company's compliance with the set regulations is not compromised.

CASE STUDY: VOLTAGE MOTORS AND THE CHALLENGE OF GDPR COMPLIANCE

Returning to VoltAge Motors, they recently expanded its operations into Europe. With this growth, the company saw an influx of European customers engaging with its Web site, app, and dealership network. While the expansion offered tremendous opportunities, it also introduced new challenges—chief among them, ensuring compliance with the GDPR.

The Current State

VoltAge's existing MarTech stack included a CDP for centralizing customer data and a CMP to capture user preferences. These systems were primarily configured for regions with less stringent privacy regulations, such as the United States, however. For example, their CMP displayed generic cookie banners but lacked the functionality to handle the granular consent options required by the GDPR, such as allowing users to opt out of specific types of data processing.

Additionally, their marketing team was not fully aligned with the IT and legal departments, leading to inconsistent handling of customer data. Marketing campaigns occasionally targeted European users without verifying whether they had given explicit consent, exposing VoltAge to potential non-compliance risks. Their existing data retention policies also failed to meet GDPR standards, as they lacked the automation to delete data after its intended use.

Key Issues Identified

Inadequate Consent Management

VoltAge's CMP was not designed to manage the detailed consent preferences mandated by the GDPR. For instance, while it could record whether a user had accepted cookies, it could not differentiate between consent for analytics, advertising, or personalization.

Data Localization Challenges

European customers' data was stored on US-based servers, raising questions about compliance with the GDPR's cross-border data transfer rules.

Disjointed Processes Across Teams

The lack of coordination between the marketing, IT, and legal teams meant that compliance gaps were not always identified or addressed promptly.

Steps Toward Compliance

To address these challenges, VoltAge initiated a comprehensive overhaul of its data privacy and consent management processes.

Upgrading the CMP

VoltAge implemented a GDPR-compliant CMP capable of handling granular consent options. The new platform allowed users to specify their preferences for analytics, advertising, and functional cookies, ensuring transparency and control. It also provided detailed audit trails to document when and how consent was obtained, a key GDPR requirement.

Improving Data Localization

The company partnered with a cloud provider that offered EU-based data centers, ensuring customer data collected in Europe remained within the region. Additionally, they updated their data transfer agreements to include SCCs, further solidifying compliance.

Establishing Cross-Functional Workflows

VoltAge created a privacy task force involving representatives from marketing, IT, and legal. This team implemented regular audits to identify compliance gaps and developed clear guidelines for handling customer data across campaigns. Training sessions ensured all departments understood the GDPR's requirements and their role in maintaining compliance.

Automating Data Retention Policies

Using their CDP, VoltAge configured automated processes to delete customer data after a specified retention period. For example, inactive customer profiles were purged after 18 months unless the customer re-engaged.

The Path Forward

With these changes, VoltAge Motors has significantly reduced its compliance risks and positioned itself as a trustworthy brand in the European market. Their upgraded CMP ensures that customers feel empowered to control their data, while improved data localization and retention policies align with the GDPR's requirements.

As VoltAge continues to grow, its commitment to data privacy will remain a cornerstone of its strategy, helping it build trust with customers while navigating the complex regulatory environment. Future initiatives include exploring AI-driven privacy tools to further automate compliance and maintain their competitive edge in an increasingly privacy-conscious world.

CONCLUSION

With customer data playing such a key role in all aspects of marketing, consumer data privacy and compliance have become inseparable from the function of marketing. As businesses strive to engage customers in meaningful and personalized ways, they must also respect and protect the data entrusted to them. Failing to address these concerns not only exposes organizations to legal and financial risks but also undermines the trust that is essential for long-term success.

MarTech platforms play a key role in bridging the gap between regulatory requirements and customer expectations. Tools such as CMPs, CDPs, and analytics systems provide the infrastructure needed to handle data transparently, securely, and ethically. These platforms empower marketers to align their campaigns with privacy standards while delivering engaging personalized experiences that build loyalty and aim to increase customer lifetime value.

Adopting a privacy-first mindset is not just a reactive measure to comply with laws—it is also a proactive strategy for ensuring sustainable growth amidst increasing competition and disruption. By investing in privacy-focused technologies, building collaborative workflows, and staying ahead of regulatory trends, organizations can safeguard their operations while building deeper, more trustworthy relationships with their customers. A privacy-first approach isn't a limitation—it's an opportunity to lead with integrity in an increasingly data-driven world.

NOTES

1. Pew Research Center. (2023, October 18). *How Americans view data privacy*. Pew Research Center. https://www.pewresearch.org/internet/2023/10/18/how-americans-view-data-privacy/

2. European Union. (2016). *Regulation (EU) 2016/679 of the European Parliament and of the Council of 27 April 2016 on the protection of natural persons with regard to the processing of personal data and on the free movement of such data (General Data Protection Regulation)*. EUR-Lex. https://eur-lex.europa.eu/legal-content/EN/TXT/PDF/?uri=CELEX:02016R0679-20160504

3. California Office of the Attorney General. (2018, June 28). *California Consumer Privacy Act (CCPA)*. California Department of Justice. https://oag.ca.gov/privacy/ccpa

4. Presidência da República do Brasil. (2018). *Lei nº 13.709, de 14 de agosto de 2018: Lei Geral de Proteção de Dados Pessoais (LGPD)*. http://www.planalto.gov.br/ccivil_03/_ato2015-2018/2018/lei/L13709compilado.htm

5. Innovation, Science and Economic Development Canada. (n.d.). *Consumer Privacy Protection Act*. Government of Canada. https://ised-isde.canada.ca/site/innovation-better-canada/en/consumer-privacy-protection-act

6. U.S. Department of Health and Human Services. (2013). *Health Insurance Portability and Accountability Act (HIPAA) simplification*. U.S. Department of Health and Human Services. https://www.hhs.gov/sites/default/files/hipaa-simplification-201303.pdf

7. Federal Trade Commission. (n.d.). *Gramm-Leach-Bliley Act*. Federal Trade Commission. https://www.ftc.gov/business-guidance/privacy-security/gramm-leach-bliley-act

8. Digital Advertising Alliance. (n.d.). *Application of the self-regulatory principles to the mobile environment: Best practices for connected devices*. Digital Advertising Alliance. https://digitaladvertisingalliance.org/sites/aboutads/files/DAA_files/Connected_Devices_Best_Practices.pdf

PART 3

CREATION, WORKFLOW, AND OPERATIONS

4 Categories of MarTech Platforms

Customer Data
Helping brands understand their customers

Content, Campaign, and Multichannel Delivery
Serving customers with content, offers, and experiences across the journey

Creation, Workflow, and Operations
Empowering teams to be more efficient in content and campaign creation

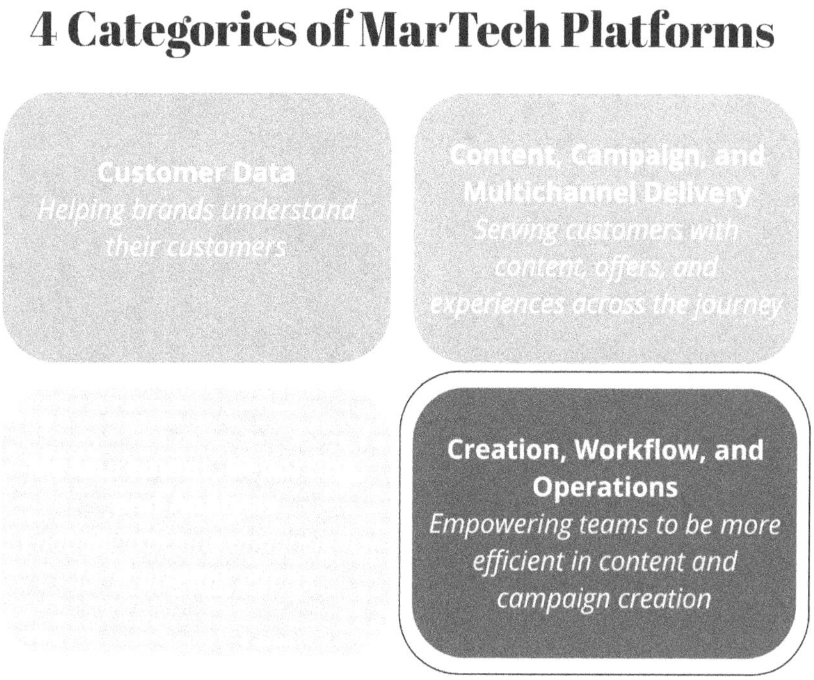

FIGURE P3.1 Creation, Workflow, and Operations

The next part of this book will explore MarTech platforms that enable teams to more efficiently create content and marketing campaigns. These platforms include project management tools, content management platforms, and workflow automation tools.

Building on the customer data foundation explored in the previous part, these platforms enable marketing teams to create content and manage the process of putting together small-scale or large-scale marketing campaigns across multiple channels.

CHAPTER 12

Key Considerations with Creation, Workflow, and Operations

A critical component of marketing is the creation of content and other assets to be used in marketing campaigns and other branded assets from Web sites, mobile apps, in-store retail displays, social media videos, or direct mail pieces, to name just a few. Thus, *how* these materials are created forms the second category of MarTech platforms.

These platforms help marketing teams produce high-quality content at scale while maintaining brand consistency and, in the best cases, they foster greater collaboration amongst the increasingly diverse teams that need to work together to create multi-channel marketing campaigns.

This chapter explores some of the key considerations that go into choosing the best and most appropriate MarTech platforms to enable teams to create compelling, on-brand content efficiently and effectively, at the scale that multi-channel marketing efforts demand.

STRATEGY AND THE CREATIVE PROCESS

While the creative output of marketing teams was traditionally in the form of hand-drawn sketches and advertising copy written using manual typewriters, the modern marketing team is heavily reliant on computers at every stage in the creative process, from strategy to execution.

This dependence on technology includes initial audience research, as well as the need for content planning and ideation tools.

Content Planning and Ideation Tools

Successful creative marketing strategies begin by pinpointing suitable ideas and developing them into workable action plans. While these plans have been created in a variety of methods in the past, from group brainstorming to sketching ideas for later consideration, AI-assisted brainstorming platforms accelerate the ideation process through theme and angle recommendations based on large data collections. These solutions incorporate audience data including demographic preferences and engagement patterns to produce content that connects with specific audience segments. When marketing teams embed data-driven decision-making into their creative process, they create campaigns that meet real audience requirements resulting in stronger outcomes and fewer planning mistakes.

Asset Management

Efficient digital asset management (DAM) streamlines collaboration between teams of all sizes and becomes necessary once ideas reach their final form, where they may need to be accessed by a variety of internal and external teams involved in the planning and execution of marketing efforts. Throughout the content creation process, marketing team members need to store images, videos, copy, and other content elements in an organized manner that allows quick locating, updating, and repurposing. DAM systems provide centralized control over content versions, which helps maintain consistency and currency across all channels. Marketing teams who organize their assets through clear naming conventions and metadata tags will prevent redundant work and save time while preserving their brand image.

Aligning Creative Efforts with Marketing Objectives

Creative work requires connection to overarching marketing goals, including brand awareness, the potential for lead generation, customer acquisition, and other end goals, since it cannot exist independently and serve its function.

Thus, creative teams that support the marketing function must check whether each content item supports the overall strategy before assets receive approval or deployment. The strategic alignment of creative outputs maximizes their ability to produce measurable results while reinforcing brand positioning. Marketers following clear guidelines and objectives during their processes can prioritize ideas while crafting targeted messages and maintaining a unified narrative that produces business results.

Multi-Channel Needs

The discipline of marketing is wide-reaching, spanning social media, Web sites, email campaigns, television ads, and much more. Each of these channels introduces specific creative and content requirements for format, tone, and targeting. Coordinating all of them effectively requires a clear strategy and diligent oversight, which is often augmented by software platforms that store, track, and provide access to this diverse set of assets. Since teams must track various timelines and dependencies, as well as manage a wide range of assets, this can strain both creative resources and project management capacity without the right tools to support them.

Channel-Specific Optimization

An additional consideration for marketers is the unique needs of each endpoint that their content reaches customers through. Not all marketing channels treat content in the same way, nor are they engaged with in the same manner by consumers. Therefore, a message that works on Instagram may fail to engage on LinkedIn or in an email newsletter. Each platform demands unique formats, from short-form posts to long-form articles or video scripts.

Additionally, in regulated industries, certain channels carry additional compliance guidelines that marketers must respect. Accounting for each channel's nuances leads to better audience engagement and helps minimize legal and reputational risks.

Repurposing Content and Assets

Creating entirely original content for every channel can quickly deplete time and budgets. Repurposing allows marketing teams to adapt core material for different audiences or contexts. A well-researched blog post can be condensed into social media updates, transformed into an email newsletter, or turned into a short video. This approach ensures consistency while maximizing the value of each asset produced. Providing centralized access to content and assets with helpful tagging taxonomies and effective search mechanisms can provide advantages to teams that need to make a single piece of content translate across many different channels.

Automation and AI-Driven Recommendations

Automation tools simplify channel management by scheduling posts, tracking performance metrics, and suggesting content adjustments in real time. More recently AI-based systems or generative AI features within existing platforms can analyze user behavior, forecast what resonates with particular demographics, and pinpoint the best times to publish. These insights help teams personalize messaging for each platform, reduce manual oversight, and maintain relevance in fast-paced, multi-channel environments.

Efficiency

Inefficient workflows can delay campaign launches, waste creative resources, and increase the likelihood of errors. When marketing teams lack unified processes or rely on outdated methods, they often spend unnecessary time locating assets or redoing work. MarTech platforms address this inefficiency by providing centralized dashboards, real-time collaboration features, and clear task ownership to keep projects on schedule.

Project and Task Management Tools

Software platforms such as Asana, Monday.com, and Wrike act as command centers for content creation, enabling task assignment, deadline tracking, and progress visualization. These systems help teams break larger projects into manageable chunks and clearly outline responsibilities. By integrating with other marketing tools—such as content repositories or CRM systems—task management platforms ensure that everyone has immediate access to the latest assets, data, and guidance.

While not specifically created for marketers, many organizations adopt enterprise-wide tools such as Jira, which can also be adopted by marketing teams to assist in their management of campaigns and initiatives.

Workflow Automation

Automation tools reduce manual tasks across multiple stages of content production and help marketing teams avoid the risks of relying on manual checks and balances, or one-off communication via Slack, Teams, or emails. AI-driven approvals can flag potential issues or noncompliant language, while automated content distribution pushes materials to various channels at optimal times. Automated processes can also apply pre-built templates, ensuring consistent design and messaging from the outset. These features not only save time but also help teams maintain a steady flow of well-coordinated content.

Version Control and Approval Processes

Marketing teams that rely on file-naming conventions to be their guide to which version of a document is most up to date are often plagued with issues, particularly when timing and accuracy are most critical. Thus, maintaining version control is essential for brand consistency, legal compliance, and timely distribution, placement, or publication.

Many MarTech platforms include built-in review and approval workflows. This functionality keeps track of edits, logs changes, and clarifies who is responsible for sign-offs at each stage. By combining these features with automated alerts, teams can prevent the duplication of work and confirm that final assets meet brand and regulatory standards.

Collaboration

Writers, designers, video editors, and strategists all bring specialized skill sets that must align with broader marketing goals. Without clear workflows, projects can lose momentum, and teams can misinterpret each other's deliverables. MarTech platforms address these risks by offering centralized dashboards and communication channels, ensuring tasks and deadlines remain visible to everyone involved.

Content Collaboration Tools

Platforms such as Google Workspace, Notion, and Airtable create real-time, shared work environments where team members can edit, comment on, and review content together in a virtual, yet collaborative environment. These tools reduce version confusion that can often arise from emailing drafts of documents back and forth, provide immediate feedback loops, and often integrate with popular project management or analytics systems. By keeping discussions, documents, and data in a single location, they help maintain transparency and speedy decision-making.

Cross-Team Coordination

Effective collaboration goes beyond the immediate creative team. Marketing, sales, product, and customer experience teams each hold vital pieces of customer data and feedback. By leveraging platforms that connect these departments—such as CRMs or shared analytics dashboards—organizations can ensure consistent messaging and informed campaign planning. Close alignment also helps reduce rework, as every team understands the project's objectives and timelines from the start.

Remote and Global Teams

Working across time zones and regions requires cloud-based tools that remain accessible at all hours. Modern collaboration platforms enable document editing, asset sharing, and asynchronous communication, allowing geographically dispersed teams to stay productive without waiting on live meetings. This flexibility broadens the talent pool and boosts overall efficiency, as contributors can update or review materials whenever they are available.

AI'S GROWING ROLE IN CONTENT CREATION

Within a relatively short period of time since the widespread adoption of generative AI tools such as ChatGPT, Claude, and Jasper, the content development cycle now depends heavily on AI technology. AI tools assist content teams from idea generation through initial copywriting by decreasing turnaround times, which allows employees to dedicate their efforts to more complex tasks. AI systems analyze extensive datasets to deliver personalized content suggestions for distinct audience segments, which keeps content relevant and up to date.

Both detailed strategies and explicit protocols are essential for integrating AI into existing workflows. Marketers need to learn how to utilize these tools for conducting research and performing drafting, editing, and quality assurance tasks. Education for users should include understanding AI's limitations, which encompass the potential risks of using generic or unverified information. Teams need to recognize scenarios where human oversight is crucial, such as when sensitive data verification occurs or when preserving a brand's unique voice.

Leaders need to develop best practices that specify the appropriate roles and situations for AI engagement. Automated content generation proves effective for basic outlines or short summaries, but high-stakes copy and complex narratives require human expertise for creative brand elements. Organizations can benefit from AI capabilities yet prevent automation-related problems through appropriate balance.

CASE STUDY: VOLTAGE MOTORS' QUEST TO MANAGE CONTENT CREATION

At VoltAge Motors, the marketing team is finding it increasingly difficult to manage content creation across the many marketing channels they use in their customer acquisition and retention efforts. Marketing, product, and sales teams often work in silos, leading to slow approval cycles, inconsistent messaging, and repeated work on similar assets. Version control is a constant headache, resulting in confusion about which files were the latest and who had final editing rights.

A Disconnected Set of Tools

Before deciding to make any changes, VoltAge Motors relied on a patchwork of disconnected tools for marketing content storage and coordination. Different marketing teams used separate project management apps depending on the channels they were assigned to, while sales teams tracked materials in a basic file-sharing tool that many in the marketing department were not even able to access. Thus, important updates to files, often relating to time-critical ad campaigns and car model launches, failed to sync between platforms, causing duplications and conflicting campaign details. As a result, misaligned messaging sometimes reached customers, reflecting poorly on the brand's professional image.

The Solution

Recognizing the need for a more unified approach, VoltAge Motors made the decision to consider the implementation of a centralized content management and workflow automation platform. This system would provide a single hub for drafting, storing, and reviewing assets, along with built-in approvals and version control features. Marketing and sales teams would gain access to shared calendars and real-time asset updates, simplifying cross-department communication. Automation tools would be able to streamline distribution, allowing the company to push approved content to multiple channels without manual duplication.

Next Steps, Results, and Lessons Learned

While the needs were clearly articulated, the VoltAge Motors team also acknowledged that the symptoms they were experiencing in the area of content creation were similar in some ways to the disconnects they experienced in other marketing and campaign-related tasks.

Thus, instead of immediate action, they continued to collect requirements from other teams about their MarTech infrastructure needs.

Despite temporary frustration at having defined a solution without immediately trying to solve it, this approach would pay off later. Then, the VoltAge team was tasked with doing a more comprehensive evaluation of the MarTech infrastructure and creating a plan to solve several wide-reaching challenges.

CONCLUSION

Marketing functions within an organization rely heavily on creation, workflow, and operations platforms for its foundational support. These platforms centralize tasks while coordinating team efforts to enable smooth asset flow from initial concepts to final launch. Technology solutions that oversee processes from brainstorming to approvals enable organizations to establish consistent operations while eliminating redundant tasks and upholding accountability. Multi-channel environments require these systems because different content formats and audience segments need tailored delivery.

The next chapter will take a look at the first of three types of platforms that enable marketing teams to produce better content more efficiently and effectively: content marketing platforms (CMPs).

CHAPTER 13

Content Marketing Platforms

The scope of content marketing extends well past traditional blog posts and social media updates. In today's crowded digital world, brands need content that goes beyond creativity to include strategic planning and data analysis for consistency. Content marketing platforms (CMPs) fulfill the need for strategic content management in modern marketing. CMPs organize content strategy processes through the centralization of planning and creation tasks while tracking performance to establish structured marketing approaches that align teams and maximize impact.

A CMP goes beyond basic content storage or publishing functions to serve as a complete content marketing hub with integrated modules for complete content management. Marketers can orchestrate multi-channel content campaigns through a single platform that integrates editorial calendars with asset management tools, workflow automation features, distribution tools, and analytics capabilities. CMPs optimize content management throughout its entire lifecycle by eliminating inefficiency and boosting team collaboration while delivering insights based on data analysis.

A CMP provides a foundational structure for efficient and results-oriented content operations while helping marketing teams manage large volumes of high-quality content. This chapter will illustrate how CMPs integrate within a comprehensive MarTech stack and guide them through selecting the right platform to enhance marketing campaign effectiveness.

How a Content Marketing Platform Helps Marketers

A CMP functions as a strategic center that boosts operational performance by improving collaboration and efficiency throughout all content marketing activities. CMPs bring together planning, creation, distribution, and analytics in a unified platform to help marketers make better decisions and improve resource use for optimal results. The following are several key contributions CMPs make to contemporary marketing strategies.

Streamlining the Content Process

Content teams often handle various spreadsheets and email messages while using separate tools to keep projects organized when there is no established system. These issues generate operational inefficiency alongside miscommunication and repeated work effort. A CMP integrates content planning and teamwork, which makes sure that all participants, including writers and designers as well as editors and strategists, follow the same structure. Using shared editorial calendars along with workflow automation and content libraries eliminates duplication and maintains efficient project advancement. A unified platform enables teams to synchronize their work processes, which helps them achieve consistent results while meeting deadlines with greater reliability.

Improving the Quality of Content

A brand's credibility suffers when its content remains unreviewed or inconsistent due to rushed production. Through structured review and approval processes, a CMP makes sure all content reaches a polished state before publication. Teams can access content assets within the platform while simultaneously tracking revisions and providing feedback to uphold quality standards. CMPs are often equipped with content scoring tools and readability assessments alongside compliance checks, which empower marketers to fine-tune messages and maximize content effectiveness. Brands can both improve their content quality and ensure voice consistency throughout all channels by embedding these features into their workflow processes.

Enhancing Personalization Efforts

Current market consumers anticipate receiving content that aligns with their distinct needs along with their preferences and behaviors. Marketers use CMPs to develop personalized content for targeted audience segments by utilizing demographic information along with data from past interactions and browsing patterns. Certain platforms make use of AI algorithms to provide recommendations that help content creators customize messages according to the needs of distinct audience groups. This personalization goes beyond solely text modifications, as CMPs often have the ability to allow teams to modify images and even layouts as well as language to fit specific personas that enhance engagement and relevance for each reader.

Optimizing Content Distribution

Although producing high-quality content is essential it remains incomplete until you deliver it to the right audience at the optimal moment. Through content performance tracking across various channels, CMPs enable marketers to discover which content formats yield the best results on specific platforms. A CMP enables teams to modify their content strategy by transforming long-form blog posts into shorter social media snippets when LinkedIn success contrasts with Twitter challenges. Content performance management systems integrate with SEO platforms and email marketing tools as well as paid advertising systems to enhance content amplification and maximize reach.

CMPs provide essential support to contemporary marketing teams through their structured workflows and automation capabilities, which deliver data-driven insights for enhanced content performance. Organizations can create personalized, high-quality content that reaches wide audiences effectively by utilizing these platforms to drive business growth through improved audience engagement.

WHO ARE THE USERS OF A CMP?

The CMP acts as the fundamental support system for all teams responsible for content planning, along with content production and distribution. It operates as a collaborative environment that unites input from strategists, creators, and analysts to align content and execute it efficiently while maintaining its impact.

Content Strategists and Marketing Managers

These professionals handle content strategy and guarantee that all activities support business objectives and specific marketing campaigns while reaching target audiences. They establish the content calendar while prioritizing content tasks and evaluating performance results.

Let's look at some typical tasks and a workflow for content strategists and marketing managers:

- Operate the platform to create and control the editorial calendar.
- Distribute content topics to writers and designers according to the objectives of each campaign.
- Employ audience segmentation tools to create detailed target persona profiles.

- Ensure brand messaging alignment by approving and refining content drafts.
- Track content performance statistics to inform strategic adjustments.

Writers, Editors, and Content Creators

Writers and editors create compelling written content, while designers and video creators produce visual assets that support campaign initiatives.

Let's look at some typical tasks and a workflow for writers, editors, and content creators:

- Marketing managers assign tasks and distribute content briefs to team members.
- Retrieve brand assets, templates, and guidelines that have been pre-approved from the CMP storage system.
- The platform allows real-time collaboration, where users can add drafts and receive immediate feedback.
- Submit content for internal review and approval.
- Leverage built-in analytics and SEO tools to enhance search engine visibility and engage your audience with optimized content.

SEO Specialists and Content Optimizers

Search Engine Optimization (SEO) specialists work to optimize content so that it achieves better search visibility and drives audience engagement. These team members evaluate trends in keywords and backlinks while examining on-page elements to enhance search rankings.

Let's look at some typical tasks and a workflow for SEO specialists and content optimizers:

- Identify appropriate keywords through research and apply them to content briefs.
- Utilize SEO tools within CMP to evaluate your content's readability and its relevance to search engines.
- Monitor search engine performance data and offer recommendations to improve content rankings.
- Determine which successful content pieces should be adapted or refreshed for new use.

Social Media and Email Teams

The social media teams distribute content through social channels and email marketing, alongside other distribution platforms. Their objective is to deliver content to appropriate audiences through effective methods.

Let's look at some typical tasks and a workflow for social media and email teams:

- Repurpose platform content for distribution across multiple channels.
- Create a social media posting schedule and enable automation for all channels.
- Team members need to work together with designers to make visual adjustments suited for each specific distribution channel.
- Track user interaction with content and adjust distribution methods based on these insights.

Project Managers

These team members ensure that content production remains on schedule and meets required deadlines. They coordinate tasks, approvals, and content deployment.

Let's look at some typical tasks and a workflow for project managers:

- Within the platform, operators distribute tasks and control the progression through workflow stages.
- Monitor workflow progress while identifying any delays or bottlenecks within the process.
- Oversee the approval process to confirm content review completion prior to publication.
- Leverage workflow automation tools to make content requests and stakeholder feedback processes more efficient.

Analytics and Performance Teams

These teams how well content performs by measuring success through key metrics, including engagement levels, conversion rates, and return on investment (ROI).

Let's look at some typical tasks and a workflow for analytics and performance teams:

- Gather insights from built-in reporting dashboards.
- Evaluate content performance by comparing results across multiple formats alongside various channels and different audience segments.
- The analytics team should deliver guidance on enhancing content effectiveness along with strategic directions for future development.
- Identify underperforming content and suggest improvements.

As demonstrated, a CMP accommodates a diverse amount of marketing-related roles and their unique needs. A central place to work can often facilitate better teamwork and workflow efficiency while providing transparency into content performance results. As will be demonstrated in the following chapters and sections, there are also some overlaps between what a CMP does and what some other MarTech platforms can do.

KEY FEATURES

CMPs can function as robust systems that manage content workflows while maintaining strategic goal alignment with tools to foster audience engagement. While the actual features of any CMP will vary, the following are some of the most common features.

Content Planning and Strategy

Executing effective content marketing successfully requires a detailed strategy, which a CMP implements through its structure for seamless deployment. Content calendars enable teams to plan and arrange content well ahead of time, which helps maintain campaign consistency while supporting broader marketing goals. Content calendars enable teams to track production deadlines and publishing schedules to maintain their marketing timelines.

Marketers find content opportunities through search demand and industry trends with the help of keyword research and competitor analysis tools in a robust CMP. Analyzing competitor content enables businesses to improve their strategies by identifying gaps and creating unique messaging. Marketing professionals utilize campaign management capabilities to manage multichannel initiatives and track performance across different touchpoints while maintaining business goal alignment.

Content Creation Tools

CMPs provide built-in content editors and templates that enable writers, designers, and videographers to create brand-consistent content through streamlined production processes. The editors provide various functions, including text formatting and multimedia embedding, together with structured content layouts that help keep all content within company guidelines.

The search visibility of content is improved through SEO optimization tools in many CMPs that examine readability metrics and keyword density while inspecting metadata as content is created. The tools provide structured content organization which maximizes organic reach. Certain platforms integrate AI-driven content creation support which suggests headlines and sentence structures along with automated creation based on target audience analysis. Content production speeds up with AI features, which maintain a high-quality standard.

Collaboration Tools

Team collaboration is fostered through a CMP's workflow management systems, which handle task assignments and deadline settings while tracking progression. These workflows create a seamless content development process that spans ideation through creation and review to final publication.

Within the platform, real-time collaboration tools give teams the ability to work jointly by enabling writers, designers, and strategists to provide comments and make immediate edits and changes. The system's integrated feedback mechanism eliminates dependencies on outside communication tools while centralizing all feedback responses. The version control and approval systems enable teams to record document modifications while preserving an audit trail and eliminating the risk of publishing antiquated information.

Content Distribution

Many CMPs enable teams to simultaneously publish content on Web sites along with social media platforms, email campaigns, and digital advertising networks without having to use separate tools.

Using automated scheduling and posting features, marketers can organize content distribution ahead of time, which guarantees that posts appear at the best possible moments. Certain platforms offer content personalization features that modify messaging and formats to suit audience preferences while adapting images based on location or prior user interactions.

Also note that CMPs are often separate from social media management and other channel-specific platforms as well, and these will be discussed in their respective chapters.

Analytics and Performance Tracking

While there are many tools that allow marketers to assess content effectiveness, CMPs often deliver built-in performance data, including page views, engagement rates, social shares, and time spent on content.

Marketing professionals gain insights into audience behavior and engagement statistics through these platforms, which reveal how different segments respond to content. Through the use of CMPs marketers can determine which blog topics lead to maximum conversions as well as identify video formats that result in higher engagement rates. When integrated throughout the customer lifecycle, ROI measurement tools help businesses monitor how their content marketing strategies generate revenue and evaluate investment effectiveness.

Asset Management

Marketing teams can quickly access necessary materials when digital assets are stored and managed in a centralized location. In some cases, CMPs offer digital asset management (DAM) capabilities that efficiently sort and maintain images, videos, templates, and brand collateral for consistent use and minimized duplication. Note that DAM platforms are also often standalone, as we will see in Chapter 14.

The implementation of content tagging and organization systems allows marketers to efficiently search for assets by using keywords, metadata, or campaign labels. This feature helps teams avoid time spent looking for files while guaranteeing access to current content.

Integration Capabilities

The best performance of a CMP occurs when it integrates flawlessly with different components of the MarTech stack, as these platforms rarely stand on their own. Marketers can unify their data through integrations with CRM systems and email marketing platforms as well as analytics tools, which ensures that content supports overall business objectives.

API access for custom integrations provides businesses with unique requirements for deeper connectivity between their proprietary systems and third-party applications. Content marketing initiatives maintain synchronization with enterprise digital strategies through flexible integration capabilities.

Brand Management

Brand consistency remains vital to establish credibility and recognition. CMPs provide brand management tools that maintain style guidelines and ensure tone and message consistency for teams operating in different locations.

Brand guidelines combined with style enforcement features make certain that all content follows established standards while minimizing off-brand communication risks. Enterprises benefit significantly from these tools when they have several content creators or decentralized marketing departments.

Through the utilization of most or all of these capabilities, marketing teams are able to achieve higher operational efficiency while maintaining uniform brand messaging along with amplifying content effectiveness throughout their communication channels. Selecting the appropriate CMP platform requires a thorough analysis of the business requirements along with the scalability potential and integration capabilities.

HOW TO EVALUATE A CMP

The exact needs of an organization will depend on many factors, however, there are some core criteria that can be used to choose a suitable CMP. The following are five such criteria.

Usability and User Experience

The primary purpose of a CMP should be to boost productivity without creating extra complexity. All team members including strategists, writers, designers, and analysts should find the platform intuitive to use.

Let's take a look at the evaluation criteria:

- *Interface simplicity*: The platform provides a simple dashboard design that enables both technical experts and non-technical staff to navigate it effortlessly.

- *Customization options*: Does your team have the ability to modify workflows and content templates as well as organize dashboards according to their specific requirements?

- *Onboarding and training support*: Does the vendor provide comprehensive documentation, as well as tutorials and customer support, to enable fast team onboarding and training?

An intuitive design for a CMP helps decrease the training time and boosts team adoption, which results in smoother operational processes.

Integration Capabilities

A CMP requires seamless integration with existing marketing and business tools to function effectively. Integration facilitates better workflow efficiency while simultaneously minimizing data silos.

Let's take a look at the evaluation criteria:

- *Pre-built integrations*: Does the CMP establish seamless connections with fundamental systems, including CRM solutions such as Salesforce and HubSpot, as well as email marketing platforms, DAMs, and analytics suites?
- *API and customization flexibility*: Is the system equipped with API capabilities to enable custom integrations when pre-built connectors are not available?
- *Cross-platform data flow*: The platform must manage data transfer seamlessly with existing MarTech solutions to facilitate the sharing of insights and assets throughout the whole ecosystem.

Strong integration features in a CMP prevent operational bottlenecks while aligning content marketing strategies with wider corporate goals.

Content Management and Workflow Automation

The main function of a CMP should be to make the process of content planning and distribution more straightforward instead of adding complexity. Workflow automation, together with asset organization and approval streamlining, is a fundamental requirement.

Let's take a look at the evaluation criteria:

- *Editorial and campaign planning*: Does the CMP feature content calendars, along with scheduling tools and campaign tracking capabilities for editorial and campaign planning?

- *Workflow automation*: Do teams have the ability to automate content assignments together with approvals and publishing while maintaining version tracking to boost efficiency?

- *Collaboration features*: Does the platform support real-time collaboration, which helps teams move away from relying on external communication tools such as email and Slack?

Structured workflows within a platform deliver quicker content production and minimize redundancy while improving team handoffs.

Analytics and Performance Tracking

A CMP enables teams to refine their strategies through data-driven insights into content performance instead of relying on intuition. While teams may have other tools to evaluate either channel-specific or journey-based analytics, often a CMP provides a focused view into targeted content performance.

Let's take a look at the evaluation criteria:

- *Granularity of reporting*: A comprehensive CMP delivers precise performance metrics, including audience engagement levels along with conversion rates and the ability to track multi-channel attribution.

- *Real-time insights*: Marketers need the ability to track content performance in real time to make immediate strategy adjustments.

- *Predictive analytics and AI*: Does the platform use AI insights and content recommendations to boost future content effectiveness?

Marketers struggle to make informed decisions without access to powerful analytical tools. Data-driven decision-making becomes feasible through a strong CMP that guarantees that content investments produce measurable outcomes.

Scalability and Future-Proofing

As content requirements develop businesses need a CMP that can grow alongside the organization instead of becoming outdated.

Let's take a look at the evaluation criteria:

- *Support for various content formats*: The platform must support diverse content types including blogs, videos, podcasts, and interactive media while marketing strategies continue to develop.

- *Enterprise-readiness*: Does the CMP enable simultaneous use by multiple users and global teams while managing large content repositories for enterprise-level operations?
- *Adaptability to emerging trends*: The platform receives regular updates to keep pace with the latest developments in content marketing trends as well as privacy regulations and AI innovations.

A scalable CMP prevents future expensive platform migrations while enabling businesses to meet new marketing demands without needing to change platforms.

Selecting an appropriate CMP for your organization, like any component of a MarTech stack, requires a long-term investment perspective as well as a clear understanding of the current and potential future needs of the teams using them. Marketers need to evaluate what matters most to them, test platforms using demos or trials, select a system that improves their capacity to create, distribute, and optimize high-quality content instead of obstructing it.

RELATED PLATFORMS

These tools offer support for multiple elements of content strategy development as well as content creation and audience interaction.

Content Management Systems (CMSs) and Digital Experience Platforms (DXPs)

Optimizely, Sitecore, Drupal, and WordPress are platforms that deliver content management and publishing capabilities that enable content delivery through Web sites and digital experiences. DXPs and CMSs specialize in presenting content effectively while optimizing it for user experience, whereas CMPs focus on workflow and strategic processes. The book will cover these platforms in more detail in Chapter 17.

Focused Content Creation and Collaboration Tools

The platforms Contently, Kapost, and DivvyHQ focus on helping users generate content ideas and collaborate editorially during content creation. Content production workflows become more efficient through these platforms, which enable teams to delegate tasks while tracking progress and managing approvals. Dedicated content creation tools specialize in the writing and creative process and avoid full content lifecycle management, unlike some CMPs that offer similar functions.

Content Curation Platforms

Scoop.it and Curata provide marketers with the means to identify valuable third-party content to supplement their original material through discovery and curation processes. For brands that need ongoing industry-specific content, these tools allow continuous updates without complete original content creation. These platforms work with CMPs to develop content strategies that extend past-owned assets.

Video Content Marketing Platforms

As video content becomes increasingly popular and effective at driving consumer engagement, specialized platforms have developed to support video creation and distribution as well as management processes. These tools facilitate scriptwriting while supporting editing and publishing functions together with performance tracking to optimize video content engagement throughout YouTube social media channels as well as brand Web sites.

INFLUENCER MARKETING PLATFORMS

Influencer marketing platforms enable brands to discover content creators and develop as well as manage collaborative relationships with them. The software delivers detailed analytics for influencer reach and engagement while simplifying campaign teamwork and monitoring content performance. The growing importance of influencer content in marketing strategies enables these platforms to enhance CMPs by merging influencer content into comprehensive marketing campaigns.

These related platforms are often used in conjunction with CMPs and fulfill niche content creation and management needs.

CONCLUSION

CMPs can have a broad range of functions, collectively helping marketers plan, create, distribute, optimize, and measure the performance of their content marketing efforts. The specific tools used may vary depending on the organization's needs and the complexity of their content marketing operations.

In the next chapter, DAM systems will be explored.

CHAPTER 14

DIGITAL ASSET MANAGEMENT (DAM) SYSTEMS

Businesses today generate such large amounts of digital content that a systematic method is necessary for organizing the storage and distribution of assets. A digital asset management (DAM) system functions as a centralized storage space where organizations can oversee and distribute digital assets such as images, videos, documents, brand materials, and multimedia content. Content management platforms allow teams to access necessary assets with speed while enforcing version control and brand uniformity throughout marketing and communication activities.

DAM systems developed in response to the growing use of digital media. Businesses began their digital content organization efforts using simple file storage systems along with shared drives and hard-coded repositories. Traditional storage solutions became inefficient when digital marketing and omnichannel strategies grew more complex because they caused duplication issues and mismanaged content alongside workflow bottlenecks. Structured platforms known as DAM systems started to appear in the early 2000s and they included metadata tagging and categorization features along with search capabilities, which improved asset location for teams.

The exponential growth in content production drove modern DAM platforms to adopt cloud-based storage and AI-powered tagging while introducing automated workflows and access control to boost security and usability. Modern brands managing extensive content libraries depend on these systems to ensure proper governance and compliance while maintaining operational efficiency throughout their global teams. The upcoming sections provide insights into essential features and evaluation standards while presenting best practices for DAM system implementation that meet business requirements.

How a DAM Helps Marketers

Content serves as the fundamental element of brand storytelling, customer engagement, and experience for marketing teams. The management and optimization of content assets become increasingly complex when organizations expand their content production operations. DAM systems enable marketing teams to effectively manage digital assets throughout their campaigns and channels by providing streamlined access and organization.

WHO ARE THE USERS OF A DAM?

Marketing teams form the core user base for DAM systems, but creatives and sales teams, along with IT and legal departments and external partners, also utilize DAM systems to handle digital assets efficiently.

Marketing Teams

Marketing teams depend on DAM systems to store digital content and deploy it across various channels. Marketing teams utilize DAM systems to maintain brand uniformity while managing the large amounts of assets required for campaigns and product launches. This includes the following uses:

- Organize and store-branded assets such as logos, graphics, videos, and promotional materials using a DAM system.
- Access approved marketing assets for use in social media posts, email campaigns, and advertising initiatives.
- Distribute content seamlessly across multi-channel marketing initiatives.
- Ensure that teams use up-to-date, brand-compliant materials.
- Monitor content effectiveness through analysis of asset usage information.

Marketing teams can concentrate on their strategic and execution tasks since a DAM removes obstacles related to retrieving assets and managing files.

Creative Teams (Designers, Videographers, and Copywriters)

Designers alongside videographers and copywriters develop and improve digital assets, which are maintained in a DAM. Creative teams require a centralized system they can use to upload content and perform editing tasks while tracking different content versions with efficiency. This includes the following uses:

- Creative teams can store both raw and completed design files together with videos and branded templates in the DAM.
- Maintain version control to make sure distribution includes solely the most recent approved files.
- Creative teams must work together with marketing departments to approve assets while maintaining the correct brand alignment.
- Ensure that high-resolution photography and video clips are stored alongside appropriate metadata to simplify search functions within your DAM.

Creative teams that use a DAM eliminate duplicated work while maintaining consistent outputs and achieve smoother collaboration across departments.

Sales Teams

Sales teams must have constant access to current sales collateral and customer-facing materials including product presentations, and they can often spend a lot of time searching for the correct sales deck or brochure version without a DAM in place. Their usage includes the following:

- Sales teams can rapidly access product sheets and case studies along with pitch decks.
- Real-time access to approved sales collateral helps reduce customer interaction delays.
- Maintain consistency between sales materials and current marketing messages.
- Deliver customized content to prospects that addresses their specific interests and requirements.

Sales reps can extract appropriate content from DAM systems while engaging with customers by integrating DAM systems with CRM platforms such as Salesforce.

Product Management Teams

Product teams rely on both accurate and current assets to effectively manage product launches and updates while conducting internal training. A DAM provides seamless access to product specifications as well as feature documentation and launch materials. This includes the following uses:

- Product teams should save demonstration videos and store feature comparison documents alongside product documentation.
- Work with marketing groups to verify that product communications maintain their accuracy.
- Establish a repository for past product versions and assets so teams can access them as needed.

Product teams benefit from DAM systems because they establish an organized approach to handling product updates and customer education content.

Legal and Compliance Teams

Amidst rising issues about intellectual property and data privacy together with content rights legal teams require access to information about digital asset storage and usage. This includes the following uses:

- Keep an eye on digital rights management (DRM) to stop unauthorized usage of copyrighted materials.
- Assets must adhere to their licensing terms and track their expiration dates to ensure compliance.
- Create a compliance tracking system for asset usage that focuses on regulated sectors such as finance and healthcare.

DAM systems help organizations lower legal risks by maintaining compliance between digital content usage and both industry standards and internal corporate guidelines.

IT and Data Teams

The IT team works to merge the DAM with current technological setups while data teams utilize the DAM system to measure content effectiveness and user activity. This includes the following uses:

- Protect sensitive data through proper management of user access permissions and security settings.
- Make sure that the DAM system interfaces properly with CMS platforms and CRM systems along with marketing automation tools.

- Enhance performance in storage and cloud environments to enable scalable and accessible systems.
- Employ DAM analytics tools to monitor engagement levels and efficiency statistics of digital assets.

A DAM system maintains security and scalability while offering seamless integration with enterprise tools through IT supervision.

External Agencies and Partners

Organizations collaborate with external agencies and freelancers who require regulated access to their brand materials. A DAM enables secure sharing of approved materials through structured methods that eliminate email and public cloud storage risks. This includes the following uses:

- Access approved brand materials to help support advertising, PR initiatives, and media campaigns.
- Keep all creative assets under version control when uploading new ones.
- Use a self-serve portal to access assets rather than reaching out to internal teams.

Organizations decrease unnecessary communication cycles with agencies through controlled access while guaranteeing external teams use only approved assets.

A DAM system serves as an enterprise-wide asset that integrates teams while enhancing collaboration efficiency. It provides improved asset retrieval and better content governance while eliminating inefficiency in the marketing, sales, IT, and legal departments.

KEY FEATURES OF A DAM PLATFORM

As with any type of software platform, the exact features—as well as the strengths and weaknesses—will vary from platform to platform. The following, however, are generally core features of a DAM.

Key Features of a DAM System

A DAM provides a structured and efficient way to store, organize, manage, retrieve, and distribute digital assets. The following features ensure scalability, security, and seamless collaboration, making these systems essential for businesses managing large volumes of content.

Centralized Asset Library

A DAM serves as a single source of truth, consolidating all digital assets—logos, images, videos, marketing collateral, and documents—into a central repository. This ensures that teams across different departments, locations, and time zones can easily access the latest approved files, eliminating redundant storage and version confusion.

Example: A global retail brand with multiple product lines stores all product photography, promotional videos, and campaign assets in a DAM. The marketing, sales, and design teams can pull assets directly from the system rather than requesting files from different teams.

Advanced Search and Filters

With thousands of assets in storage, finding the right file quickly is crucial. DAMs provide AI-powered search, keyword tagging, and filters to help users refine results by file type, category, upload date, or usage rights. Some platforms even provide image recognition and text extraction to enhance searchability.

Example: A social media team searching for a product image can filter by campaign name, color scheme, or file size, ensuring they retrieve the correct high-resolution image for Instagram while avoiding outdated versions.

Metadata and Tag Management

Metadata enhances searchability by associating descriptive tags, categories, and attributes with each asset. Custom metadata fields allow organizations to structure their content library based on specific industry or business needs.

Example: A film studio organizes its digital content by tagging assets with actors' names, shooting locations, production year, and licensing rights, making it easy for editors and marketing teams to find relevant footage.

Access Controls and Rights Management

DAMs enforce security and compliance by offering granular permissions, user roles, and expiration controls on digital assets. This ensures that only authorized users can edit, download, or share specific assets, reducing the risk of misuse or unauthorized distribution.

Example: A legal team at a pharmaceutical company configures the DAM to restrict access to confidential medical research files, ensuring that only approved teams can retrieve and modify them.

File Format Support and Conversion

A DAM supports a wide range of file formats, including images (JPG, PNG), videos (MP4, MOV), audio (MP3, WAV), and design files (PSD, AI, PDF). Some platforms allow users to convert assets into different formats or resize them on the fly, eliminating the need for external tools.

Example: A design team creates a high-resolution poster in Adobe Illustrator, but the social media manager automatically converts it to a PNG file within the DAM for use in an Instagram ad.

Version Control

Marketing and creative teams frequently update assets, leading to multiple versions of the same file. DAMs track all revisions, approvals, and past edits, ensuring teams use the latest approved version while maintaining access to historical versions.

Example: A product packaging team updates a nutrition label design and uploads the new version to the DAM. The system archives the previous versions, preventing outdated labels from being mistakenly used in future campaigns.

Collaboration Tools

DAMs streamline content creation workflows by allowing teams to comment on assets, approve drafts, and track project progress. Some platforms integrate with project management tools to align asset approvals with broader marketing timelines.

Example: A brand manager working on a commercial campaign leaves feedback directly on a video file in the DAM, prompting the video editor to make the necessary changes before final approval.

Brand Management

A DAM safeguards brand integrity by storing pre-approved logos, fonts, and design templates, ensuring consistent branding across all marketing materials. It can also restrict unauthorized modifications to brand assets.

Example: A multinational corporation provides regional marketing teams with access to localized brand assets, preventing the use of incorrect fonts or outdated logos in international campaigns.

Asset Analytics

Understanding how, when, and where digital assets are used helps teams optimize content strategies. DAMs provide insights into asset downloads, views, engagement rates, and performance metrics, allowing teams to measure which assets drive the most impact.

Example: A content team discovers that a series of product demo videos is underperforming, prompting them to update the content and optimize distribution.

Integrations

A DAM should integrate seamlessly with content management systems (CMSs), CRM platforms, marketing automation tools, and design software. This allows direct publishing, automated asset updates, and streamlined workflows.

Example: An e-commerce company integrates its DAM with Shopify, ensuring product images update automatically when a new collection launches.

User-Friendly Interface

A DAM should have an intuitive, customizable interface that allows users to drag and drop files, set up asset collections, and access frequently used content quickly.

Example: A nonprofit organization with volunteers and employees of varying technical skills relies on a DAM with simple navigation and onboarding tutorials to ensure easy adoption.

Asset Sharing

DAMs provide secure, trackable methods for sharing files internally and externally. Features such as password-protected links, expiration dates, and restricted access prevent unauthorized downloads and misuse.

Example: A fashion brand shares high-resolution product images with an advertising agency using a branded, time-limited download link instead of email attachments.

Scalability

Much like any platform, DAM must grow alongside an organization's needs, accommodating increasing volumes of assets, users, and integrations. Enterprise-level DAMs offer multi-user access, multilingual support, and extended storage.

Example: A technology company expanding into new global markets upgrades its DAM to support regional marketing teams, additional storage, and AI-powered content localization.

AI and Machine Learning Capabilities

Many modern DAMs leverage AI-powered features to enhance searchability, automate metadata tagging, and improve asset recommendations. AI capabilities streamline asset organization and reduce manual data entry.

Example: A sportswear company uploads thousands of product images, and the DAM's AI engine automatically tags images based on product type, color, and brand logo visibility.

A DAM system is far more than a storage solution—it is a content powerhouse that enables teams to organize, collaborate, distribute, and measure asset performance effectively. Whether supporting marketing campaigns, creative teams, or enterprise-wide operations, a DAM ensures efficient asset usage, brand consistency, and long-term content management success.

These features collectively enable organizations to efficiently store, organize, manage, retrieve, and distribute their digital assets, improving productivity, brand consistency, and overall content management.

HOW TO EVALUATE A DAM

Organizations that handle extensive digital content collections must choose their DAM system wisely, and thus need to conduct thorough evaluations to select the platform that best fits their requirements. The following section outlines five essential evaluation criteria and additional considerations for DAM selection.

Asset Organization and Searchability

A DAM primarily enables quick storage, organization, and retrieval of assets. Effective DAM platforms need advanced search and tagging features to allow teams to find files quickly instead of spending time searching.

Let's look at the evaluation criteria:

- *Metadata and tagging features*: Does the DAM system allow users to create custom metadata fields while supporting AI-generated tags and structured taxonomies for enhanced asset organization?
- *Advanced search and filtering*: Do users have access to advanced search capabilities that enable rapid asset retrieval through keyword search along with filters and AI technologies such as image recognition and facial recognition?
- *Folder structures and asset relationships*: Can users create groups that link related assets within the system, such as multiple versions of the same file and collections dedicated to specific campaigns?

A DAM system that has strong asset organization capabilities enables teams to access relevant content rapidly and save time as well as effort.

Integration with Existing Systems

The DAM needs to connect effortlessly with the MarTech stack tools to facilitate seamless workflows that link content creation processes with marketing and distribution systems.

Let's look at the evaluation criteria:

- *CMS and marketing automation integration*: Does the DAM provide integration capabilities to connect with CMSs such as WordPress, Drupal, or DXPs for direct asset publishing?
- *Design software compatibility*: The DAM software provides integration capabilities with Adobe Creative Cloud as well as Figma and Canva to enable seamless creative workflows.
- *API access and custom integrations*: Does the DAM provide APIs and pre-existing connectors that facilitate business-specific requirements and third-party system connections?

Effective DAM integration enables teams to transfer assets across multiple platforms without needing to download or re-upload files.

User Access, Security, and Compliance

The high-value and proprietary nature of digital content assets requires DAMs to implement strong security controls and compliance functions to secure sensitive data.

Let's look at the evaluation criteria:

- *Role-based access control*: Is the system designed with granular permission controls that grant specific access levels to various user roles, including marketing, sales, and external agencies?
- *DRM*: The DAM system needs to monitor licensing agreements, track expiration dates, and enforce copyright restrictions to stop unauthorized use.
- *Compliance with industry regulations*: The DAM system maintains secure storage and sharing practices for digital assets by complying with the GDPR, the CCPA, and additional data privacy standards.

Secure digital asset management enables organizations to maintain brand integrity and effectively manage asset distribution while minimizing legal risks.

Collaboration and Workflow Automation

Asset storage should not be the sole function of DAMs because they must enable team collaboration and automate workflows for smoother content creation and publishing processes.

Let's look at the evaluation criteria:

- *Approval and review workflows*: Does the DAM enable users to assign tasks to others while requesting approvals and maintaining version history tracking?
- *Real-time collaboration*: Does the system offer real-time collaborative features that allow multiple users to annotate assets with feedback and comments?
- *Automated tagging and file organization*: The DAM leverages AI to manage metadata tagging automation, file classification systems, and content suggestion functions.

A DAM that boosts teamwork minimizes workflow obstructions while enabling efficient operations among departments and external collaborators.

Scalability, Performance, and Future-Proofing

Asset libraries grow with organizational expansion, which requires DAM systems to scale up to meet these developments. An ideal DAM solution needs to manage high storage capacities while supporting large files and adapting to new technology developments.

Let's look at the evaluation criteria:

- *Storage capacity and file type support*: Does the DAM system support terabytes of data storage while accommodating high-resolution video files and multiple file format options?
- *Cloud and on-premises options*: The DAM solution provides scalability through cloud storage options or on-premises deployment depending on organizational requirements.
- *AI and machine learning capabilities*: The system integrates AI features that support auto-tagging functions predictive asset recommendations and smart search enhancements.

With its scalable capabilities, a DAM system allows businesses to handle rising content demands without sacrificing system performance and accessibility.

The selection process for a DAM system demands rigorous assessments of organizational compatibility, integration capabilities, security measures, collaboration features, and scalability options. An appropriate solution takes all of these factors into account, as well as the current and projected future state of the organization's needs.

CONCLUSION

DAM system enhances organizational efficiency and workflow continuity while safeguarding assets for marketing and creative teams as well as operational units. Organizations need to evaluate both existing and anticipated requirements through testing and demonstrations to identify a DAM solution that will sustain content management achievements over the long term.

In the next chapter, project management systems will be explored, completing this review of creation, workflow, and operations platforms.

CHAPTER 15

Project Management Platforms

Marketing teams function in dynamic settings where they must simultaneously handle numerous projects and campaigns. Marketing operations need seamless collaboration along with precise scheduling and efficient resource allocation, whether launching new products, managing content production, or coordinating multi-channel advertising campaigns. Project management approaches using spreadsheets, email chains, and separate communication tools typically result in operational inefficiency and missed deadlines while reducing visibility into project advancement. Marketing teams now rely on dedicated project management platforms to enhance their workflow coordination and accountability alongside strategic alignment.

Project management platforms enable marketing teams to centrally direct tasks and timelines while managing budgets and facilitating team cooperation to grant stakeholders real-time access to updates and project status. Through their integration with content management systems (CMSs), customer relationship management (CRM) tools, and digital asset management (DAM) platforms, these solutions create streamlined processes and destroy silos to boost productivity. Project management platforms that include automation features alongside reporting dashboards and task prioritization capabilities enable teams to concentrate on execution tasks instead of administrative burdens.

This chapter examines the main advantages of using project management platforms, along with the diverse user types that depend on them and the distinct requirements of marketing teams that change with organizational size and campaign complexity. It also provides a comprehensive analysis of necessary platform features alongside a systematic method for choosing the most suitable tool. Efficient project management becomes essential to marketing teams that scale up to manage more complex campaigns for achieving success.

HOW DOES PROJECT MANAGEMENT SOFTWARE HELP MARKETERS?

To execute marketing effectively you need not just creative ideas and strategic planning but also accurate coordination efforts and organized teamwork processes. Project management software delivers solutions for marketing teams by offering a unified system to track tasks while managing resources and maintaining campaign schedules. The platforms give marketers the ability to arrange campaigns and allocate duties while monitoring progress and assessing results with full visibility amongst teams and stakeholders.

Streamlined Workflows

Marketing projects require coordination across various components including content creation and design approvals as well as campaign execution and performance tracking. Tasks lose organization when workflows lack structure, which results in missed deadlines and process inefficiency. Project management platforms enable marketers to optimize their workflows by implementing standardized project structures, automating repetitive tasks, and making sure every step is tracked.

A project management system for product launch campaigns generates predefined workflows to assign specific tasks to designers, copywriters, ad buyers, and social media managers so the process remains fully comprehensive. Automated notifications, together with deadline tracking, maintain team alignment while stopping bottlenecks and last-minute rushes.

Improved Collaboration

Successful marketing campaigns depend on collaboration between cross-functional teams composed of content creators, designers, developers, data analysts, and external agencies. The lack of a central communication platform results in fragmented discussions that generate misaligned expectations and duplicated work while delaying approval processes.

Project management platforms create a unified working space that enables stakeholders to monitor progress while giving feedback and collaborating instantly. Marketing teams gain better control and clarity by commenting on tasks directly and using built-in workflows to approve content while providing full visibility into all active projects. This approach eliminates departmental barriers while boosting operational effectiveness throughout creative production processes and the execution as well as analysis of marketing campaigns.

Resource Allocation and Time Management

When marketing teams handle several campaigns at the same time they must balance workloads and allocate resources wisely to avoid burnout. Teams without a structured system often face imbalanced workloads with some becoming overloaded while others stay underused, which results in productivity gaps and operational inefficiencies.

Project management platforms include tools that help teams distribute workloads while managing project timelines and tracking budgetary expenditures. Through platform tools, marketers can distribute tasks according to team capabilities while watching progress in real time to detect bottlenecks before they obstruct campaign timelines. Project management tools enable global enterprises with diverse product lines to focus on high-priority campaigns while maintaining proper support for current initiatives.

Visibility and Transparency

Marketing project execution faces significant obstacles due to insufficient insight into campaign status and potential roadblocks throughout the progress. Leadership finds it difficult to determine which projects are progressing as planned while identifying those that need improvement and pinpointing necessary adjustments.

Real-time dashboards within project management platforms display comprehensive views of marketing activities, including their deadlines and approval requirements and any existing project dependencies. Marketing managers and executives can monitor important performance indicators and spot trouble areas while making strategic changes in advance. Real-time insights provide quicker decision-making and agile campaign management options compared to static reports that become outdated before review.

Scalability and Adaptability

The scope of marketing operations ranges from small, focused teams to large enterprises handling global multi-channel campaigns. An effective project management platform provides the flexibility to accommodate diverse team configurations, operational methods, and intricate campaign demands.

Small teams benefit from basic Kanban boards and task lists, whereas enterprise marketing teams need complex features, including advanced automation capabilities and MarTech platform integrations. Project management software must develop with the organization to enable team productivity during growth phases, structural changes, or campaign strategy adjustments.

Project management software enables marketing teams to execute campaigns more effectively by streamlining workflows and enhancing collaboration together with optimized resource allocation and increased visibility while supporting scalability. Through these platforms, teams can prioritize strategic execution over administrative disorder, which allows their marketing efforts to produce quantifiable business results.

WHO ARE THE USERS OF PROJECT MANAGEMENT SOFTWARE FOR MARKETING?

Project management software for marketing helps teams work together efficiently while automating workflows and enabling flawless campaign execution. The primary users of marketing platforms, such as those for project management, are marketing teams, but several other departments also utilize these platforms to synchronize activities and manage resources within compliance guidelines. Distinct teams utilize project management software for various objectives that span high-level strategic oversight to detailed task execution.

Marketing Leadership

Marketing executives and directors rely on project management platforms to gain visibility into active campaigns, monitor team productivity, and align marketing efforts with broader business objectives.

Project management platforms enable the following for marketing leadership:

- Track high-level progress across multiple projects, ensuring that initiatives stay on schedule and within budget.
- Allocate resources effectively based on campaign priorities, business goals, and available team capacity.
- Use performance tracking features to analyze campaign effectiveness and adjust strategies accordingly.

- Ensure cross-team collaboration by integrating marketing initiatives with sales, product development, and executive leadership priorities.

For example, a CMO at a global enterprise may use project management dashboards to review campaign progress, identify underperforming initiatives, and reallocate budgets to higher-impact activities.

Campaign and Content Teams

Marketing campaign managers, content strategists, and digital marketers use project management software to organize, track, and execute content-driven campaigns. Since modern marketing campaigns involve multiple assets, channels, and approvals, these tools help to maintain efficiency and consistency across deliverables.

Project management platforms enable these teams to do the following:

- Use task management tools to assign roles, set deadlines, and ensure smooth handoffs between teams.
- Automate workflows for content creation, review, and approvals to eliminate bottlenecks.
- Track progress across multiple content types—blog posts, social media updates, video production, ad creatives, and email campaigns.
- Coordinate with designers, writers, and strategists to ensure branding and messaging consistency.

For instance, a content marketing team managing a product launch can use the platform to schedule blog posts, coordinate email promotions, and track social media content rollouts in a unified calendar.

Creative and Design Teams

Designers, video producers, branding designers, and other creatives need structured workflows to manage creative requests, track revisions, and ensure visual assets align with campaign goals. Without a centralized project management system, design teams often struggle with last-minute changes, unclear feedback, and missed deadlines.

Project management platforms enable creative and design teams to do the following:

- Manage incoming creative requests from content, social media, and advertising teams.

- Use approval workflows to streamline design revisions, ensuring that stakeholders review and approve creative assets efficiently.
- Track production timelines for branding materials, video content, and advertising graphics to align with campaign deadlines.
- Maintain asset version control, ensuring that the correct designs are used across various marketing channels.

For example, a graphic designer working on an ad campaign may use the platform to collaborate with copywriters, receive feedback from marketing managers, and track revisions before finalizing assets for deployment.

Marketing Operations and Analytics Teams

Marketing operations and analytics teams use project management tools to oversee data collection, campaign performance tracking, and process optimization. Their role is to connect insights with execution, ensuring that marketing strategies are data-driven and continuously optimized.

Project management platforms allow these teams to do the following:

- Monitor key performance metrics (KPIs) in real time, providing actionable insights to campaign teams.
- Coordinate A/B testing and experimental marketing initiatives, tracking performance against benchmarks.
- Integrate project management tools with marketing analytics platforms to align execution with performance data.
- Ensure data accuracy and compliance in performance reporting, preventing discrepancies between marketing efforts and business intelligence insights.

For instance, a marketing analytics team supporting a paid advertising campaign can use the platform to track ad performance, collaborate with content teams on optimizing messaging, and ensure timely reporting to executives.

Cross-Functional Teams (Sales, Product, IT, and Legal)

While marketing teams drive campaign execution, successful marketing initiatives often require alignment with other departments, including sales, product management, IT, and legal teams. Project management software provides a centralized space for cross-functional collaboration, ensuring that marketing efforts align with broader business objectives and compliance requirements.

This enables the following:

- Sales teams to collaborate on lead generation and campaign alignment, ensuring that marketing messaging supports sales outreach
- Product teams to coordinate product launches and feature releases, ensuring that marketing materials accurately reflect new offerings
- IT teams to support integration efforts, ensuring that marketing automation tools, CRM systems, and analytics platforms work seamlessly together
- Legal teams to review marketing materials for regulatory compliance, reducing risks associated with misleading claims, data privacy violations, or industry-specific guidelines

For example, during a new product launch, the marketing team must coordinate with sales to ensure lead handoffs, IT to enable new landing pages, and legal to approve promotional messaging. A project management platform facilitates these interactions in a structured, trackable manner.

Varying Needs Depending on Teams

The operational methods of marketing teams differ, and their demands for project management software change substantially depending on team size, complexity levels, and organizational structure. A lean marketing team at a startup benefits from lightweight task management software, but enterprise marketing departments that handle global campaigns and cross-functional collaboration need advanced structured project management tools that scale. The choice of project management tool should be based on team size, workflow complexity, and compatibility requirements with additional marketing systems.

Small Marketing Teams and Startups

Small marketing teams, such as startups, small agencies, or individual marketing consultants, need simple, flexible tools that allow them to stay organized without excessive administrative overhead.

Smaller marketing teams benefit from the following aspects of project management platforms:

- Lightweight task management tools such as Trello, Asana, and ClickUp help small teams visualize tasks and prioritize deadlines using Kanban boards or simple to-do lists.

- Content calendars ensure that social media posts, blog articles, and email campaigns are published on schedule without requiring a dedicated operations team.

- Ease of use is a priority, as small teams do not have the resources for extensive onboarding, training, or system customization.

- These teams may not need advanced reporting, budgeting tools, or resource management features, instead focusing on quick adoption and immediate usability.

For example, a startup launching a new product may use Trello to track campaign tasks, ensuring that social media posts, Web site updates, and email newsletters are executed without the complexity of enterprise-level workflow approvals.

Mid-Sized Marketing Teams

As marketing teams grow, so does the complexity of their campaigns, content production workflows, and cross-team dependencies. Mid-sized teams require greater structure, workflow automation, and integration capabilities to keep projects on track.

For these mid-sized teams, the following features of a project management platform may be particularly beneficial:

- Workflow automation and approval tracking become essential for streamlining content production and campaign management.

- More structured project roadmaps and campaign calendars allow mid-sized teams to handle multiple concurrent initiatives across SEO, paid media, email marketing, and content strategy.

- Integration with CRM, CMS, and marketing automation platforms ensures seamless coordination between marketing activities and customer data.

- Teams may use platforms such as Wrike, Asana, or Monday.com, which offer scalability without the heavy customization required by large enterprises.

For example, a mid-sized e-commerce company running seasonal campaigns across email, social, and digital ads may rely on Monday.com to track marketing workflows and align campaign timelines with inventory availability.

Enterprise-Level Marketing Departments

Large enterprise marketing teams managing global campaigns, multiple product lines, and distributed teams need a fully scalable, feature-rich project management platform. These organizations prioritize cross-functional collaboration, security compliance, and detailed resource planning.

Enterprise marketing teams can benefit from the following project management platform features:

- Multi-team collaboration and role-based access control ensure that stakeholders across marketing, sales, IT, and legal can contribute efficiently.
- Advanced reporting and budget tracking help leadership monitor performance and allocate resources effectively.
- Multi-tiered approval processes and compliance tracking ensure that campaigns meet brand, legal, and regulatory standards before launch.
- Enterprise-grade solutions such as Workfront, Smartsheet, and Adobe Workfront offer custom workflows, integration with enterprise-wide systems (e.g., Salesforce, SAP), and AI-driven task automation.

For example, a platform such as Adobe Workfront can enable global consumer goods companies to handle marketing assets while coordinating across time zones and tracking multiple campaign budgets during multi-region product launches.

Marketing project management requires tailored solutions rather than a universal approach. Small teams need straightforward agile tools for rapid task completion, whereas mid-sized teams achieve operational scaling through automation and integrations. Enterprises require unified platforms with flexible customization capabilities to enable collaboration among extensive teams. Organizations can effectively streamline their marketing execution by understanding their team structure, workflow patterns, and growth trajectory when choosing suitable project management software.

KEY FEATURES OF PROJECT MANAGEMENT PLATFORMS

Project management platforms enable marketing teams to handle complex campaigns through efficient coordination and streamlined workflows while maintaining collaboration among multiple stakeholders. Teams can manage tasks and track deadlines while allocating resources efficiently through a well-equipped platform that also integrates essential marketing tools. Marketing operations become disorganized without proper capabilities, which results in missed deadlines along with inefficient processes and campaigns that are poorly executed.

The following criteria represent essential features marketing teams need to consider when choosing a project management platform.

Task and Workflow Management

The primary features of any project management platform are task assignment and tracking capabilities while ensuring task prioritization in marketing projects. The complexity inherent in marketing campaign projects that involve content creators, designers, strategists, and analysts demands an efficient organizational system to manage their tasks:

- Through task assignment and tracking, team members gain clarity on their responsibilities and deadlines.
- Customizable workflows make sure projects progress through structured phases such as content approvals and creative briefs as well as ad campaign execution.
- Teams can prevent work stoppages and maintain smooth workflow progress with the help of automated alerts and task dependencies.

A content marketing team that generates blog series creates a workflow structure where writers submit drafts that editors review, followed by designers creating visuals, until managers approve final versions before publication.

Campaign Planning and Calendar Management

Marketing teams need to manage several campaigns that span various channels as well as different timelines and deadlines. Teams can use visual dashboards and campaign calendars to plan content schedules and track dependencies while aligning all components of their projects:

- Marketing-specific timelines show all ongoing and future campaigns, which helps teams maintain their planned schedule.
- Teams use editorial and campaign calendars to organize and control their publishing schedules for blogs and various online marketing efforts.
- Teams can understand how a single delayed task affects the entire campaign launch through dependency mapping.

- Teams can plan holiday promotional campaigns months ahead with established milestones for ad creative development and content production as well as email sequences and social media promotions.

Team Collaboration and Communication Tools

The successful execution of a marketing campaign requires teamwork from different departments along with external agencies and remote teams. The absence of a centralized communication system leads to vital information getting buried in endless email chains and dispersed throughout multiple platforms:

- The platform's integrated messaging and comment threads keep feedback about campaign materials contained within its system.
- Teams can upload their assets to one location where they can review and make revisions using file-sharing features.
- Connectivity with Slack, Microsoft Teams, and Google Workspace establishes seamless communication between different tool-using teams.

Social media marketing teams collaborating with external design agencies can maintain all their communications and design updates on their project management platform, which prevents lengthy email correspondence.

Resource Allocation and Time Tracking

Project management platforms enable marketing leaders to distribute tasks efficiently, which prevents team members from becoming overloaded while others stay underused:

- Teams use time-tracking tools to measure task completion durations, which facilitates project timeline optimization.
- Managers can use workload-balancing dashboards to distribute tasks to team members who have the capacity to complete them.
- Budget-tracking features enable project management teams to maintain adherence to their designated marketing budgets.
- An email marketing team that manages multiple nurture campaigns can use time-tracking and resource-allocation tools to distribute work among content writers, designers, and automation specialists evenly across the campaigns.

Integration with Other Marketing Tools

Marketing projects require integration with fundamental MarTech platforms, including CRM and CMS systems, along with email marketing platforms and analytics tools:

- Marketing teams achieve real-time synchronization of content along with customer information and performance analytics through the seamless integration of marketing automation platforms with CRM and CMS systems.
- Businesses can connect their project management software to platforms such as HubSpot through custom APIs and pre-built connectors.
- Data synchronization provides marketing teams with up-to-the-minute access to their project performance details as well as engagement metrics and ROI figures.

A performance marketing team can monitor paid advertising campaign execution by integrating its project management software with its analytics dashboard to track both campaign performance and execution milestones.

Approval Workflows and Version Control

Marketing teams that manage extensive creative assets face risks of publishing outdated content and misusing branding materials when they lack structured approval processes that ensure compliance with regulations:

- Approval workflows require every piece of marketing content to undergo a complete review before it is published.
- Version control systems track all file modifications to eliminate confusion from multiple edits while guaranteeing that users work with the latest versions of files.
- Audit trails maintain records of all approvals, which proves essential within regulated sectors such as finance and healthcare.

Product marketing teams could implement an approval workflow that requires legal teams to review compliance concerns, followed by designers who finalize branding before marketing directors grant their final approval before the product is launched.

Analytics and Reporting

Project management software serves multiple functions as it delivers execution capabilities while offering detailed insights into project efficiency and marketing impact:

- Project tracking dashboards deliver real-time visibility into project progress, along with identification of bottlenecks and completion rates.
- Marketing leaders are able to evaluate campaign performance using customizable reporting tools that measure against KPIs.
- Analyzing data-driven insights enables the refinement of future projects by distinguishing successful elements from failures.

Project management software enables content marketing teams to monitor blog article publication timelines and performance metrics related to engagement and conversions.

Beyond organizing marketing teams, a top-tier project management platform boosts both collaboration and strategic execution while improving efficiency. These tools that supply task management solutions along with workflow automation options and team communication channels combined with resource allocation capabilities, seamless integrations, structured approval processes, and data-driven insights allow marketing teams to execute impactful campaigns successfully within both time and budget constraints. Marketing teams achieve peak operational efficiency and future scalability by choosing platforms with optimal feature combinations.

HOW TO EVALUATE PROJECT MANAGEMENT SOFTWARE FOR MARKETING TEAMS

Marketing teams must choose their project management software carefully because selecting an inappropriate tool can cause operational inefficiency and create both bottlenecks and frustration among team members. A platform should match team workflows while scaling to meet changing requirements and achieve flawless integration with existing MarTech. Project management tools should be evaluated based on their ability to support daily operations and enable marketing teams to perform their tasks with precision while promoting collaboration.

Five essential criteria for evaluating project management platforms for marketing teams are outlined below, together with specific considerations for each category.

Ease of Use and Adoption

The effectiveness of a project management platform depends entirely on how widely it is adopted by team members. When software complexity or navigation challenges arise, marketers will turn back to spreadsheets and separate tools while undermining the software's efficiency.

- *User-friendliness*: Does the platform provide an intuitive experience for marketing teams that have different levels of technical knowledge? Does it have a clean, easy-to-use interface?
- *Pre-built templates*: Are there pre-existing templates available on the platform to support typical marketing processes, including campaign planning and creative production?
- *Onboarding and training*: What is the speed at which teams can master the platform and integrate it into their everyday workflows? What training resources, knowledge bases or live support options does the provider offer for new users?

For example, a fast-growing content marketing team should use a platform that includes drag-and-drop task management combined with pre-built editorial calendars to bring new employees up to speed without a steep learning curve.

Customization and Scalability

Marketing teams change frequently as they expand or contract and their organizational structure becomes more intricate. A project management system needs to accommodate small teams and large enterprises through adaptable customization options for specific workflows.

- *Adaptability*: How does the system ensure scalability for an increasing number of users, teams, and projects?
- *Custom workflows and permissions*: Is it possible for teams to develop custom workflows while establishing approval protocols and assigning permissions based on user roles to provide appropriate project access?

- *Support for multiple teams and brands*: The software needs to support cross-functional collaboration between teams while managing multiple brands and global marketing campaigns without generating data silos.

For example, a multinational company needs a project management tool that provides multilingual capabilities and region-specific campaign workflows along with brand-specific permissions to allow local teams to function efficiently while meeting global marketing goals.

Collaboration and Communication Features

Marketing campaigns necessitate seamless collaboration between internal departments and external agencies and partners. Teams face misalignment risks and duplication of work along with inefficient approval processes when they lack powerful communication tools:

- *Cross-team collaboration*: Can the software effectively enable teamwork among content, design, analytics, and product marketing teams within the organization?
- *Integration with communication tools*: Is platform integration available with communication tools such as Slack, Microsoft Teams, Google Workspace, and email, to enable teams to receive updates and notifications directly in their current work processes?
- *External collaboration*: Agencies and third-party vendors can participate in secure collaboration efforts without having access to internal projects or files.

For instance, a social media marketing team collaborating with a freelance design agency needs secure external access to asset approvals but must retain limited project visibility for partners.

Automation and Workflow Efficiency

Marketing teams manage repetitive tasks that require multi-level approval processes and operate according to strict deadlines. The platform needs to simplify process flows and automate repetitive tasks as well as reduce the necessity for manual follow-ups:

- *Task automation*: Does the software provide automatic notifications together with task dependencies and recurring task scheduling functions to maintain project momentum?

- *Approval workflows and version control*: The platform should demonstrate effective management of approval processes and asset revisions while maintaining proper storage of finalized materials.

- *Process optimization*: Does the software enable automation for repeated marketing operations such as email approvals, campaign rollouts, and asset review cycles?

For instance, a B2B SaaS marketing team handling multiple marketing channels requires automated notifications for content reviews and scheduled dependencies between tasks to complete product launch materials timely.

Reporting and Performance Tracking

Marketing leaders require real-time insights into project progress along with team performance and campaign effectiveness. An effective project management tool delivers live monitoring capabilities that track both campaign execution and operational efficiency:

- *Project status dashboards*: The software provides live tracking of task progress to help identify which tasks meet deadlines, face delays, or are at risk.

- *Custom reporting*: Marketing leaders can produce reports that analyze task completion rates along with bottlenecks and workload distribution for better resource allocation.

- *Marketing ROI integration*: Is the tool designed to link with analytics platforms such as Google Analytics, HubSpot, and Salesforce for real-time evaluation of marketing campaign performance?

A team handling marketing operations for campaign performance analysis needs a project management tool that integrates analytics to track both campaign progress and marketing KPIs.

Selecting project management software requires more than just looking for comprehensive features because the best choice is a system that supports marketing workflows and grows with team size while boosting collaboration and efficiency.

CONCLUSION

Project management software serves as an essential tool for marketing teams as it changes campaign planning and execution processes while enabling optimization. These platforms enable marketing teams to reduce inefficiency and improve accountability while keeping campaigns on track by offering a structured framework for task management together with collaborative streamlining and progress tracking.

The best tools provide intuitive design for quick learning and have flexible features to adapt to various workflows while delivering strong automation capabilities and insightful data analysis. Assessing usability, customization features, automation capabilities, and reporting functions enables marketing teams to select a solution that enhances daily operations while contributing to long-term business expansion.

PART 4

Content, Campaign, and Multi-Channel Delivery

4 Categories of MarTech Platforms

Customer Data
Helping brands understand their customers

Content, Campaign, and Multichannel Delivery
Serving customers with content, offers, and experiences across the journey

Creation, Workflow, and Operations
Empowering teams to be more efficient in content and campaign creation

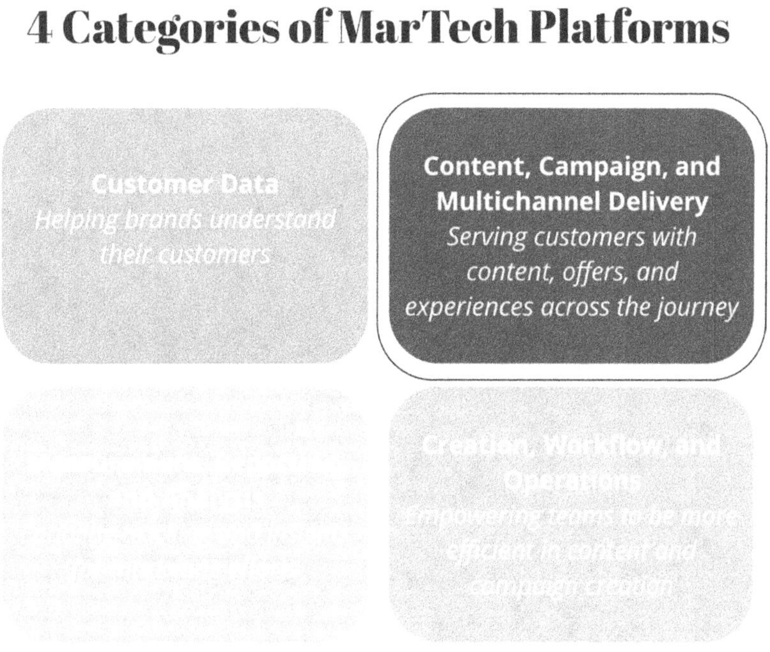

FIGURE P4.1 Content, Campaign, and Multi-Channel Delivery

The next part of the book discusses a critical set of MarTech applications that serve customers with content, offers, and experiences, and can include platforms such as Digital Experience Platforms (DXPs), Content

Management Systems (CMSs), email marketing, marketing automation platforms, customer journey orchestration, and many more.

These platforms are critical because they are often direct interfaces to end customers and provide the messages, offers, and digital experiences that have the opportunity to drive customers' engagement and loyalty.

CHAPTER 16

KEY CONSIDERATIONS FOR CONTENT, CAMPAIGN, AND MULTICHANNEL DELIVERY

Marketing teams encounter both new possibilities and complex challenges due to the proliferation of marketing channels and digital touchpoints that their customers engage on. Because of this, marketers need to explore and employ increasingly sophisticated content and campaign delivery platforms that support multichannel marketing operations to satisfy growing customer demands for personalized and seamless experiences. These technologies allow advanced engagement possibilities and automation functions but lead to integration difficulties, scalability issues, and data consistency problems.

The evolution of marketing content and campaign delivery platforms over the last ten years has expanded their scope beyond traditional Web site content management systems (CMSs) and email platforms to include digital experience platforms (DXPs), marketing automation solutions, customer journey orchestration tools, and AI-powered personalization features. These platforms assist marketers in producing and distributing content and campaigns across various channels with efficiency while maintaining message consistency and timeliness. The growing number of tools within the marketing stack leads to increased complexity in workflow coordination and data integrity maintenance while striving for true omnichannel experiences.

This chapter identifies major obstacles marketers encounter during the deployment and administration of content delivery systems alongside campaign and multichannel platforms.

Marketers who grasp the nature of these challenges and their consequences can improve their evaluation and implementation of content delivery platforms, which allows optimized campaign execution, leading to more fluid and data-informed customer engagement experiences.

THE MARKETING LIFECYCLE AND ITS IMPACT ON PLATFORM SELECTION

MarTech platforms should support every phase of the marketing lifecycle while integrating content creation with campaign execution across multiple channels. Choosing appropriate platforms requires more than evaluating features and vendor credibility because successful tools must enable support for all customer journey stages including awareness, engagement, conversion, and retention. Without a unifying strategy, different platforms serving isolated functions create inefficiency and data silos that result in disconnected customer experiences.

Aligning Platforms with the Marketing Lifecycle

Every stage of the marketing lifecycle requires specific technologies and capabilities to support its objectives. The following outlines where these platforms can fit, from the very early stages of campaign planning to assessing the results of a campaign after it is complete:

- *Planning and strategy*: Content planning alongside workflow management tools and team collaboration platforms help marketing teams synchronize their messaging strategies with creative assets and campaign objectives.
- *Content and asset creation*: Marketing teams benefit from CMS or DAM platforms, which streamline the creation, storage, and retrieval of brand-approved assets.
- *Campaign execution and delivery*: Through marketing automation, email marketing, and customer journey orchestration and DXP campaigns, marketing teams successfully connect with their intended audiences at optimal moments.
- *Engagement and personalization*: Through the utilization of AI personalization tools and customer data platforms (CDPs), businesses craft content that aligns with each customer's unique tastes.

- *Measurement and optimization*: Analytics platforms together with attribution models and performance tracking tools enable continuous improvement of marketing strategies through real-time data insights.

The proper organization of MarTech components enables all elements to function together seamlessly instead of remaining segregated in separate silos.

Platform Needs Across the Customer Journey

In addition to the internal marketing planning and execution lifecycle, customer journeys need to be accounted for as well. As marketers know, consumers do not follow a straight path, which means marketing platforms need to be flexible for effective customer engagement throughout various stages of the journey:

- *Awareness stage*: To draw in new audiences, brands have to utilize advertising platforms together with content marketing tools and organic search optimization. CMS systems, alongside DXPs and social media management tools, serve as fundamental elements for the delivery of educational and engaging content.
- *Consideration stage*: Customers now evaluate their choices and need marketing platforms for email campaigns and marketing automation to nurture leads properly.
- *Conversion stage*: Conversion rates rely heavily on personalization techniques combined with e-commerce integrations, along with landing page builders and AI-driven recommendations.
- *Retention and loyalty*: Brands utilize customer journey orchestration together with loyalty management tools and post-purchase engagement platforms to establish enduring relationships and maintain customer engagement.

The absence of an integrated MarTech stack creates potential messaging inconsistency and response time delays that lead to missed customer engagement opportunities.

Balancing Long-Term Brand-Building and Short-Term Performance Marketing

Marketers regularly struggle to find an equilibrium between their efforts to build enduring brand value and their often high-stakes objectives for immediate marketing performance. Some platforms prioritize immediate data optimization, while others focus on long-term user engagement and narrative development. MarTech investments must meet dual objectives without introducing operational disturbances.

Successful brand-building needs platforms that enable content creation and thought leadership while reaching audiences organically through engagement. The success measurement of performance marketing goals depends on automating processes and utilizing real-time analytics alongside AI optimization methods and conversion tracking.

Scalability, adaptability, and cross-channel orchestration are also essential features of a well-designed MarTech stack to enable different marketing approaches to function together effectively instead of competing for resources.

Choosing appropriate content and campaign delivery platforms requires marketers to thoroughly understand how each platform functions within the entire marketing lifecycle. Strategic platform choices boost operational efficiency while maintaining customer journey consistency and ensuring sustainable marketing success.

The Challenges of Planning, Creation, Management, and Delivery

In the race to beat stiff competition and meet customer demands, marketing teams face increasing pressure to produce high-quality content, execute personalized campaigns, and manage multichannel delivery at scale. The complexity of modern marketing operations, however, introduces challenges related to fragmentation, agility, and alignment. Without seamless collaboration, efficient workflows, and flexible tools, marketers risk losing speed, consistency, and strategic focus.

Fragmentation Across Teams and Workflows

Marketing organizations include several specialized teams that manage content creation tasks as well as campaign execution and customer engagement and analyze performance. These teams function as standalone entities that rely on separate systems and methods causing operational inefficiency together with strategic misalignment.

Some symptoms of this fragmentation include the following:

- *Lack of integration between platforms*: Marketing teams use CMSs, digital asset management (DAM), customer relationship management (CRM), and analytics tools, but these systems don't always communicate effectively.

- *Inconsistent messaging and content duplication*: Without a unified content strategy, multiple teams may create similar materials, leading to wasted effort and brand inconsistency.

- *Delayed approvals and bottlenecks*: Creative teams often face slow review cycles due to disconnected approval workflows, causing campaign delays.

- *Difficulty tracking project ownership*: Unclear responsibilities can lead to missed deadlines, redundant work, and poor visibility into progress.

For example, a global marketing team launching a product campaign may struggle if regional teams create their own assets, social media teams draft separate messaging, and analytics teams use different reporting tools, resulting in misaligned execution, fragmented customer experiences, and inefficient resource use.

Marketing teams need solutions that support real-time collaborative efforts while optimizing workflow automation and consolidating content approval mechanisms. Built-in integrations in a centralized project management system enable seamless cooperation between content teams, creative teams, and analytics teams, which ensures marketing activities maintain alignment.

Maintaining Agility Without Compromising Strategy

The digital marketing environment requires marketers to operate quickly and adaptively, but they must maintain consistent strategic intentions. Marketers face the challenge of executing quickly while maintaining long-term brand-building and personalization objectives.

The need for quick content creation requires marketers to respond to trends and customer activities but without structured processes. This haste can damage messaging and compromise content quality. Brands aim to deliver customer-specific personalized content but find that too much customization impacts their execution speed and scalability.

The inflexible structures of certain marketing platforms prevent businesses from adapting to changes in audience behavior and technological advancements. Marketers operating in regulated sectors such as finance and healthcare should execute campaigns with agility while meeting all legal requirements.

For instance, retail brands operating omnichannel holiday campaigns face obstacles when attempting rapid messaging updates across email, social media, and paid ads because strict approval processes combined with platform restrictions cause delays in market response times.

Marketing platforms need to provide teams with pre-built templates and automation tools for efficient operations while supporting customization abilities to maintain personalization and brand consistency. Marketing tools that provide agility in content delivery enable marketers to swiftly shift strategies while preserving organizational direction.

The need for MarTech platforms becomes apparent when fragmented workflows and the necessity to balance agility with strategic planning require tools that enable collaboration and streamlined execution while supporting rapid iteration without quality loss. Marketing teams achieve efficient operations and precise strategic execution across content creation and campaign delivery through proper tool integration and workflow optimization.

A GROWING NUMBER OF MARKETING CHANNELS

The marketing landscape today demonstrates unprecedented complexity due to consumers interacting with brands through multiple digital touchpoints and devices. Brands need to synchronize their content delivery across various formats to meet audience expectations and implement effective engagement strategies. The increase in available channels results in growing difficulties surrounding platform selection and brand consistency alongside scaling content efficiently.

The Proliferation of Digital Touchpoints

Today's marketing strategies extend beyond traditional platforms such as Web sites and email to include social media and other digital touchpoints. Brands need to connect with customers through multiple digital touchpoints that continue to grow. These channels seemingly grow daily and include the following:

- *Owned channels*: The spectrum of owned channels comprises Web sites, blogs, mobile applications, customer portals, and email marketing initiatives.
- *Paid channels*: Paid marketing channels consist of digital ads together with social media promotions and influencer partnerships, as well as programmatic advertising.

- *Conversational channels*: Chatbots, SMS, in-app messaging, and voice assistants.

- *Experiential and emerging channels*: The experiential and emerging digital touchpoints include virtual reality (VR), augmented reality (AR), connected TV (CTV), and the Metaverse.

Distinct technical needs, audience expectations, and content demands define each channel. Managing them effectively requires platforms that do the following:

- Ensure consistent brand messaging while delivering content through various platforms.

- Integrated campaign execution requires platforms that enable simultaneous support for paid media strategies together with owned media approaches.

- Create scalable personalization solutions that integrate data across multiple platforms instead of maintaining siloed information.

Marketers face significant risks when they overcommit resources to numerous platforms. Marketers who adopt every emerging technology might end up with redundant tools and underused features along with operational inefficiency. Businesses often end up with an overloaded MarTech stack because they struggle to integrate platforms that fail to communicate effectively or match their strategic goals.

For example, retail brands that use separate tools for their Web site management while scheduling social media content and running email marketing campaigns along with digital ads often find these platforms fail to integrate properly, which results in mixed messaging, repeated work, and ineffective performance tracking.

To be successful amidst the growing set of possibilities, marketers need to thoroughly assess which platforms will best support their primary channels and customer engagement tactics. To achieve optimal results, teams need to direct their energy toward enhancing integration capabilities and automating workflows within a unified MarTech system.

Adapting Content Across Channels

Copying and pasting identical content across multiple channels will not result in maximum engagement because each platform demands specific content strategies. Every digital platform demands specific format guidelines and engagement standards that cater to distinct audience behavior. An effective Instagram campaign might not succeed on LinkedIn and desktop emails may fail to convert mobile users.

Similarly, there are key differences among other platforms and their content requirements:

- Social media audiences expect visually engaging, bite-sized content, whereas email subscribers may prefer in-depth messaging with personalized recommendations.
- Search-optimized blog content differs from conversational chatbot responses, requiring platforms that help adapt messaging accordingly.
- Video-based storytelling may thrive on TikTok and YouTube, while static infographics perform better on B2B blogs and whitepapers.

The challenge is maintaining brand consistency while ensuring that messaging feels native to each platform. Because of this, marketers often struggle with the following:

- Repackaging long-form content into digestible, channel-specific formats (e.g., turning a whitepaper into a series of LinkedIn posts)
- Balancing automation with human oversight—how much can AI-generated content be trusted to maintain brand integrity?
- Scaling content adaptation efficiently without compromising quality

As an example, a B2B software company launching a product update may create a detailed blog post for SEO visibility, an email campaign for existing customers with personalized use cases, short-form LinkedIn and Twitter posts to engage industry professionals, plus a targeted YouTube explainer video demonstrating key features. All of these have their unique content needs and requirements, meaning that the marketing team must take these into account when planning, writing, editing, and publishing each piece of content on each specific channel. In some cases, this may be done by a single team, and in others, there may be a team assigned to a single marketing channel.

Creating multiple variations of content manually is unsustainable at scale. AI-driven content automation tools can help by generating channel-specific adaptations using AI-powered copywriting tools, automating video subtitles, social captions, and blog summaries based on existing content, and personalizing messaging dynamically based on audience behavior and engagement data. Marketers must invest in platforms that facilitate scalable content adaptation—balancing automation with human creativity. DXPs, AI-powered personalization tools, and marketing automation software can help ensure that content resonates across every channel while maintaining brand voice and strategic intent.

With dozens of marketing channels available, brands must prioritize efficiency, integration, and scalability when selecting platforms. A growing channel landscape should enhance marketing reach—not create silos, inefficiency, or diluted messaging. By leveraging strategic platform selection, AI-driven automation, and integrated content workflows, marketers can successfully orchestrate campaigns across multiple digital touchpoints while maintaining consistency and relevance.

Customer Channel Switching and Evolving Preferences

Consumer interactions with brands have evolved beyond a one-directional path. Consumers now navigate across various channels including Web sites, social media platforms, emails, SMS messages, applications, physical store interactions, and voice assistant technology while seeking a unified experience. The necessity for continuous channel switching creates substantial obstacles for marketers because most platforms remain built for single-channel execution instead of genuine omnichannel engagement.

Marketing teams can preserve their competitive position only through continuous touchpoint management while adapting to customer behavior changes and choosing flexible platforms. This section explores two key challenges: the ability to move smoothly between channels and consumer behavior shifts driving platform choices presents major challenges for marketers.

The Expectation of Seamless Transitions Between Channels

Consumers expect the brands they support to recall their personal details and previous interactions no matter which channel they choose to connect with next. Many more traditional MarTech platforms fail to monitor customer interactions across various channels simultaneously, which has resulted in fractured customer experiences and lost engagement opportunities due to repeated messaging.

Consumers anticipate seamless transitions between digital platforms and physical locations while shopping, which includes exploring products online followed by social media reviews and retargeting ads before visiting a store and completing the transaction via a mobile application. When there is friction within this process it produces lost conversions and the abandonment of potential sales.

The siloed operation of numerous MarTech platforms poses challenges for brands when they try to monitor customer interactions across different channels. This lack of cohesive platform communication leads customers to receive irrelevant recommendations and redundant emails without recognition of their previous interactions.

When platforms don't synchronize customer data customers have to repeat information and restart conversations across channels, which causes frustration and leads to them abandoning their interactions.

For instance, when a bank markets a credit card through various channels it may overlook a customer who accessed more information on their Web site after clicking a social media ad but has already completed a live chat application—resulting in unnecessary email prompts for application submission.

Marketing teams need to adopt platforms that support instant data sharing and interaction tracking between different channels. Unified CDPs, customer journey orchestration tools, and AI-driven personalization engines ensure that customers receive appropriate messaging at the correct time irrespective of their chosen communication channel.

Changing Consumer Behavior and Platform Selection

The market landscape changes continuously as customer preferences evolve, making today's effective methods obsolete tomorrow. Marketers must frequently reevaluate their MarTech investments because new social platforms, messaging channels, and engagement trends appear at a fast pace.

Marketers face challenges when trying to allocate funds between established marketing channels and exploring opportunities in emerging platforms. Companies that fail to adapt quickly may give competitors an advantage but premature investment in trending technologies can result in waste on tools that provide no return on investment.

For instance, retail brands testing SMS marketing face challenges when incorporating this channel into their email automation workflows, which necessitates separate management for campaigns, tracking for analytics, and compliance procedures, and results in additional complexity.

Marketers require technology stacks that enable experimentation while preventing large-scale disruptions to keep up with changing customer behaviors. Technology platforms that combine modular designs and API accessibility with multichannel support and low-code/no-code integration enable organizations to assess and expand new communication channels without modifying current systems.

While customers demand cohesive personalized experiences throughout multiple channels, MarTech platforms mostly operate in isolated segments. Marketers who do not address these gaps risk alienating customers and losing engagement chances while new channels become dominant.

Brands that choose flexible integrated MarTech platforms will achieve omnichannel consistency while keeping up with changing consumer patterns and securing their marketing strategy against future changes.

Balancing Channel-Specific Expertise with Omnichannel Excellence

Marketing today demands a presence across multiple channels, but simply having access to those channels is not enough. Effectively managing each one requires deep expertise, as platforms have unique capabilities, audience behaviors, and technical requirements. While omnichannel marketing aims to create a seamless customer experience, the reality is that most organizations lack comprehensive expertise across all channels, leading to inefficiency, misalignment, and underutilized tools.

Complexity of Managing Multiple Channels Effectively

It's easy to assume that omnichannel marketing means being everywhere at once, but each channel operates differently, requiring specific content strategies, engagement tactics, and measurement approaches. Many organizations attempt to centralize marketing functions but lack specialized knowledge across every channel, leading to inefficiency and missed opportunities.

Some of the requirements of this multi-channel complexity are as follows:

- Deep expertise is required to manage different platforms effectively. An email marketing team must know deliverability best practices, segmentation strategies, and compliance regulations, while a paid media team must understand bidding algorithms, audience targeting, and real-time ad optimizations.

- Many platforms require dedicated specialists, creating silos. Social media management platforms, search marketing tools, and programmatic ad platforms all demand unique skill sets. Organizations without experts in each area risk underperforming campaigns due to a lack of knowledge.
- Keeping up with evolving platforms is a constant challenge. Algorithms change, privacy laws shift, and consumer behavior evolves, meaning marketers must continuously adapt strategies or risk falling behind.

For example, a retail company using an omnichannel approach may struggle if its paid media team runs high-performing ad campaigns, but its email team lacks the segmentation skills needed to nurture leads post-click—resulting in wasted ad spend and lost conversions.

Organizations need to balance specialization with cross-functional knowledge sharing. While dedicated teams may own specific platforms, marketing leaders must ensure strategic alignment across all channels. A unified CDP can help bridge knowledge gaps by centralizing insights across teams, ensuring that channel-specific expertise does not exist in isolation.

TRAINING AND ADOPTION CHALLENGES

Despite their expense and complexity, advanced marketing platforms often remain underused or improperly operated by organizations because they lack adequate training for their teams. Teams often barely scratch the surface of available features and miss the potential benefits of automation, personalization, and analytics capabilities.

Some of these training and adoption challenges are as follows:

- Platform complexity leads to slow adoption across marketing groups. Teams tend to use just basic functions of multi-feature tools when they lack proper training, which results in poor campaign performance.
- A limited group of skilled users becomes a bottleneck in operations. When just a few employees have the expertise to operate a platform properly they become critical bottlenecks which impede execution speed and organizational scalability.
- Ongoing education is necessary, butis often neglected. While many companies conduct one-time training sessions during onboarding, teams need persistent educational support and practical learning opportunities to handle regular software updates and new features.

For instance, when a B2B company adopts a customer journey orchestration platform, only a few marketing operations specialists know how to create multi-step automation sequences. The lack of full understanding leads other marketing teams to revert to manual operations instead of maximizing platform capabilities.

Marketing organizations need to maintain consistent platform training while establishing governance procedures and documenting best practices. Key strategies include the following:

- Establishing internal "MarTech Champions" who will provide training for other teams about the platform's capabilities and best practices.
- Collaborating with vendors to obtain certifications and provide ongoing training sessions as well as continuous education programs.
- Developing robust platform governance structures to ensure teams utilize tools effectively and consistently throughout the organization.

Successful implementation of an omnichannel marketing strategy needs both the appropriate technology and specialized expertise. Marketing teams without specialized knowledge across channels will probably struggle to manage campaigns properly and fail to use platform capabilities while losing engagement opportunities.

Organizations that focus on specialized talent acquisition while promoting inter-team collaboration and continuous training will maximize their MarTech investments and successfully run powerful omnichannel marketing campaigns.

Integration and Coordination Between Channels

A genuine omnichannel marketing strategy involves more than simply existing across multiple platforms since it demands smooth coordination and integration between all channels. MarTech platforms specialize in distinct functions because one system handles email, while another manages social media, and yet another supports customer service, so creating unified cross-channel strategies becomes complicated. The lack of proper integration between marketing channels leads to disjointed customer experiences, so brands have problems maintaining consistent personalized messaging throughout the customer journey.

Providing a Cohesive Omnichannel Experience

Customers expect brands to identify them across all the channels they utilize throughout their customer journey and maintain engagement by understanding their preferences at all times. The lack of integration between various marketing platforms leads to difficulties in monitoring customer interactions and delivering smooth user experiences.

There are several areas where some MarTech platforms fail to deliver a cohesive omnichannel experience:

- *Most marketing tools are built for isolated functions*: Email marketing platforms, social media schedulers, Web site CMSs, and customer support tools all serve different purposes, often with little to no communication between them.

- *Inconsistent messaging results from disconnected platforms*: A customer who abandons a shopping cart online may still receive irrelevant promotional emails, while another who contacts support for an issue may continue receiving sales offers without acknowledgment of their inquiry.

- *The risk of broken customer experiences is high when interactions aren't shared*: Without real-time data synchronization, customers may receive repetitive, conflicting, or untimely messages, leading to frustration and disengagement.

For example, a travel company running marketing campaigns across email, mobile apps, and paid ads may fail to recognize that a customer already booked a flight through their Web site, resulting in continued retargeting ads urging them to book the same trip.

Brands must connect their marketing platforms through unified data architectures, ensuring that customer interactions are tracked, shared, and acted upon in real time. CDPs, marketing automation tools, and journey orchestration platforms play a critical role in enabling smooth omnichannel experiences.

Personalizing Across Multiple Channels

Effective personalization requires real-time access to customer data, behavioral insights, and capabilities for dynamic content adaptation across multiple channels. Many marketing teams, however, lack the infrastructure to deliver relevant, personalized experiences consistently.

Siloed platforms prevent real-time personalization—if an email marketing platform doesn't integrate with a CRM or CDP, customer data remains trapped within separate systems, limiting context-aware messaging.

Timing is everything in personalization—a relevant offer or recommendation must be delivered at the right moment. Many brands struggle with delayed data processing, resulting in stale or irrelevant messaging. AI-driven automation can help, but only if systems are properly integrated—AI-powered recommendation engines, predictive analytics, and journey orchestration tools are only as effective as the data they have access to.

For example, a B2B software company using email, LinkedIn ads, and sales outreach may find that its sales team isn't aware of which marketing emails a prospect has engaged with, leading to redundant conversations and missed opportunities to tailor the sales pitch.

Marketers should invest in platforms that integrate customer data seamlessly—particularly CDPs, marketing automation software, and AI-powered personalization engines. This ensures that customer interactions are continuously updated across all platforms, enabling timely, context-aware engagement.

Creating Personalized Content at Scale

While consumers expect personalized experiences tailored to their preferences, marketing teams struggle to create unique content for every customer segment, device, and channel, as this is unsustainable at scale. The challenge lies in delivering customized experiences efficiently without overwhelming content production teams.

Other challenges related to creating this personalized content at scale include the following:

- Scaling personalization is difficult when content creation is manual—traditional content production methods require extensive time and resources, limiting the ability to customize content for every audience segment.

- AI-generated content and dynamic creative optimization (DCO) are emerging as solutions—AI-powered tools can generate text, images, and videos in real time, automatically tailoring content based on audience data.

- Modular content frameworks allow for scalable customization—by breaking down content into reusable components (e.g., headlines, images, CTAs), marketers can assemble personalized messages dynamically without having to create entirely new assets for every variation.

For example, an e-commerce company running an email campaign may use AI-driven dynamic content blocks to automatically swap out product recommendations based on past browsing behavior, cart history, and customer preferences—without requiring manual intervention.

Marketers need to implement AI-driven content creation tools along with automated design platforms and modular content systems to expand personalized messaging operations effectively. Dynamic content platforms enable teams to offer personalized experiences without adding more manual tasks.

Brands will achieve success in omnichannel marketing by transitioning from standalone campaigns to comprehensive strategies that connect customer engagement across all channels. When marketing platforms operate independently without coordination their efforts split into disconnected pieces that fail to connect with genuine customer experiences.

Brands can achieve seamless multi-channel engagement by combining marketing technologies with real-time personalization and scalable content solutions to deliver precise messages to the right audience at the right moment.

VOLTAGE MOTORS' MULTICHANNEL MARKETING CHALLENGES

VoltAge Motors has established itself as a leader in innovative design while focusing on customer needs throughout its history. The firm's MarTech strategy fell short due to a fragmented ecosystem of platforms and processes that failed to support rapid business growth and meet expanding customer expectations. The expansion of VoltAge into new regions combined with the launch of more advanced models and services such as subscription-based software upgrades and charging solutions led to increased complexity in its marketing operations. VoltAge possessed numerous tools yet faced difficulties in providing seamless personalized customer interactions throughout various points of contact.

The car company's primary challenge lay in the absence of integration across its marketing platforms. VoltAge invested in different systems for managing content as well as email marketing with an additional system for CRM and social media scheduling. The content department, together with creative teams, regional marketing, and analytics, functioned as separate entities, each using distinct tools and processes.

Regional teams across Europe and North America often created duplicate content since they lacked a centralized system to share assets. The separation of customer engagement data across CRM systems and marketing automation tools led to disjointed customer journeys. Due to poor coordination systems, customers sometimes receive vehicle upgrade emails after purchase or duplicate messages across various channels.

VoltAge faced difficulties adapting promptly to evolving customer behavior and their preferred communication channels. The fluctuation of consumers between online research platforms and dealership visits along with social media interactions created difficulties for VoltAge's marketing teams, who had trouble keeping their messaging consistent. VoltAge struggled to modify campaigns instantly because its fixed content management systems and campaign execution processes created obstacles for real-time adjustments. VoltAge faced deployment delays for its flash promotion offering free charging credits because of approval bottlenecks and slow coordination among email, Web, and mobile teams. The campaign launch coincided with a significant decline in urgency and customer interest.

VoltAge Motors started an extensive assessment of its MarTech stack to address current issues with specific attention to DXPs. The marketing leadership team found that a DXP could function as a single platform to integrate content management along with customer data and personalization engines for campaign orchestration. The goal was to establish an omnichannel experience that provided real-time content tailoring and messaging for each individual customer across all channels. VoltAge examined additional technologies to support their strategy, which included CDPs for creating unified customer profiles together with AI-powered personalization tools for dynamic content delivery and customer journey orchestration platforms that help to streamline cross-channel engagement.

The company plans to eliminate departmental silos and speed up campaign launch times while delivering unified customer experiences throughout Web interfaces, email communications, social media platforms, mobile applications, and in-vehicle digital displays through their adoption of integrated DXP systems and revised content delivery strategies. Marketing teams will adjust their workflows by implementing standardized approval methods and utilizing automation to speed up relevant content delivery.

Through its strategic overhaul progression VoltAge positions itself to provide customer-first experiences that correspond to the innovative quality of its electric vehicles. The next chapters will provide additional insights into their journey to create a marketing ecosystem that is both unified and agile.

CONCLUSION

The management of content and campaign execution across multiple channels continues to grow in complexity. Marketing professionals must successfully manage expanding digital platforms while adapting to changing consumer behavior and meeting demands for personalized experiences. Numerous organizations face difficulties caused by disconnected technology systems along with isolated data repositories, which makes delivering uniform customer experiences across various channels particularly challenging. Marketing teams face reduced efficiency and missed engagement opportunities when they lack the proper tools and strategies to deliver the expected customer personalization levels.

CHAPTER 17

DXPs and CMS

Modern Web sites and apps have evolved from static platforms to become dynamic and data-based touchpoints that deliver personalized experiences influencing customer engagement. Digital Experience Platforms (DXPs) and Content Management Systems (CMSs) serve as the essential tools for brands to manage and distribute their Web-based content while optimizing it across multiple channels to meet increasing demands for personalized content, offers, and experiences. DXPs function as fundamental infrastructures that enable organizations to deliver interactive and scalable digital experiences that maintain seamless customer engagement across Web sites, mobile apps, and new digital interfaces.

Web content management platforms have evolved significantly since the days of basic Web site builders. Whereas initial Web content management systems focused only on static content publishing, modern DXPs and CMS solutions provide advanced features such as real-time personalization through AI automation omnichannel distribution and deep integrations with customer data and marketing tools. The latest technological developments enable brands to customize user experiences while enhancing touchpoint engagement and scaling their content operations more effectively.

This chapter examines how DXPs and CMS platforms function in digital marketing by focusing on their major differences and vital features while providing guidance for business evaluation. This exploration will begin by defining what a DXP and CMS and then will look at their similarities and differences.

CMSs

CMSs allow users to create Web site content and manage publication processes, mostly without needing coding or Web development skills. CMS platforms help businesses, bloggers, e-commerce stores, and publishers build and maintain Web sites with streamlined processes for organizing, editing, and distributing content.

CMS platforms serve as the essential infrastructure for digital publishing across corporate Web sites, e-commerce stores, news portals, and personal blogs. Web site creation required custom development and manual coding, which presented significant challenges for non-technical users to manage content. A CMS enables users to modify text, images, videos, and layouts through an easy-to-use interface that decreases dependency on developers while enhancing content update processes.

Common use cases for a CMS include the following:

- *Blogging and news sites*: News organizations and personal bloggers rely on WordPress and Drupal to manage their content thanks to features for scheduling posts and multimedia handling alongside SEO tools.
- *Corporate Web sites*: Companies employ CMS platforms to keep their informational Web sites active and share company information without relying on an in-house IT team.
- *E-commerce*: Brands can manage online storefronts and sell products through CMS platforms that integrate with online shopping systems such as Shopify and WooCommerce.
- *Marketing and lead generation*: Marketers can build landing pages, gated content, and digital campaigns using a CMS to capture leads through integrated forms and analytical tools.

How a CMS Empowers Non-Technical Users

One of the biggest barriers to entry for a company to have a robust Web presence is the technical and design skills needed to create and maintain a professional Web site presence.

A CMS provides essential functionality by allowing users with minimal coding skills to edit and manage digital content. A CMS eliminates the need for HTML, CSS, and JavaScript skills by providing a graphical user interface (GUI) and often has drag and drop capabilities to create layouts. Other features of a typical CMS include the following:

- They utilize a What-You-See-Is-What-You-Get (WYSIWYG) editor to create and modify content in an interface similar to a word processor.
- The media library feature enables users to upload and manage their images, videos, and documents through the CMS platform.
- Create custom Web site templates and layouts through a user-friendly interface without direct code editing.
- Control user roles and permissions to enable team member contributions to content creation while maintaining system security.

Marketing teams can easily update product pages, publish blog posts, or add customer testimonials independently of IT support, which demonstrates the crucial role CMS platforms play in agile content management.

Examples of CMS Platforms

Numerous CMS platforms exist to serve various requirements and complexity levels. The following are some of the more popular examples.

WordPress

WordPress stands as the most widely used CMS platform responsible for managing more than 40% of Web sites worldwide[1]. The platform allows comprehensive customization options along with thousands of plugins and maintains an easy-to-use interface.

Drupal

Drupal serves as an advanced CMS platform designed to handle complex data-driven Web sites that require both high scalability and robust security measures. As one of the more technically complex open-source CMS platforms, it is used by some large enterprise organizations in place of more expensive proprietary DXPs.

Contentful

Contentful delivers an API-based headless CMS solution that businesses can use to deliver content across various channels beyond traditional Web site platforms.

Webflow

Webflow provides both design and visual development tools so marketers and other non-technical users can maintain full creative control over Web site layouts without the need for engineers.

There are many other types of Web site CMS platforms, including those aimed at consumers or "prosumers" that have small businesses or hobby businesses. Two of these platforms are Squarespace and Wix, platforms that while sometimes used by larger companies are geared towards non-technical users that favor ease of use to robust technical capabilities.

A CMS provides essential digital content management infrastructure that allows organizations to manage Web site publishing and updates without requiring advanced technical skills. Growing businesses seeking dynamic omnichannel solutions often need capabilities beyond those of traditional CMS systems. DXPs step in to deliver enhanced personalization options alongside comprehensive integration and multi-channel content management solutions.

DXPs

Many large company Web sites are built on DXPs, which do much more than simply host Web content. DXPs represent comprehensive software solutions that enable the creation, management, and optimization of digital experiences across numerous customer touchpoints, including desktop and mobile Web sites. DXPs go beyond traditional CMS by providing a comprehensive framework that incorporates content management with data integration and customer engagement across various digital touchpoints including Web sites and mobile applications.

The Shift from Traditional CMS to DXPs

Businesses required advanced digital solutions that delivered personalized and integrated experiences across multiple platforms when digital engagement moved past single Web sites. Original CMS platforms served static content needs until growing customer expectations pushed organizations to seek systems that provided real-time personalization and AI-driven insights alongside smooth CRM, analytics, and commerce platform integrations.

DXPs were developed as the latest advancement in content management systems to solve several critical challenges:

- *Omnichannel engagement*: The ability of customers to use various devices and platforms requires organizations to implement a content delivery system that works uniformly.
- *AI-powered personalization*: DXPs connect with customer data platforms (CDPs) alongside AI-driven recommendation engines and behavioral analytics to generate customized content that meets user needs.
- *Customer journey management*: DXPs enable marketers to manage and enhance customer journeys through multiple interactions between brand touchpoints to deliver consistent brand experiences across all channels.
- *SaaS and on-prem versions*: Increasingly, DXPs are offering a cloud-based or software as a service (SaaS) model for hosting that simplifies the hosting infrastructure needs.

DXPs and a Multi-channel Focus

DXPs function as the core engine that manages content and orchestrates customer interactions alongside commerce activities and system integrations.

Content Management and Distribution

DXPs enable brands to produce content and distribute it through multiple digital channels while maintaining a consistent brand identity.

DXPs combine customer data and AI-driven segmentation tools with behavioral analytics to immediately deliver personalized content and offer recommendations.

E-Commerce Integration

DXPs establish seamless connections with e-commerce platforms to support personalized shopping experiences through product recommendations and transactional interactions.

API-First and Headless Capabilities

DXPs support headless content management systems that enable the delivery of content through APIs to diverse digital interfaces such as mobile apps and IoT devices instead of restricting themselves to Web site presentation. As more companies consider and adopt headless and composable approaches to their marketing technology infrastructure, these features become more compelling.

Marketing and Analytics Connectivity

DXPs connect to marketing automation systems, analytics platforms, CRM software, and AI optimization tools, which enables marketers to monitor customer journeys and optimize engagement methods through testing different experiences.

Examples of DXPs

A few dominant enterprise-level DXPs control the market with their advanced content-driven digital experience capabilities. Also note that in Chapter 21, multi-function platforms will be discussed, and several of the below DXPs are also part of those larger and more cross-functional systems.

Adobe Experience Manager

Adobe Experience Manager (AEM) is built on a broader platform (Adobe Experience Platform, or AEP), which includes many other functions, such as a CDP, multi-channel journey-based analytics, e-commerce functionality, and more. AEM integrates content management with personalized experiences and AI analytics to create immersive digital environments.

Optimizely

Optimizely's current DXP is the product of their original testing and personalization platform (called Optimizely), as well as their acquisitions of both Episerver and Ektron. Optimizely provides a DXP solution combining content management with experimentation tools and AI-based optimization capabilities to help brands create data-driven experiences. The Optimizely DXP is provided as both an SaaS cloud version as well as a hosted version.

Sitecore

Sitecore delivers real-time personalization together with comprehensive customer data management and strong marketing and e-commerce integrations. The platform is provided as both a cloud-hosted and self-hosted version.

Acquia

Acquia offers scalable DXP solutions with robust content and commerce features built on Drupal. This offering is unique amongst those mentioned because Drupal is the only platform of the four that is open source.

DXPs extend beyond basic content hubs to function as complete ecosystems that enable businesses to provide highly personalized experiences across multiple channels. Organizations aiming to deliver seamless and data-driven customer interactions across various digital touchpoints will adopt DXPs to consolidate their content management with marketing and commerce strategies.

THE DIFFERENCES BETWEEN A DXP AND A CMS

It might seem to be difficult to understand the differences between DXPs and CMSs since both systems manage and deliver digital content. These platforms accomplish different objectives and deliver distinct ranges of capabilities. A CMS concentrates on managing Web site content, while a DXP extends its capabilities by integrating various digital touchpoints alongside personalization and enhanced customer engagement features. Businesses must understand these differences to determine which platform most effectively meets their requirements.

Scope and Capabilities

The fundamental difference between a CMS and a DXP stems from the varied scope and purpose they serve.

A CMS functions to build and maintain Web content and publish it for Web site use. The platform provides editing capabilities for Web pages alongside organizational tools to update content without needing extensive technical expertise.

A DXP offers a unified platform to control digital customer experiences across various touchpoints, such as Web, mobile, email, social media, chatbots, and apps.

A small business operating a corporate Web site or blog should consider a CMS, while a global retail brand should adopt a DXP to manage personalized customer interactions across Web platforms and mobile applications.

Integration and Extensibility

DXPs enable businesses to connect their customer data systems with marketing automation tools and CRM platforms while integrating e-commerce and analytics solutions into one cohesive platform.

Most CMS platforms concentrate on Web site content management yet certain CMS platforms can extend their functionality through plugins to connect with additional tools. CMS platforms lack the inherent extensive connectivity features found in DXPs.

DXPs act as central hubs that facilitate the seamless exchange of data among systems to enable personalized data-driven customer interactions.

An enterprise brand could utilize a DXP platform to combine real-time customer data from its CRM system with personalized Web site content driven by past purchases and automated marketing campaigns in one unified system. Without extensive customization and third-party integrations, a CMS would lack these capabilities.

Personalization and AI Capabilities

Modern customer engagement demands personalization, and this is another major area where DXPs outshine traditional CMS platforms.

CMS platforms typically provide basic personalization features, such as displaying different content to logged-in users, but require third-party tools for deeper behavioral personalization.

DXPs incorporate AI-driven personalization natively, using customer segmentation, behavioral analytics, and machine learning to deliver customized content in real time.

For example, a CMS-powered e-commerce site may show generic homepage banners to all visitors, whereas a DXP-powered e-commerce site could analyze visitor behavior and dynamically serve personalized product recommendations, custom offers, and targeted messaging.

TABLE 17.1 Comparison of a DXP and a CMS

Capability/Characteristic	DXP	CMS
Primary Purpose	Delivers personalized digital experiences across multiple channels	Primarily used for creating and managing Web content
Content Management Scope	Manages content across Web, mobile, apps, kiosks, and other digital touchpoints	Manages content primarily for Web sites and blogs
Personalization	Advanced AI-driven personalization and customer segmentation	Basic personalization features, usually requiring third-party tools
Omnichannel Support	Designed for seamless customer journeys across multiple platforms	Limited to Web content, with some extending to mobile or email
Integration with Other Systems	Deep integration with CRMs, marketing automation, commerce, and analytics tools	Integrates with some third-party tools, but not as deeply as DXPs
Data and Analytics	Robust data analytics, customer insights, and journey tracking	Basic reporting on content performance and Web site traffic
User Roles	Used by marketing, sales, IT, customer experience, and e-commerce teams	Used mainly by content creators, editors, and marketing teams
Scalability	Built for large enterprises with complex digital experience needs	Suitable for businesses of all sizes, from small to enterprise
Flexibility and Customization	Highly customizable with modular architecture, supports composable approaches	Customizable via themes, plugins, and APIs, but less modular than DXPs
Ideal Use Cases	Enterprise brands with multi-channel experiences, e-commerce, and customer portals	Businesses needing Web site management, blogs, and corporate content hubs

Target Users and Use Cases

Both CMSs and DXPs are widely used across industries, yet they serve different optimal functions when deployed successfully.

CMS platforms serve businesses regardless of their size when they need a straightforward and user-friendly system for content publishing. Bloggers along with media companies and smaller e-commerce sites alongside corporate Web sites often use CMS platforms for content-focused Web sites that do not need advanced personalization features or omnichannel capabilities.

DXPs are designed for enterprise organizations that need solutions to manage their digital interactions throughout various touchpoints. The retail, finance, and healthcare industries and major digital publishers experience advantages by merging customer experiences across Web platforms, mobile applications, physical stores, and other digital touchpoints.

Here are some examples:

- Small businesses that maintain both a Web site and a blog platform can utilize WordPress as their CMS to effortlessly publish content and administer their Web site.
- To deliver consistent and personalized experiences across all customer touchpoints, a global bank with an integrated mobile app, Web site, customer support portal, and AI-driven chatbot needs a DXP.

Businesses needing cross-channel engagement and personalization as well as deeper marketing and commerce platform integration can achieve better results through a DXP beyond basic Web site management with a CMS. Organizations must decide based on their specific operational necessities along with their digital strategy and long-term goals for customer engagement.

A CMS is the cost-effective solution that businesses need to efficiently manage Web site content. Organizations aiming to build integrated digital experiences using data across multiple channels need the sophisticated capabilities of a DXP to thrive in the modern digital marketplace.

Primary Users of DXPs and CMSs

Just as Web sites support a wide range of external audiences, a diverse group of team members utilize DXPs and CMS platforms. Properly deployed CMS or DXP platforms enable seamless interdepartmental collaboration, which enhances efficiency and maintains consistency in digital content delivery.

Marketers

Marketers utilize CMS and DXP platforms as their main tools to manage Web site content without needing to rely on software engineers to make changes for them. Marketing teams can rapidly build and distribute digital content for the Web, mobile apps, and other channels, such as email, through these platforms and preserve consistent branding throughout all channels.

Marketing teams create landing pages and product pages along with blog posts by utilizing pre-approved templates and drag-and-drop editors.

Personalization tools and functionality within a DXP can be used to deliver customized experiences for distinct audience segments. Additionally, marketing automation platforms connect with DXPs to enable real-time campaign management alongside customer journey coordination, and a retail company's marketing team can launch seasonal campaigns through their DXP platform to update Web site banners and blog posts and send personalized email promotions from one unified interface.

Content Creators and Editors

Writers, along with editors and designers, use CMS and DXPs to produce structured content efficiently. These users require systems that support seamless collaboration, manage approval workflows efficiently, and distribute content across multiple channels.

Web content writers use rich text or WYSIWYG editors and SEO optimization tools along with metadata fields to create and format their content and position it for optimal consumption by human readers and search engines.

Editors handle content approval procedures to confirm that every piece matches brand standards prior to being published.

CMS and DXP workflow automation facilitates collaborative processes between writers, editors, and legal teams to achieve compliance before publishing content.

A news organization that operates with CMS platforms such as WordPress or Drupal provides journalists with a submission system while enabling editors to review and approve content and designers to incorporate multimedia elements within the same platform.

Developers, Engineers, and IT Teams

The primary function of CMS platforms is to enable non-technical users to create Web-based content, but developers and IT teams remain essential in maintaining platform customization and security. The transition from traditional CMS systems to advanced DXPs results in more responsibilities for organizations in many cases.

Developers implement integration features between CMS/DPX systems and CRM systems, together with e-commerce solutions, marketing automation platforms, and analytics tools.

Enterprises are adopting headless CMS solutions and API-based DXPs, which make developers responsible for providing dynamic content delivery to Web platforms, mobile applications, and IoT devices.

Security teams keep track of user access rights alongside data protection measures and adherence to privacy standards to protect customer information.

Even in a headless or composable environment, these technology teams are still vital to the success of the platforms. A global enterprise implementing a headless DXP such as Contentful needs its developers to construct custom APIs for distributing content to various digital platforms and preserving one unified content source.

UX and Design Teams

The UX and design teams work to enhance site architecture and layout while improving digital experiences within CMS and DXP platforms. Their work guarantees that content functions properly while remaining engaging and accessible to users on various devices.

Site-mapping tools help UX designers establish customer navigation paths and build intuitive structures for Web site navigation.

Design teams develop unique themes alongside branding elements and templates to maintain visual consistency and branding fidelity across all content.

The A/B testing tools included in many DXPs enable UX teams to test various layouts alongside headlines and CTAs to enhance user engagement.

An e-commerce brand can use A/B testing inside a DXP to test how a new home page layout affects conversion rates, which helps the brand make data-driven design decisions.

A CMS or DXP can impact many teams within an organization. The technical team handles platform scalability and security while developers build upon the work of marketers and content creators who focus on publishing personalized content. The UX and design teams work to improve digital interactions while making sure customer journeys remain seamless and optimized for better engagement and conversion results. These platforms reach their potential when teams work together to manage content properly while integrating it into an overarching digital strategy driven by data.

EVALUATING DXP AND CMS PLATFORMS

While they share many characteristics, DXP and CMS platforms also have unique features and focus areas that require their own sets of evaluations.

An organization that is choosing one platform or the other should keep these considerations in mind.

DXPs

DXPs enable businesses to provide personalized and data-driven experiences through multiple customer interaction points at once. DXPs expand beyond Web site content by connecting with multiple MarTech tools such as CRMs, e-commerce systems, and analytics platforms to deliver integrated customer experiences.

Businesses that need real-time personalization capabilities and AI content delivery across multiple channels will find more advantages in DXPs compared to traditional CMS platforms.

Key Features of a DXP

AI-Powered Personalization and Customer Journey Orchestration

DXPs deliver personalized experiences by dynamically adjusting content according to user actions while using AI to segment audiences and automate interactions in real time.

Multi-Channel Content Delivery

DXPs deliver content across multiple touchpoints, including Web interfaces, mobile applications, digital kiosks, IoT devices, social platforms, and chatbots, unlike traditional CMS platforms, which focus solely on Web sites.

Deep Integration with the MarTech Stack

DXPs establish native connections to CRM systems, marketing automation tools, e-commerce solutions, and CDPs that enable smooth customer interactions throughout various platforms.

How to Evaluate a DXP

First, let's look at how to evaluate a DXP for scalability and performance:

- Does the platform support enterprise content distribution alongside personalization and traffic spike management?
- Is the system designed to handle scaling in the cloud during peak demand times?

Next, let's look at the DXP's integration capabilities:

- What level of integration capability does the system offer with current CRM systems, as well as e-commerce platforms and marketing automation and analytics tools?

- Does the platform deliver an API-first design to enable flexible headless implementations?

You'll need to evaluate the personalization and AI-driven content delivery capabilities:

- What level of sophistication does it provide in its AI-powered segmentation processes and customer behavior analysis, along with recommendation systems?

- Can the system deliver immediate modifications to digital experiences through user interaction data?

Depending on the industry, specific marketing requirements, and other factors that vary based on the company, there may be other criteria that it will be beneficial to evaluate. For instance, a healthcare company evaluating DXPs will want to ensure that patient data privacy regulations are being handled in a way that meets regional or local regulations.

Additionally, a retailer with a prominent e-commerce business will want to ensure their DXP supports their needs to feature products on the Web site.

CMSs

A CMS functions mainly to create and administer Web site content before publishing it online. CMS platforms with multi-channel distribution capabilities fail to deliver the advanced AI personalization and customer journey management features found in DXPs. A CMS can be a more cost-effective solution as compared to a DXP, simplifying business operations for companies that manage Web sites and publish content and blogs compared to a DXP.

Key Features of a CMS

User-Friendly Content Editing Interfaces

The majority of CMS platforms provide WYSIWYG editors that enable non-technical users to format pages and publish content while also adding images.

SEO and Performance Optimization Tools

The integrated SEO tools within CMSs enable content optimization for search engines, which leads to better visibility and improved search rankings.

Modern CMS platforms provide performance improvements through caching capabilities alongside lazy loading features and mobile-responsive designs.

Plugin and Extension Support

Users can expand the capabilities of WordPress and Drupal CMS platforms by implementing plugins that support e-commerce and additional tools for SEO optimization and marketing automation along with analytics.

The size of the development community that writes and supports these plugins can also be a selling point for a particular CMS.

How to Evaluate a CMS

First, let's look at ease of use:

- Does the CMS provide a sufficient level of intuitiveness that allows non-technical users to create and manage content?
- Does the platform include drag-and-drop functionality along with template options and WYSIWYG editors?

Now let's look at flexibility and customization:

- Does the system allow expansion by adding themes along with plugins and external integrations?
- Does the system support custom development capabilities for advanced functionality?

Next, let's evaluate security and compliance:

- Does the platform include built-in security measures such as role-based access control, SSL support, and ongoing security updates?
- Does the CMS meet the requirements of data protection laws such as the GDPR and CCPA and/or are plugins available to support these needs?

While seemingly simple in functionality, a CMS can provide a robust set of Web site functionality, and many large organizations rely on them over

DXPs. It is important to make the effort to understand what will be the best long-term fit for the organization.

Differing Approaches to Managing Content

The selection of a CMS or DXP offers organizations several alternatives, each featuring unique methods for content management and delivery along with customization capabilities. Certain businesses focus on SaaS solutions for their ease of use and scalability, while other organizations choose headless or composable architectures to achieve enhanced flexibility. Selecting between proprietary and open-source choices affects a business's expenses, its security measures, and how well the solution can be customized. Businesses must understand platform differences to choose the appropriate system for their content management needs and technical specifications.

SaaS DXP and CMS

SaaS CMS and DXP solutions eliminate the need for on-premise hosting and manual updates by providing a fully managed cloud-based platform. These platforms provide automatic updates, scalability, and lower maintenance costs, making them an attractive choice for organizations that want to reduce their IT overhead.

These are the advantages of SaaS CMS/DXP:

- No need for internal hosting or server management
- Automatic security patches and software updates
- Built-in scalability to handle traffic spikes and content growth

Here are some examples of SaaS CMS and DXP platforms:

- *HubSpot CMS*: An all-in-one marketing CMS with integrated SEO, automation, and analytics
- *Contentful*: A cloud-based headless CMS that enables API-driven content management
- *Adobe Experience Cloud*: A DXP with advanced personalization, AI-driven recommendations, and deep MarTech integrations

SaaS solutions are ideal for organizations that want a managed service with low IT complexity but may not be as customizable as self-hosted or open-source alternatives.

Headless CMS

A headless CMS separates content management (back-end storage and organization) from content delivery (front-end display and user experience). This API-first approach allows businesses to deliver content dynamically across multiple digital touchpoints, including Web sites, mobile apps, IoT devices, and even voice assistants.

How Headless CMS Works

In a headless CMS system, the structured database holds content functions independently from its presentation layer. Content creators gain full control to manage updates without being restricted by particular front-end designs.

APIs function as the link between the CMS and multiple front-end frameworks which enable dynamic content delivery to Web platforms as well as mobile apps, kiosks, chatbots, and IoT devices. A headless CMS enables organizations to deliver uniform brand experiences throughout multiple digital channels.

Developers have total freedom to construct bespoke digital experiences through front-end frameworks including React, Vue.js, and Angular. Traditional CMS platforms provide pre-made templates, whereas headless CMS systems enable front-end teams to develop bespoke user experiences that meet specific business requirements.

Benefits of Headless CMS

A headless CMS enables businesses to deliver their content across all channels as one of its most significant benefits. Businesses can distribute content across multiple platforms from a single publication point, which maintains consistency and reduces redundant work.

Developers gain increased flexibility because they can operate outside of pre-built themes and templating systems. Developers possess the ability to create optimized performance experiences that support business objectives.

Future-proofing is another key benefit. Through the use of a headless CMS, organizations can rapidly adjust to new digital interfaces such as smartwatches and both AR and VR systems while maintaining content accessibility through technological advancements.

Here are some examples of headless CMS platforms:

- *Contentful*: API-driven CMS for multi-channel content distribution
- *Strapi*: Open-source headless CMS with extensive customization options
- *Sanity*: A developer-friendly CMS with real-time collaboration features

Headless CMS solutions are best suited for companies that prioritize multi-channel content delivery, require deep customization, and have development resources to build front-end applications.

Composable CMSs vs. Headless CMSs

Composable CMSs and headless CMSs are similar, but they serve different strategic needs, as shown in Table 17.2.

TABLE 17.2 A comparison of headless and composable CMS

Feature	Headless CMS	Composable CMS
Focus	Separates content storage from display	Modular architecture with interchangeable services
Content Delivery	API-first, delivers content anywhere	API-first but integrates multiple services
Flexibility	Requires a custom front-end framework	Allows swapping of components such as analytics, search, and AI
Best For	Multi-channel content delivery	Organizations needing a highly adaptable, evolving tech stack

Composable CMSs take a modular, API-first approach that allow organizations to assemble various content, analytics, and experience-building tools into a unified system.

Headless CMS platforms are back-end content repositories that strictly focus on managing and delivering content through APIs.

A composable CMS provides greater flexibility, allowing businesses to integrate and swap services (e.g., AI-driven personalization, e-commerce, or search capabilities) without needing to overhaul the entire platform.

For example, a large enterprise using a composable CMS could integrate its preferred search engine, digital asset management (DAM) system, and AI-powered personalization tool, whereas a headless CMS would require custom development to achieve similar functionality.

Proprietary vs. Open Source

Organizations must also decide between proprietary (licensed) such as Optimizely, Sitecore, or AEM, or open-source CMS/DXP platforms such as WordPress or Drupal, as each approach has distinct benefits and trade-offs.

Proprietary

Proprietary platforms require licensing fees and are typically maintained by a single vendor, offering enterprise support, reliability, and built-in security. They have the following advantages:

- Dedicated customer support and enterprise security
- Seamless integration with other proprietary tools (e.g., Adobe Experience Manager with Adobe Analytics)
- Faster deployment with pre-built functionality

They also have the following challenges:

- Higher costs due to subscription fees and licensing
- Less customization compared to open-source alternatives

Open Source

Open-source platforms provide complete customization but require in-house expertise for setup, maintenance, and security. They have the following advantages:

- No licensing fees, making them more cost-effective
- Highly flexible and customizable, allowing tailored digital experiences
- Large developer communities, offering plugins, themes, and community support

They also have the following challenges:

- They require internal IT or developer resources for ongoing maintenance.
- Security vulnerabilities may arise if not properly maintained.

Businesses that need a large amount of control, customization, and cost-efficiency may prefer an open-source CMS, while those requiring enterprise support, security, and seamless integrations may opt for a proprietary DXP.

Organizations have numerous options for managing digital content, from SaaS-based platforms and headless CMS solutions to fully composable DXPs. Several factors will influence the best choice and outcome for an organization, including the technical resources available for support, scalability requirements, the need for customization, as well as the need for personalization for customer experiences.

CONCLUSION

When choosing between CMS and DXP platforms, marketers must make important decisions that involve evaluating business requirements and scalability while developing their content strategy. A traditional CMS efficiently handles Web site content management, but DXPs provide more extensive solutions by incorporating personalization tools, omnichannel delivery systems, and AI analytics. Content management evolves through headless and composable architectures, which enable brands to deliver flexible multi-platform digital experiences that keep up with changing consumer expectations.

NOTE

1. https://w3techs.com/technologies/details/cm-wordpress

CHAPTER 18

Marketing Automation, Multichannel Personalization, and Journey Orchestration

As this book has covered so far, traditional marketing methods with highly scheduled methods have, in many cases, given way to a more advanced, always-on marketing that engages customers when and where they are choosing to interact with a company's marketing efforts. Thus, companies must provide customers with streamlined personalized communication across multiple touchpoints as real-time engagement and automation are vital for reaching consumers with near-limitless options. Consumers' demands for real-time engagement have led to the adoption of marketing automation alongside multichannel personalization and journey orchestration platforms, which has revolutionized brand management of customer interactions while enabling marketers to scale their engagement and optimize their workflows for better efficiency.

Marketing automation has played a large role in this, and its share of the overall MarTech market is considerable. As of 2022, the market size of marketing automation software was valued at just over $5.2 billion, and it is predicted to rise to $13.5 billion by 2030[1]. This growth demonstrates the practical application that the marketing automation platform (MAP) provides to marketers, and while email remains the primary channel used by marketers on these platforms (58%), other channels are being used as well, including social media management (49%) and SMS marketing (30%)[2].

Through automated email sequences and AI-driven content recommendations, together with dynamic journey orchestration, these tools deliver marketing that remains both timely and data-driven while maintaining relevance. Businesses face the risk of providing disjointed customer interactions and inefficient promotional activities, which lead to missed revenue possibilities if they don't employ these systems.

COMPARING THE TYPES OF PLATFORMS

Marketing teams depend on three main types of automation and personalization platforms for managing customer engagement at a large scale. Despite their shared functionalities, these platforms operate with specific roles within a MarTech stack, as illustrated in Table 18.1 and the following list:

- MAPs automate and optimize repetitive marketing activities, including email campaigns, lead nurturing, and scoring processes. The platforms maintain a system where potential customers get personalized outreach delivered promptly according to established rules and workflows.

- Multichannel personalization platforms enable the real-time customization of content depending on user behavior across various digital platforms. Dynamic content and messaging adjustments on these platforms enhance engagement rates and conversion success.

- CJO platforms enable businesses to document customer paths while streamlining interactions through automated processes. CJOs alter messaging and engagement strategies in real time based on active customer interactions across multiple channels, which is different from MAPs and personalization platforms.

TABLE 18.1 A comparison of platforms

Feature	Marketing Automation (MAP)	Multichannel Personalization	Customer Journey Orchestration (CJO)
Primary Focus	Automating campaigns and workflows	Delivering tailored experiences across channels	Managing customer interactions across multiple touchpoints
Core Functionality	Email automation, lead nurturing, segmentation, CRM integration	AI-driven content customization, behavioral tracking, dynamic messaging	Journey mapping, real-time decision-making, cross-channel engagement
Best For	Scaling email marketing, automating lead management, managing large customer lists	Personalizing Web site, email, ads, and content experiences	Coordinating interactions across digital and offline experiences

When to Use Each Type of Platform

Businesses can scale their email marketing campaigns and nurture leads effectively while automating communication workflows for prospects and customers with MAPs.

Brands rely on multichannel personalization platforms to enhance engagement across Web sites and messaging channels by delivering targeted content based on user behavior in real time.

Customer journey orchestration (CJO) platforms unify interactions across various touchpoints to deliver seamless and relevant customer experiences that respond quickly to real-time actions.

The selection of appropriate platforms or platform combinations relies on marketing objectives and the complexity of the audience as well as how much automation is necessary to connect with customers effectively throughout their journey.

How Do These Platforms Help Marketers?

Marketing automation, multichannel personalization, and CJO platforms play a critical role in optimizing how businesses engage with their audiences. The individual functions of these platforms combine to enable marketers to achieve greater efficiency while delivering personalized experiences and enhancing customer engagement. An introductory overview of their advantages appears here, while further detailed explanations follow in subsequent sections.

Efficiency Gains

Marketing teams need to handle multiple-channel campaign execution as they also deal with expanding customer data volumes. By automating key functions, including email marketing and customer segmentation, these platforms save time from executing campaigns and scoring leads. Automation facilitates the smooth operation of repetitive processes by minimizing human intervention and enhancing both consistency and scalability. Tools that schedule emails and trigger personalized experiences enable marketers to concentrate on strategic planning instead of operational tasks.

Personalized Customer Engagement

One-size-fits-all messaging no longer meets customer expectations. MarTech platforms apply AI and automation to behavioral data to create customized content and interactions. Personalization platforms automatically modify the content of Web sites and email messages together with advertising experiences according to user actions. Journey orchestration tools make customer interactions relevant to their buying process stage while MAPs target leads with precise messages at appropriate times.

Improved Customer Journeys

Marketing professionals must establish links among customer interactions that occur both online and offline. These platforms enable marketers to design customer journeys that optimize engagement across multiple touchpoints. When a visitor looks at a product page without buying, they might receive an automated email or targeted advertisements to encourage purchase completion. Journey orchestration tools enable real-time adaptation to customer behavior to deliver relevant communications that accompany customers during their brand interaction journey.

Data-Driven Decision-Making

Without data, marketing decisions rely on guesswork. These platforms deliver analytical insights about campaign success while monitoring customer actions and journey results. Marketing automation tools assess engagement metrics such as open rates and conversion rates, while personalization platforms determine which content performs best with audiences and journey orchestration platforms map out the entire customer experience. Marketers who apply these insights can optimize their strategies and budget allocations while enhancing customer engagement.

While each has its strengths and core areas of focus, these platforms work together to create marketing strategies that are efficient, personalized, and driven by data. Sections that follow will provide an in-depth look at marketing automation and multichannel personalization platforms alongside CJO platforms to reveal how each system tackles these challenges differently.

WHO USES THESE PLATFORMS?

As discussed, these platforms provide support to different teams throughout the organization. Sales and customer experience teams, together with IT and data departments, work alongside marketing teams to maximize the effectiveness of these platforms. Every team engages with these technologies through unique approaches, which results in consistent customer interaction across all platforms. The following section will explore how the platforms affect different teams within an organization.

Marketing Teams

Marketing departments are the main users who utilize these platforms to automate operations, deliver personalized experiences, and manage customer journeys.

Demand generation teams use MAPs to develop leads, along with running automated email sequences and guiding potential customers through the sales funnel. Automated sequences deliver timely relevant information to prospects, which helps to boost conversion rates.

With multichannel personalization tools, *customer engagement teams* deliver customized content experiences through Web sites, emails, social media, and other digital touchpoints. AI-powered personalization delivers content that matches customer preferences and behaviors.

Retention and loyalty teams manage interactions with existing customers through CJO platforms. These tools enable businesses to keep customers engaged and loyal by sending personalized renewal reminders as well as upselling relevant products and sending proactive support messages.

Customer Experience Teams

To provide a seamless customer experience, businesses must synchronize their marketing, sales, and support interactions.

Experience teams utilize CJO platforms to connect departmental touchpoints, which creates consistent and contextual customer engagement. The CJO system maintains conversation continuity for support interactions initiated after promotional emails by incorporating previous customer interactions.

Sales Teams

Sales teams achieve improved visibility of prospect and customer behavior through the use of marketing automation and journey orchestration systems.

Marketing automation platforms enable sales teams to detect top prospects by analyzing their engagement levels along with Web site visits and email activity through lead scoring and prioritization. Sales teams can maximize their efficiency by directing their efforts toward leads that show the highest potential to convert.

Sales enablement tools track potential customer content interactions before sales outreach actions. Sales representatives can customize their discussions according to customers' preferences which boosts their capacity to finalize sales agreements.

IT and Data Teams

Marketers create strategies that utilize these platforms, while IT and data teams maintain technical infrastructure to enable seamless execution and adherence to governance and regulatory requirements.

Successful management of platform integrations is essential for making marketing automation tools operate together with personalization systems and journey orchestration solutions in the MarTech stack. IT teams maintain efficient data transfer between CRM systems and email marketing solutions along with analytics platforms and other digital platforms.

Ensuring compliance efforts becomes more important because of regulations such as the GDPR and CCPA. Data teams enforce privacy policies and consent tracking while ensuring responsible customer data handling.

These teams collaborate to make MarTech platforms deliver personalized and scalable customer experiences that use data-driven approaches. The upcoming sections will detail the operational processes of each platform type and provide methods for organizations to assess their functionality.

MAPs

MAPs simplify repetitive tasks in marketing and facilitate lead nurturing while improving customer engagement through automated processes. These platforms enable marketers to expand their reach while maintaining prompt communication and refining customer engagement through the analysis of behavioral patterns.

Key Features

Email marketing automation tools manage drip campaigns along with triggered emails and A/B testing to enhance messaging and improve customer engagement results. The system delivers timely content directly to customers without any need for human input.

The lead scoring and nurturing system calculates scores from customer actions such as email opens or Web site visits to identify top prospects for sales follow-up.

CRM and sales integration guarantees smooth transitions between marketing and sales teams through synchronized customer engagement data on CRM platforms. Sales representatives can respond to live data about prospect actions and behaviors.

Real-time dashboards and analytics deliver information regarding campaign success and marketing ROI alongside lead conversion rates.

Example MAPs

The following are a few popular examples of MAPs that businesses can use to scale their operations and personalize customer interactions more effectively:

- HubSpot Marketing Hub gives businesses an accessible platform to automate email marketing and lead scoring while integrating CRM for improved inbound marketing outcomes.
- Adobe Marketois an enterprise-grade marketing automation solution that specializes in B2B lead management while enabling audience segmentation and multi-channel marketing operations.
- Salesforce Pardot functions as a B2B marketing automation platform that focuses on lead nurturing and scoring while integrating seamlessly with Salesforce CRM to facilitate alignment between sales and marketing teams.

Strengths

Marketing automation enhances efficiency by streamlining tasks such as email follow-ups and lead qualification, which allows marketers to concentrate on strategic initiatives.

Businesses can nurture large lead volumes automatically, which guarantees that prospects receive the appropriate personalized messages at the correct times.

Behavioral data informs the customization of messaging and outreach, which enhances both engagement levels and conversion rates.

Weaknesses

With excessive use, automation produces robotic communications that lack personalization and may drive customers away.

MAPs function through established workflows instead of real-time adjustments to customer behavior, which multichannel personalization platforms accommodate.

Older MAPs face integration difficulties with contemporary CRM systems and analytics platforms, which necessitates additional configuration work or specialized development.

How to Evaluate MAPs

While every organization has unique needs and goals to achieve, MAPs should be evaluated for those specific requirements and according to more common needs, such as the following;

- *Ease of use*: The platform should enable marketers to set up and manage campaigns with minimal learning curve. Do marketers need advanced technical skills to set up workflows or is the configuration process straightforward?
- *Scalability*: The platform needs to support expansion by handling larger audience segments and more complex automation sequences along with increasing data volumes.
- *Integration capabilities*: The system provides seamless integration with CRM systems as well as analytics and content delivery platforms for comprehensive customer engagement.

MAPs serve as fundamental elements in contemporary marketing by enabling business automation of outreach and lead conversion while boosting engagement efforts.

MULTICHANNEL PERSONALIZATION PLATFORMS

Multichannel personalization platforms allow businesses to provide tailored content and messaging across multiple touchpoints including Web sites, email, mobile apps, and advertisements. Marketers use AI-based features, together with customer segmentation and real-time behavioral tracking on these platforms, to modify marketing actions according to individual user behavior thereby enabling brands to deliver relevant and seamless experiences.

Key Features

The platforms in this area are likely to include a wide variety of features, depending on the channels that they support. For instance, some may incorporate SMS marketing, while others focus more on solely digital and visual channels. There are several key features that most share, however:

- *AI-driven content recommendations* utilize machine learning techniques to evaluate browsing patterns while reviewing past interactions and purchase histories to provide tailored products, articles, and promotional offers. Businesses gain the ability to deliver personalized experiences to each individual user through this functionality.

- *Dynamic content personalization* enables the customization of messaging elements together with visual components and CTAs instantaneously across digital platforms to suit audience segments and behavioral triggers.

- *Customer segmentation* allows the creation of dynamic customer segments by analyzing both real-time behavioral signals such as abandoned carts and browsing history with historical data, including past purchases and loyalty status.

- *A/B and multivariate testing* enable constant experience improvements by examining various content versions and layout combinations to find the optimal fit for diverse audience groups.

Example Multichannel Personalization Platforms

The main function of these platforms is to provide personalized user experiences across various digital channels by utilizing behavior-driven customization methods. A few examples are as follows:

- *Dynamic Yield* employs AI algorithms to create personalized experiences across Web, mobile, and email platforms by analyzing real-time user behavior to enhance business engagement and conversion rates.

- *Optimizely* provides A/B testing and AI-driven personalization alongside content recommendations for Web sites and mobile apps, which enables brands to improve digital experiences through data analysis. Optimizely's personalization features are available alongside several other features on their Optimizely ONE platform, including a DXP and DAM.

- *Monetate* provides real-time personalization solutions for e-commerce and retail brands by dynamically modifying content and product suggestions according to individual customer behavior.

Strengths

Personalization has been proven to increase customer engagement and interactions, and these platforms customize user experiences so that users get relevant content based on their individual behavior and previous interactions.

Brands often achieve reduced bounce rates and increased sales conversions through the delivery of personalized content.

Real-time operations enable marketing personalization, which adapts immediately based on current customer interactions, which in turn enhances responsiveness and effectiveness and can reduce wasted interactions, increasing click-through rates.

Weaknesses

The implementation of these platforms requires high-quality data input along with proper integrations and a robust strategy to deliver meaningful and effective personalization.

AI models and personalization algorithms require ongoing refinement and monitoring to maintain the delivery of relevant recommendations.

Personalized marketing strategies depend on monitoring user behavior but generate data privacy concerns and pose compliance and consent challenges.

How to Evaluate Multichannel Personalization Platforms

Evaluating personalization platforms should center around the platform's ability to analyze customer data and deliver accurate recommendations, relying on its effectiveness in predicting user preferences:

- Does the platform synchronize customer data to deliver personalized experiences across Web sites, email, mobile devices, and paid media platforms?
- What is the time frame for the system to adapt content delivery and offer presentations in response to customer behavior?

Multichannel personalization platforms enable marketers to generate precise data-based experiences at various points of contact instead of using generic communications.

CJO PLATFORMS

CJO platforms enable businesses to effectively oversee customer interactions at different stages while delivering a smooth personalized customer experience across their entire lifecycle. CJO platforms differ from traditional automation tools because they combine data from multiple sources to create multi-step customer journeys while dynamically tailoring messaging according to user behavior and engagement history.

Key Features

CJO platforms often vary by the amount that they enable their users to dictate the exact steps a customer takes, as opposed to enabling a more dynamic journey. That said, there are several key features that most platforms share:

- *Cross-channel automation*: This feature enables managing customer communications and interactions across email systems, SMS services, social media platforms, Web sites, and customer service points. These platforms eliminate siloed campaign management by delivering a unified customer journey across mobile, Web, and offline channels.

- *Journey mapping and predictive analytics*: These are tools that are used to create visual customer paths and predict future actions with AI-based insights. CJO platforms analyze behavioral patterns to help brands direct customers toward conversion and retention or re-engagement.

- *Adaptive workflows*: These enable CJO platforms to change their processes dynamically according to real-time user interactions while static automation sequences remain unchanged. When a customer leaves items in their shopping cart, the platform initiates a journey that begins with an email reminder and proceeds to show targeted advertisements before sending an SMS with an exclusive time-limited discount. Customer responses develop through their unique interactions at each stage of the journey.

- *Integrations with channel platforms*: CJO platforms establish connections with current MarTech solutions, allowing customer data to move smoothly across marketing departments, sales teams, and support staff. Brands benefit from this integration because it allows them to maintain message consistency between departments while providing customers with uniform experiences.

Example CJO Platforms

These CJO platforms connect customer interactions across digital and offline channels to create seamless, data-driven journeys:

- *CSG XPonent (formerly Kitewheel)*: Provides real-time journey orchestration, helping businesses map and personalize customer interactions across multiple channels
- *Salesforce Interaction Studio (part of Salesforce Marketing Cloud)*: Enables brands to track, analyze, and optimize customer engagement through journey orchestration
- *Braze*: Specializes in cross-channel engagement, using real-time data to personalize messaging across push notifications, email, in-app messaging, and more

Strengths

CJO platforms eliminate departmental barriers between marketing, sales, and customer service by establishing a single centralized customer view. They guarantee consistent messaging and engagement strategies across all customer interactions, including ad viewings, support calls, and email campaign receptions. Businesses achieve a smoother customer experience by synchronizing their teams with a unified data source.

CJO platforms respond to customer behavior and preferences in real time, while traditional automation relies on fixed sequences. When customers visit a product page without purchasing, the system responds by delivering a tailored offer through the customer's chosen communication method, such as email or SMS. The dynamic approach leads to higher customer engagement while simultaneously enhancing conversion rates.

These platforms enable brands to sustain long-term customer relationships through personalized journeys with consistent and relevant touchpoints. The system allows businesses to anticipate customer requirements by initiating engagement activities such as subscription renewal reminders and product recommendations based on previous interactions, which builds loyalty and lowers churn rates.

Weaknesses

Deploying a CJO platform requires intensive coordination between multiple teams, such as the marketing, IT, and customer experience departments. Businesses need to synchronize their strategies with data structures and business processes to fully utilize the platform's features. The implementation demands extensive time investment because it needs solid support from every department and effective coordination efforts.

CJO platforms require real-time data to initiate appropriate customer interactions, but inaccurate or incomplete data leads to irrelevant messages and journey misalignment. Businesses must maintain effective data management systems, including proper CRM and CDP maintenance, to generate dependable insights and automation.

The deployment of these platforms can often become costly for large enterprises because their advanced capabilities demand significant expenditure while managing complex customer journeys. The investment involves not only licensing fees but also covers implementation steps, customization work, staff training programs, and continuous optimization efforts. Organizations should evaluate their preparedness to realize the maximum ROI prior to launching comprehensive deployment initiatives.

How to Evaluate CJO Platforms

While exact needs may vary in the form of the types and number of channels to be connected, as well as the complexity of the journeys that a typical customer might engage with, there are several core features that a CJO platform can be evaluated on:

- *What level of detail and adaptability does the journey builder provide?* An effective CJO platform needs to provide marketers with an intuitive visual journey builder for designing complex multi-channel interaction sequences. The platform must enable adaptable branching logic that responds to user behavior patterns, along with demographic data and engagement histories. Creating dynamic pathways that respond to customer behaviors such as email opens or shopping cart abandonments delivers a more effective journey design. A strong platform delivers real-time customer journey progress insights, which enables marketers to adjust and enhance flows when necessary.

- *Is the platform capable of establishing seamless connections with existing CRM and personalization systems?* Customer journeys rely on accurate, up-to-date data. An effective CJO platform needs to integrate without difficulty into current MarTech stacks which encompass CRMs, Customer Data Platforms (CDPs), analytics tools, and personalization engines. Effective integration enables platforms to collect live customer data and monitor how customers interact across various channels while offering a consolidated view of all customer activities. Teams face difficulties with fragmented data, which results in disjointed customer experiences and inefficient automation when deep integration capabilities are missing.

- *To what extent does the system activate appropriate responses according to customer behavior patterns?* Leading CJO platforms not only automate fixed workflows but also modify customer interactions instantaneously based on real-time behavior analysis. Marketers need to assess the speed at which the system handles data to activate personalized responses. When a customer looks at a product but doesn't make a purchase, can the system instantly deliver a targeted ad while also sending an abandoned cart email or alerting a sales representative for follow-up interaction? Creating timely and relevant interactions drives engagement and conversion rates while boosting customer satisfaction.

Organizations can find a CJO platform that matches their requirements and enhances their marketing performance by concentrating on criteria that also support their customer engagement plans.

CONCLUSION

Choosing the appropriate platform in each of these categories needs strategic evaluation of business goals alongside data systems and organizational competencies. Businesses need to determine whether they require basic automation capabilities, sophisticated personalization tools, or comprehensive journey orchestration and confirm that their selected solutions operate seamlessly with their current systems.

NOTES

1. Fortune Business Insights. (n.d.). *Marketing automation software market size, share & industry analysis.* Fortune Business Insights. https://www.fortunebusinessinsights.com/marketing-automation-software-market-108852
2. Statista. (2024). *Share of marketing channels using automation worldwide.* Statista. https://www.statista.com/statistics/1269813/marketing-channels-automation/

CHAPTER 19

SINGLE-CHANNEL MARKETING PLATFORMS

Marketing has evolved to focus on omnichannel strategies, yet single-channel marketing platforms still serve as critical resources for executing single-channel tasks and audience-targeting activities on those channels. Thus, despite belonging to a broader multi- or omnichannel strategy, a single-channel platform still plays a critical role.

These platforms offer advanced functionality for specialized activities, including email, social media, SMS, and digital advertising, which broader or all-in-one MarTech systems may lack. Organizations operating in advanced marketing ecosystems that manage customer interactions across multiple touchpoints continue to depend on specialized single-channel platforms to ensure proper functionality while meeting compliance requirements and optimizing channel-specific engagement.

Different organizations utilize single-channel platforms to serve distinct purposes. Certain teams dedicate themselves to one marketing area, such as email marketing teams that utilize an ESP, while others handle multiple platforms to merge data into wider MarTech systems. Companies with high success rates in particular channels can use these tools to expand their efforts more effectively. The chapter investigates how businesses benefit from single-channel marketing platforms and their effects on marketing professionals, along with evaluation methods for platform effectiveness.

HOW DO SINGLE-CHANNEL MARKETING PLATFORMS HELP MARKETERS?

While a few large platforms continue to add specialized features, marketers benefit from single-channel platforms because they deliver specialized features that generalist platforms usually don't offer. Marketers receive access to specialized analytics and optimization tools that adapt to channel-specific requirements that help brands achieve peak engagement results.

Single-channel platforms enable marketers to adjust campaigns through sophisticated features that are specifically designed for each medium. Email marketing platforms support A/B testing of subject lines, whereas paid media platforms provide precise audience targeting capabilities and bidding strategies.

These platforms focus their resources on one channel, which results in streamlined campaign execution and automated workflows while minimizing manual work. Using a single interface, marketers schedule posts and initiate automated sequences while optimizing content distribution without needing to manage multiple systems.

Single-channel platforms deliver granular analytics, which enables marketers to better refine strategies based on specific engagement metrics compared to multi-channel tools that only provide aggregated reporting. Social media platforms deliver comprehensive audience demographics reports along with sentiment analysis data and top-performing content information.

Strict regulations control email, SMS, and paid advertising communications across many industries. Marketers can follow regulatory guidelines on single-channel platforms because these systems include features for opt-in management, ad policy compliance, and data privacy protection.

These platforms target one channel but usually connect with Customer Data Platforms (CDPs), analytics tools, and marketing automation systems. Businesses can include insights from their single-channel operations into broader customer journey strategies through this process.

EXAMPLES OF SINGLE-CHANNEL PLATFORMS

With 14,106 products characterized as MarTech and a 27.8% year-on-year growth rate in the category according to Chiefmartech[1], it would be impossible to catalog all of them, but the next few sections will look at some of the most commonly used and popular single-channel platforms.

Email Marketing Platforms

With an estimated 361.6 billion emails sent in 2024, email remains one of the most-used communication platforms, and despite some reports of younger generations' lessened use of the channel, email continues to grow year over year, and, despite other channels like messaging apps growing in usage, is still estimated to reach over 400 billion by 2027[2]. Brands achieve effective customer engagement and lead nurturing along with driving conversions through the direct power of email communication. The core infrastructure for managing extensive email campaigns through automated workflows and personalized messages based on customer actions is provided by email marketing platforms. While MarTech ecosystems often provide email features, specialized email marketing platforms provide more advanced tools designed specifically to enhance deliverability and optimize audience targeting and campaign performance.

These platforms enable marketers to deliver messages to targeted audiences at optimal times while providing relevant content beyond just sending emails. Advanced email marketing solutions connect with CDPs and CRMs to enable businesses to create highly targeted marketing campaigns that utilize purchase history, browsing behavior, and engagement signals.

Example Platforms

A few examples of email marketing platforms are as follows:

- *Mailchimp*: A widely used email marketing platform known for its user-friendly interface, automation features, and built-in analytics. It provides customizable templates, audience segmentation, and integrations with various CRM and e-commerce platforms, making it ideal for small to mid-sized businesses.
- *Klaviyo*: An email and SMS marketing platform designed for e-commerce brands. Klaviyo specializes in advanced automation, dynamic personalization, and AI-driven predictive analytics, allowing businesses to create targeted campaigns based on customer behavior and purchase history.
- *HubSpot Marketing Hub*: A comprehensive marketing automation platform that includes email marketing, CRM integration, and customer

journey tracking. HubSpot enables businesses to build personalized email campaigns, automate workflows, and analyze engagement data within a single ecosystem.

Key Features

These are the key features of email marketing platforms:

- *Campaign creation and management*: The platforms include intuitive campaign builders that make it easy for marketers to craft newsletters along with promotional and transactional emails. Drag-and-drop editors together with customizable templates and audience segmentation tools make setting up campaigns much more efficient.

- *Automation and personalization*: Modern email platforms provide advanced automation capabilities that allow brands to send personalized emails when customers perform specific actions. The system handles abandoned cart notifications alongside post-purchase follow-ups and customized audience-targeted re-engagement campaigns.

- *Design and customization*: Marketers can construct attractive emails through built-in design tools that meet brand standards. Most platforms provide mobile-responsive templates and dynamic content blocks together with AI-driven recommendations to boost engagement rates.

- *Deliverability optimization and compliance*: Email platforms provide businesses with tools to navigate spam filters and maintain high deliverability rates through sender authentication and IP reputation tracking while enforcing data privacy regulations such as the GDPR and CAN-SPAM.

- *Testing and analytics*: Built-in A/B testing features in most email marketing tools allow users to test different subject lines, content variations, and optimal send times. Marketers use real-time analytics to monitor email open rates alongside click-through rates and conversion rates, which allows them to adjust their email strategies as they receive new data.

Key Considerations

The introduction of automation enhances operational efficiency yet results in overly templated messages that fail to create personal connections. By using dynamic content and behavioral triggers, marketers can keep their messages fresh and appealing to their audience.

Effective email deliverability requires keeping a subscriber list that remains both clean and actively engaged. To boost engagement rates, businesses should perform regular removal of inactive subscribers and comply with opt-in/opt-out rules while segmenting audiences according to their preferences.

Email marketing platforms function optimally when they connect with CRM systems and marketing automation tools along with CDP to centralize customer information and deliver a unified cross-channel customer experience.

How to Evaluate

Ease of Use and Campaign Setup

Does the email marketing platform offer user-friendly operations through intuitive design tools and pre-built templates? Does the platform offer drag-and-drop capabilities that simplify use for people who aren't technically skilled?

Automation and Personalization Capabilities

Evaluate the sophistication of the platform's automated workflows and personalization capabilities. Does the platform feature the capability to send emails in response to real-time customer actions and dynamic customer segmentation?

Deliverability and Regulatory Compliance

Does the platform deliver email authentication protocols SPF, DKIM, and DMARC to enhance the sender's reputation? The platform needs to demonstrate the effective management of GDPR and CAN-SPAM compliance requirements, along with others.

Analytics, Optimization, and Reporting

Which reporting capabilities does the platform provide within its analytics and reporting tools? Is A/B testing along with engagement heatmaps and audience behavior insights part of the platform's features?

Pricing and Scalability

Does the platform provide scalable solutions that match the growth of the business? Does the pricing structure for the platform depend on subscriber counts or email send volumes and feature availability?

The choice of email marketing platform should match the organizational needs, which range from basic campaign handling abilities to sophisticated automation and personalization features.

Social Media Management

Successful large-scale social media management involves multiple tasks beyond simple updates, including cross-platform coordination and real-time audience engagement along with performance tracking and integration with overall marketing strategies. Social media management platforms provide marketers with a centralized system to schedule content while monitoring brand interactions and analyzing engagement trends.

As social media marketing becomes more complicated, these platforms enable businesses to maintain consistent online visibility, automate posting tasks, and analyze audience engagement for valuable insights. Social media management tools serve both small businesses and global enterprises to boost efficiency and customer engagement while maintaining brand consistency.

Example Platforms

While there are many platforms in this space, including several multi-purpose platforms that incorporate social media management into their functionality, a few social media management platforms are as follows:

- *Hootsuite* is a mature social media management platform that enables users to plan their posts, monitor brand mentions, and track engagement data while supporting teamwork across various channels. Hootsuite provides support for numerous social networks while enabling integration with CRM systems and analytics applications.

- *Sprout Social* provides businesses with robust analytics and social listening features to manage content publishing alongside customer interactions and performance tracking. Its user-friendly interface combined with powerful reporting tools makes it the preferred platform for brands that value engagement and data insights.

- *Buffer* provides a simple platform that allows users to schedule posts and analyze performance data to maximize content strategies. Buffer offers easy-to-use tools designed for managing multiple social media accounts, which makes it perfect for small businesses and content creators who want to simplify their social media operations.

Key Features

Social media management platforms can contain a wide variety of features, as illustrated in Table 19.2. While platforms will vary in the number of features they include, the features should include the ability to create, manage, schedule, and measure social media content performance.

TABLE 19.2 Key social media management platform features

Feature	Description
Multi-Channel Management	Manage multiple social media profiles from a single dashboard, supporting platforms such as Facebook, Instagram, Twitter, LinkedIn, and more
Content Creation and Scheduling	Tools for creating, editing, and optimizing posts, along with content calendars and asset libraries for efficient planning
Publishing and Automation	Automated scheduling, queueing, and AI-driven recommendations for optimizing post timing
Analytics and Reporting	Real-time tracking of engagement metrics, audience growth, and best-performing content with customizable dashboards
Social Listening and Monitoring	Monitor brand mentions, hashtags, competitor activity, and sentiment analysis to gauge audience perception
Engagement Management	Unified inbox for managing comments, messages, and mentions across platforms, with tools for customer service tracking
Team Collaboration Features	Role-based permissions, workflow automation, and internal messaging for team collaboration on content approvals and campaigns
Integration Capabilities	Seamless integrations with CRM, email marketing, and analytics tools, along with API access for custom workflows
Mobile Accessibility	Mobile apps and push notifications for managing social accounts on the go with a responsive dashboard
Audience Targeting and Segmentation	Tools for creating audience segments based on demographics and behavior, enabling personalized content delivery
Paid Social Campaign Management	Features for creating, managing, and optimizing paid ad campaigns on platforms such as Facebook, LinkedIn, and Twitter
Compliance and Security	User access controls, compliance with GDPR and CCPA regulations, and security features for protecting brand accounts.

Things to Keep in Mind

Ensure the following are top of mind when evaluating social media management platforms, as well as when determining if the platform currently being used is a good fit with organizational needs:

- *Platform-specific nuances*: While social media tools provide multi-channel management, each platform has unique engagement styles and algorithms that impact content reach. Marketers must tailor content accordingly.
- *Balancing automation with authenticity*: While automation saves time, excessive scheduling and canned responses can make brand interactions feel impersonal. A human touch is still necessary for meaningful engagement.
- *Data privacy and security*: Businesses must ensure their platform complies with security regulations and that social data is handled responsibly.
- *Cost vs. features*: Some platforms offer advanced analytics and AI-driven automation but come with enterprise-level pricing. Marketers should balance features with budget constraints.

How to Evaluate

While individual organizations and social media marketing teams will have very specific criteria to judge success, some general ways to evaluate these platforms are as follows:

- *Ease of use and UI*: Is the platform intuitive and easy to use for team members with varying levels of expertise?
- *Scheduling and automation features*: Does it provide robust scheduling, queuing, and automated posting?
- *Analytics and reporting capabilities*: How deep are the insights provided? Can it track real-time engagement, audience growth, and competitor activity?
- *Integration with other tools and platforms*: Does it connect with CRM, analytics, and marketing automation platforms for a seamless workflow?

Digital Advertising Management

Successful digital advertising management demands advanced strategies for running campaigns and targeting audiences while measuring performance.

Advertising on numerous platforms, including search engines and programmatic channels, demands tools that enable streamlining workflows while automating optimization and delivering real-time insights to advertisers. Digital advertising platforms assist marketing teams in budget allocation while testing creative variations and tracking ROI across various channels.

Digital advertising management platforms have grown substantially because AI-driven automation along with real-time bidding and sophisticated audience segmentation have become prevalent. Advertising platforms deliver accurate customer targeting and appropriate messaging at the right time while maintaining brand consistency and adhering to privacy laws. The increase in digital advertising expenditure requires companies to put money into scalable systems that deliver practical insights and integrate smoothly into their existing MarTech infrastructure.

Example Platforms

Google Ads grants marketers the capability to manage search, display, video, and shopping campaigns through real-time bidding and AI-powered targeting, all exclusively on Google's advertising network. The service delivers strong analytical capabilities alongside automated functions and cross-channel attribution instruments.

Meta Ads Manager enables advertisers to run campaigns across Facebook, Instagram, Messenger, and the Audience Network. The platform offers sophisticated targeting methods and A/B testing features together with AI-enhanced ad placements, which help to boost user engagement and conversion rates.

The Trade Desk operates as a programmatic advertising platform enabling businesses to direct and improve omnichannel campaigns across connected TV (CTV), display, native, and audio advertising formats. The platform utilizes AI bidding methods together with audience segmentation techniques to achieve accurate targeting.

Key Features

These platforms can vary widely in the features they offer, though a list of common features is provided in Table 19.3. The most common features include the ability to create, schedule, publish, and measure ads across one or more platforms.

Feature	Description
Multi-Channel Management	Manage multiple social media profiles from a single dashboard, supporting platforms such as Facebook, Instagram, Twitter, and LinkedIn
Content Creation and Scheduling	Tools for creating, editing, and optimizing posts, along with content calendars and asset libraries for efficient planning
Publishing and Automation	Automated scheduling, queueing, and AI-driven recommendations for optimizing post timing
Analytics and Reporting	Real-time tracking of engagement metrics, audience growth, and best-performing content with customizable dashboards
Social Listening and Monitoring	Monitor brand mentions, hashtags, competitor activity, and sentiment analysis to gauge audience perception
Engagement Management	Unified inbox for managing comments, messages, and mentions across platforms, with tools for customer service tracking
Team Collaboration Features	Role-based permissions, workflow automation, and internal messaging for team collaboration on content approvals and campaigns
Integration Capabilities	Seamless integrations with CRM, email marketing, and analytics tools, along with API access for custom workflows
Mobile Accessibility	Mobile apps and push notifications for managing social accounts on the go with a responsive dashboard
Audience Targeting and Segmentation	Tools for creating audience segments based on demographics and behavior, enabling personalized content delivery
Paid Social Campaign Management	Features for creating, managing, and optimizing paid ad campaigns on platforms such as Facebook, LinkedIn, and Twitter
Compliance and Security	User access controls, compliance with GDPR and CCPA regulations, and security features for protecting brand accounts

How to Evaluate

Unlike some of the other single-channel platforms in this chapter, digital advertising platforms generally fall into one of two categories and thus should be evaluated in terms of how extensive and diverse the companies' advertising needs are:

- *Single-network advertising*: The Google and Meta platforms fall into this category. For instance, it is not possible to advertise on Facebook or TikTok from Google's interface, nor is it possible to advertise on Google or Pinterest from Meta's interface. Therefore, evaluating and using Google's platform may not leave users with a lot of alternatives.

- *Multi-channel digital advertising*: The Trade Desk and other multi-channel platforms allow advertisers to access multiple networks or media types. Other platforms, such as Hubspot, allow the placement of ads across platforms such as Google and social media platforms.

Marketers should determine what approach will best serve their needs, as single-network platforms may offer more robust capabilities but require more resources to manage if multiple single-network platforms are required.

Search Engine Optimization Platforms

Search engine optimization (SEO) serves as a fundamental aspect of digital marketing because it helps Web sites and content achieve higher placement in search engine results pages (SERPs) when users perform relevant searches. The constant development of search engines such as Google and Bing requires businesses to maintain their online presence by implementing technical SEO practices and strategies for keyword research, content creation, and link building. SEO platforms enable marketers to monitor their Web site's performance while spotting improvement opportunities and verifying content alignment with search visibility best practices.

SEO is a long-term strategy that demands constant optimization efforts alongside content updates and technical modifications. SEO tools give valuable data regarding Web site organic traffic patterns and keyword placement as well as backlink profiles and competitive tactics. These tools help automate processes including site audits and metadata optimization together with structured data implementation. Businesses can improve their organic traffic numbers, strengthen domain authority, and deliver better user experiences using SEO platforms.

Example Platforms

There are several organic SEO platforms, which we will discuss in this section.

Ahrefs provides a suite for SEO that features keyword research tools together with backlink analysis capabilities, rank-tracking features, and competitor analysis options. Content strategy development and link-building campaigns rely heavily on this tool alongside technical audits.

SEMrush provides an integrated suite for SEO and digital marketing featuring keyword research tools along with site audit features and PPC competitive analysis capabilities. This platform delivers comprehensive reports alongside automation tools that boost optimization efforts.

Moz Pro provides a dedicated platform for domain authority tracking while also delivering comprehensive keyword research and link analysis services. Moz provides SEO tools that help both new users and seasoned professionals to enhance their Web site's performance and search visibility.

In addition to standalone platforms, many Web site content management systems (CMSs) and digital experience platforms (DXPs) offer some SEO functionality.

Key Features

Functionality will vary from platform to platform but core features are included in Table 19.4, and focus on research and audits, tracking search rankings, and the ability to optimize current Web site content for better search rankings.

Feature	Description
Keyword Research and Analysis	Identifies high-impact keywords and search trends for optimizing content
Site Audits and Technical SEO	Scans Web sites for technical errors, broken links, and crawlability issues
Backlink Tracking and Analysis	Monitors inbound links, evaluates domain authority, and detects spammy backlinks
Rank Tracking	Tracks keyword performance across search engines to measure SEO success
On-Page Optimization	Provides recommendations for meta tags, content structure, and schema markup
Competitor Analysis	Evaluates competitors' SEO strategies, backlink profiles, and keyword rankings
Content Optimization	Offers AI-driven recommendations for improving content relevance and readability
Local SEO and Listings Management	Helps businesses optimize for local search and manage business listings
AI-Powered Insights	Uses machine learning to predict ranking opportunities and search trends
Integration with Other Tools	Connects with Google Analytics, Google Search Console, and other platforms.

Things to Keep in Mind

SEO requires time to produce significant results, while paid advertising delivers almost immediate outcomes. Regular maintenance of content updates and technical optimizations along with link-building strategies is essential to keep search engine rankings stable. Marketers need to establish achievable expectations and dedicate themselves to continual efforts for lasting organic growth.

The continuous evolution of search engine algorithms causes previous SEO strategies to become outdated over time. Search engines such as Google and Bing along with others regularly change their ranking criteria, which requires SEO professionals to stay up to date with the latest industry best practices. Businesses that adapt to these changes can evade penalties and sustain rankings while seizing new opportunities.

The integration of SEO with comprehensive content marketing efforts and user experience (UX) strategies produces the best results. A combination of high-quality content with a seamless Web site experience results in better engagement metrics, thereby enhancing search rankings. Organizations that rely only on technical SEO factors while disregarding content relevance and Web site usability will probably face challenges in achieving sustainable business success.

How to Evaluate

Choosing the right SEO platform requires prioritizing data accuracy and valuable insights. Accurate keyword tracking combined with backlink analysis and ranking data forms the foundation for making knowledgeable decisions. Marketers need to evaluate if the platform supplies current and complete data that matches actual search trends and performance metrics.

Alongside data accuracy and insights, platforms must deliver straightforward usage and powerful reporting functions. SEO platforms need intuitive dashboards with clear visualizations and customizable reports so users can easily make data-driven decisions. Marketing teams require fast access to important insights without having to perform detailed manual analysis. A platform that makes reporting easier enables teams to optimize their strategies with enhanced efficiency.

The functionality to integrate with other systems greatly enhances the usefulness of SEO tools. Businesses need to determine if their chosen platform has integration capabilities with critical tools such as Google Analytics and Google Search Console as well as CMSs and marketing automation platforms. Through seamless integration, businesses can obtain a comprehensive understanding of their search performance and marketing impact.

Businesses operating multiple Web sites or brands need to evaluate scalability when choosing an SEO platform. The platform needs to support enterprise-level SEO requirements through multi-domain tracking capabilities, automated reporting functions, and AI-driven insights. Companies preparing for sustained growth need to select digital marketing platforms that offer scalability.

CONCLUSION

Of course, there are many other types of single-channel marketing platforms available, each with unique features and considerations. A core set of considerations for any of them will be how they integrate into a larger workflow, their ease of use, the ability to provide meaningful analytics, and technical integration capabilities with other core platforms in the MarTech stack.

NOTES

1. Brinker, S. (2024, May 6). *2024 marketing technology landscape supergraphic — 14,106 martech products (27.8% growth YoY)*. *ChiefMartec*. https://chiefmartec.com/2024/05/2024-marketing-technology-landscape-supergraphic-14106-martech-products-27-8-growth-yoy/
2. Statista. (2024). *Daily number of e-mails worldwide*. Statista. https://www.statista.com/statistics/456500/daily-number-of-e-mails-worldwide/

CHAPTER 20

CONVERSATIONAL MARKETING AND REAL-TIME COMMUNICATION PLATFORMS

Consumer expectations about communication and their patience when needing to wait have shifted dramatically. In physical retail environments, a 2024 study from Waitwhile found that customer frustration levels from waiting in lines increased 126% from 2024 to 2023[1]. On digital channels, customers become frustrated when responses are delayed and real-time engagement has now become standard, with a study from Khoros showing that 79% of customers are expecting a fast response when contacting a brand and 83% cite good customer service as the most important factor for deciding what to buy[2]. People anticipate brand availability in real time through different communication channels when they look for product advice, solve problems, or verify their orders. Marketing has evolved from delivering one-way messages to establishing real-time interactive dialogues that create customer engagement and propel business success.

Conversational marketing fundamentally transforms traditional brand communication from one-way messaging to interactive dialogue. Businesses have moved away from broadcasting messages to passive listeners to instead participate in real-time dialogues that allow personalized and context-specific customer interactions. Customer experience gets better through this shift because it enables businesses to accelerate their sales process by responding to customer questions and worries during key moments. Organizations that ignore real-time communication methods in their customer engagement approaches will probably become irrelevant because consumers now demand immediate interactions that demonstrate personalized attention and quick responsiveness.

WHAT FALLS UNDER THIS CATEGORY?

Conversational marketing and real-time communication platforms utilize various technologies to enable interactive two-way communications. Businesses can use AI-powered chatbots together with live chat systems and messaging apps to connect with customers via voice assistants and social media engagement tools. Although each technology fulfills a unique role, they work together to improve brand interactions by making them both seamless and timely.

Chatbots execute automated dialogues by processing frequent inquiries and transactional operations without requiring human input. Live chat enables representatives to deliver immediate support, which improves decision-making efficiency. Businesses can connect with their customers through messaging platforms such as WhatsApp, Facebook Messenger, and WeChat, which are environments customers frequently use on a daily basis. Amazon Alexa and Google Assistant offer hands-free interaction possibilities that function alongside smart devices. Businesses enhance real-time interaction through social media messaging by responding to customer requests and maintaining relationships in digital spaces where consumers frequently engage.

Organizations implementing these tools into their engagement approaches need to evaluate how these platforms match customer expectations and operational requirements and advance long-term business goals. Upcoming sections will explore the function of these technologies in marketing while explaining their main attributes and discussing evaluation criteria for businesses when choosing a solution.

Example Platforms

A diverse range of solutions exists within conversational marketing and real-time communication technologies to facilitate instant engagement while automating interactions and improving customer experiences. These platforms perform various functions by using either AI chatbots to independently manage inquiries or human staff to deliver direct customer support through live chat solutions. Certain platforms provide seamless interaction across different channels so that conversations maintain continuity irrespective of their starting point. This section presents the primary categories of platforms that demonstrate their unique capabilities.

AI-Powered Chatbots

Some conversations don't require human intervention. AI chatbots create a bridge between machine efficiency and meaningful customer interaction through the automated handling of repetitive inquiries and sales funnel guidance as well as problem resolution without human involvement. These systems utilize natural language processing (NLP) alongside machine learning to decipher user intent while improving their capabilities through ongoing customer engagements. Let's look at some examples:

- *Drift*: This company focuses on B2B conversational marketing through AI-powered lead qualification and meeting scheduling with sales teams.
- *Intercom*: Live chat functionality combines automated AI responses with human support by allowing bots to manage early customer interactions until they reach a point where human intervention becomes necessary.
- *Ada*: This tool automates customer support, which allows companies to expand personalized interactions without growing their workforce.

Live Chat Solutions

Some interactions require a human touch. Live chat platforms enable businesses to engage in immediate live conversations with their customers. These tools differ from AI chatbots because they enable human agents to take over whenever automated systems cannot handle inquiries effectively and thus enhance customer satisfaction while minimizing issues during purchases. Let's look at some live chat solution examples:

- *LiveChat*: Delivers businesses a quick chat solution that scales easily while connecting with CRM systems and e-commerce applications.
- *Zendesk Chat*: This tool serves as a vital component of Zendesk's support system while offering automated features for seamless customer service interactions.
- *Tidio*: This platform combines AI automation capabilities with live agent assistance and is a preferred solution for small and mid-sized business operations.

Messaging Apps and Social Media Chat

Modern customers demand that brands can be contacted through their daily-used messaging applications. Customers and brands now rely heavily on social messaging channels such as WhatsApp, Facebook Messenger, and Instagram DMs as essential platforms for both customer service delivery and marketing strategies. Businesses utilizing these platforms can connect directly with consumers at their most active locations without any barriers:

- *Meta Business Suite*: Connects business communications across Facebook Messenger, Instagram DMs, and WhatsApp while providing automation tools and team collaboration capabilities alongside analytics.
- *Twilio*: Serves as a cloud platform to help businesses weave messaging features and voice and video capabilities into their software applications.
- *WhatsApp for Business*: Delivers a focused messaging system through which companies can offer customer support services and manage large-scale automated messaging conversations.

Conversational AI and Voice Assistants

The emergence of voice search technology and smart digital assistants has broadened conversational marketing to include non-textual interactions. Companies have adopted AI voice interface technology to improve customer support services while enabling hands-free communication and operational task streamlining. These systems process vocal input to understand intent and create responses that mimic human interaction:

- *Amazon Alexa for Business*: Helps organizations incorporate voice interactions into their workplace processes, customer service operations, and commercial activities.
- *Google Dialogflow*: Serves as a conversational AI framework that builds chatbots and voice interactions across different channels.
- *IBM Watson Assistant*: AI-powered NLP technology drives the development of intelligent virtual assistants with contextual understanding and complex query resolution capabilities.

SMS Marketing and Business Texting

SMS continues to stand as one of the most potent communication channels because it achieves high open rates alongside substantial response rates even amidst the emergence of social and Web-based messaging platforms. From promotional campaigns to appointment reminders and customer support services, SMS marketing platforms enable brands to directly connect with their audiences. Let's look at some SMS marketing platforms:

- *Attentive*: This mobile messaging platform delivers personalized customer interactions utilizing both SMS and MMS platforms.

- *Podium*: Provides two-way business texting solutions for customer interactions, as well as reviews and local marketing initiatives.

- *EZ Texting*: Delivers a straightforward SMS marketing tool that includes automation capabilities along with contact segmentation and analytics features.

Successful customer engagement requires choosing tools that match business goals while meeting customer needs and operational abilities.

How Do They Help Marketers?

Through conversational marketing and real-time communication platforms, marketers can achieve superior customer engagement levels by creating valuable interactions across large audiences. Traditional marketing channels typically depend on delayed responses and asynchronous messaging, but these tools enable marketers to communicate with customers instantly in both directions. Using AI-driven automation alongside live chat services and social messaging businesses can establish stronger customer connections while boosting conversion rates and gathering critical insights from consumers.

Instant Engagement and Lead Conversion

Speed matters in marketing. When customers interact with brands through Web site chats, social messages, or SMS, they expect their queries to be answered without any delay. Delayed replies often lead to lost opportunities. Chatbots and live chat systems remove response delays by providing real-time answers to customer questions and product queries while also assisting users throughout their purchasing journey. Automated lead qualification systems detect visitors with high purchase intent and immediately forward them to sales representatives. By minimizing response times, businesses can experience substantial improvements in conversion rates.

Personalized Customer Experiences

The scope of personalization now extends past the basic practice of using customer names during interactions. AI-powered conversational platforms utilize data from previous interactions along with browsing history and behavior patterns to provide highly pertinent responses. E-commerce chatbots suggest products according to users' past buying activities and AI assistants retrieve historical customer service dialogues to deliver uninterrupted support. Businesses that use contextual awareness can create personalized interactions that drive greater customer engagement and satisfaction.

24/7 Customer Support and Automation

Today's customer demands support availability that matches their personal schedules rather than being limited by company business hours. Conversational AI tools provide businesses with round-the-clock support, which keeps them responsive at all times outside regular business hours. AI chatbots handle commonly asked questions and resolve standard problems while referring difficult inquiries to human support agents when needed. Customer service teams benefit from reduced workload while still providing quick responses to customer needs.

Omni-Channel Communication

Customers move between multiple brand touchpoints, including Web sites, social media messaging apps, and voice assistants, throughout their interaction journey. Customers become frustrated when they experience inconsistent interactions across different communication channels. Conversational marketing platforms integrate multiple channels so a chat begun on a Web site can naturally extend to WhatsApp, email, or a voice assistant. The uninterrupted nature of these marketing platforms builds stronger customer experiences while increasing trust in the brand.

Data Collection and Behavioral Insights

Each customer interaction offers a chance to gather knowledge. Through customer engagement, conversational platforms both provide interaction capabilities and capture valuable data points such as common customer questions and journey drop-off locations, along with repeated pain points. Marketers can leverage this information to enhance their messaging strategies and improve Web site content while also guiding product development decisions. These platforms' advanced analytics deliver detailed insights about user behavior patterns, sentiments, and preferences, which allows brands to precisely adjust their marketing approaches.

When marketers integrate conversational marketing with real-time communication into their MarTech stack they transition from static campaigns toward dynamic responsive engagement strategies. Subsequent sections will explain which organizational members benefit most from these platforms and provide methods for evaluating their performance to achieve maximum impact.

WHO ARE THE USERS OF THESE PLATFORMS?

This category of platform is used by a diverse set of teams and stakeholders. While it fits under the category of MarTech, multiple departments utilize conversational marketing and real-time communication platforms to support their sometimes unique business functions.

Marketing Teams

Conversational platforms function as digital campaign extensions for marketing teams that allow personalized customer interactions at large volumes. AI chatbots qualify leads through targeted questions and visitor segmentation, which directs them to appropriate content and offers. Automated messaging tools enable the delivery of promotional campaigns, together with event reminders and personalized recommendations through SMS messages, WhatsApp conversations, or Web site chat features. Marketers apply conversational insights to enhance their messaging strategies and boost campaign effectiveness.

Sales Teams

Sales professionals improve the buying experience by engaging directly with prospects during real-time interactions. Sales teams abandon traditional contact forms and slow email follow-ups in favor of live chat and AI assistants to connect instantly with customers who show strong purchasing intent. Sales representatives are able to personalize their outreach efforts because these platforms deliver detailed contextual data about prospects, including previous interactions and CRM information. Automated chat sequences function as a lead nurturing tool that prepares prospects for direct sales conversations.

Customer Support and Customer Service Teams

Support teams utilize conversational platforms to deliver customer service that is both quick and effective. Real-time customer support is delivered through live chat solutions, and AI chatbots manage standard questions about order tracking and refund policies. Support teams integrate conversational AI with knowledge bases to enable customers to access automated FAQs and guided workflows for self-service. Business operations become more cost-effective and customer satisfaction rates rise as waiting times decrease.

E-Commerce and Retail Teams

Conversational marketing is a crucial component of e-commerce businesses to increase conversions and improve shopping experiences. AI-driven chatbots help customers by suggesting products according to their browsing patterns while simultaneously addressing product inquiries and supporting checkout procedures. Messaging applications such as Facebook Messenger and SMS serve as tools for abandoned cart recovery campaigns, which remind customers of their unpurchased items and provide incentives to complete their purchases. Retail brands utilize real-time messaging systems to provide virtual shopping support while engaging customers through loyalty programs.

Technology Teams

The underlying work of technology, IT, data, and AI development teams involves integrating these solutions with pre-existing business infrastructures. Teams responsible for chatbot operations handle logic design, enforce data security measures, and ensure adherence to privacy standards. AI experts work on refining NLP models to increase chatbot accuracy, which results in more human-like and contextually appropriate responses. Technology teams supervise API connections between CRMs, analytics platforms, and customer service tools, which enable seamless omnichannel functionality.

Key Features

Real-time communication platforms, alongside conversational marketing solutions, deliver multiple capabilities that improve customer engagement while streamlining interactions and offering uninterrupted support services. Automated AI systems working alongside human customer representatives allow businesses to respond promptly to inquiries while fostering lead growth and improving customer satisfaction.

These platforms use AI-powered chatbots and advanced analytics to deliver real-time communication that is both personalized and accessible across multiple channels. Table 20.1 shows some key features of these platforms in more detail.

TABLE 20.1 Key features of conversational marketing platforms

Feature	Description	Impact on Marketing and Customer Engagement
AI Chatbot and NLP Capabilities	Uses NLP to understand user intent, context, and sentiment, enabling more natural conversations	Enhances customer interactions by making automated responses feel human-like and relevant, improving engagement and resolution rates
Live Chat and Human Handoff	Provides a seamless transition from AI chatbot to human agents when necessary, ensuring complex inquiries are handled effectively	Reduces frustration for customers by escalating issues to human representatives at the right time, enhancing satisfaction and trust
Omni-Channel Messaging Integration	Supports real-time messaging across multiple touchpoints, including Web chat, social media, SMS, and voice assistants	Creates a unified experience across all digital channels, allowing customers to engage with brands on their preferred platforms
Personalization and Customer Data Storage	Tracks past interactions, customer preferences, and purchase history to tailor responses and recommendations	Strengthens customer relationships through personalized interactions, increasing loyalty and conversion rates
Automation and Workflow Triggers	Enables automated lead routing, chatbot responses, follow-up reminders, and customer segmentation based on predefined rules	Streamlines marketing and sales processes by reducing manual intervention, accelerating response times, and improving lead management
Analytics and Performance Tracking	Measures response times, engagement rates, chatbot accuracy, and customer sentiment to assess performance	Provides actionable insights to optimize chatbot scripts, agent workflows, and messaging strategies for better results
Integration with CRM and MarTech Stack	Syncs with customer relationship management (CRM) platforms, customer data platforms (CDPs), and analytics tools	Ensures a holistic view of customer interactions across all touchpoints, allowing for data-driven decision-making and improved targeting

While exact features may vary by platform, the functionality that is core to most conversational marketing platforms enables businesses to scale real-time conversations while maintaining high levels of engagement, personalization, and operational efficiency.

HOW TO EVALUATE

Marketing teams should assess multiple factors when selecting a conversational marketing platform, and should keep in mind that other teams will probably need to weigh in. Finding a platform that meets their business goals and customer engagement requirements while supporting necessary integrations is key. The performance of platforms with AI automation and real-time messaging hinges on their ability to scale operations and personalize customer interactions while integrating with other MarTech solutions. Businesses need to evaluate several essential criteria when selecting a conversational marketing platform.

AI and NLP Capabilities

For a conversational marketing platform to be effective, it needs advanced Natural Language Processing (NLP) capabilities that enable the understanding of user intent, sentiment analysis, and contextual comprehension. The effectiveness of a chatbot or AI assistant depends on its ability to deliver appropriate responses that feel natural while avoiding customer frustration. Businesses should evaluate the following:

- The platform demonstrates proficiency in understanding user intentions and delivering responses that mimic human interaction.
- The platform needs to support various languages and dialects for diverse customer interactions.
- The system uses machine learning to evolve its performance by learning from previous interactions.

Omni-Channel Functionality

Consumers often interact with multiple platforms in a single buying journey, and they expect smooth transitions from one platform to another. A comprehensive conversational marketing platform needs to consolidate interactions across Web chat, social media channels, SMS messages, and voice assistant technologies. Key considerations include the following:

- The platform should allow customers to begin discussions on one platform and then move to another platform while maintaining context throughout the conversation.
- Messaging platforms including WhatsApp, Facebook Messenger, Instagram, as well as Web site chat systems, require support.
- Systems should support connectivity with Amazon Alexa and Google Assistant to facilitate voice-controlled interactions.

Integration Capabilities

To personalize conversations effectively, conversational tools need to operate as part of the broader MarTech stack by accessing data from CRM systems along with marketing automation and analytics tools. To ensure seamless operations, businesses should evaluate the following:

- The system must support CRM integration enabling personalized customer responses through platforms such as Salesforce and HubSpot.

- Businesses need access to APIs and existing connectors for platforms such as Marketo and Pardot to integrate with their marketing automation systems.

- The MarTech stack requires integration with analytics tools to monitor engagement levels along with chatbot performance and conversion metrics.

Scalability and Automation

The conversational marketing platform for businesses needs to expand in tandem with their growth to handle higher traffic volumes and intricate workflows. The platform maintains efficiency throughout growing engagement volumes thanks to its scalability. Factors to consider include the following:

- The platform needs to handle multiple concurrent conversations without any reduction in performance.

- The platform enables users to build advanced automation processes, including lead qualification systems and customer support distribution, while generating predictive responses.

- The learning abilities of AI systems evolve through customer interaction analysis, which results in continuous enhancement.

Customization and Personalization Features

Any powerful conversational platform needs to support extensive customization options for personalized user experiences. Businesses should assess the following:

- Businesses need to evaluate if the platform permits responses to be customized by analyzing user behavior, preferences, and past interactions.

- Branding customization options along with consistent messaging tone options.

- Real-time customer interaction data lets businesses adjust their messaging by analyzing browsing history and previous purchases.

Of course, each company has its own unique sets of requirements as well. In addition to features and functionality, teams should ensure either that the processes required to use the platforms will integrate with existing ones, or that the change management required is feasible.

CONCLUSION

Brands seeking instant connections with customers now depend on conversational marketing alongside real-time communication platforms to create meaningful engagement. AI-powered chatbots along with live chat and social messaging allow businesses to deliver personalized support and automated interactions which result in increased conversion rates. The evolution of marketing strategies toward dynamic dialogue-driven engagement enables these platforms to strengthen customer relationships and boost operational efficiency.

Selecting the right conversational platform for an organization involves evaluating AI features alongside multi-channel support and compatibility with present MarTech systems. Businesses focused on delivering seamless customer experiences and intelligent automation while maintaining real-time responsiveness will achieve a competitive advantage. Organizations that adopt conversational tools can improve customer engagement and conversion paths while developing strong customer loyalty because consumers increasingly demand instant personalized interactions.

NOTES

1. Yahoo Finance. (2024, February 6). *Study: Consumer frustration with lines skyrockets*. Yahoo Finance. https://finance.yahoo.com/news/study-consumer-frustration-lines-skyrockets-130000464.html

2. Forrester and Khoros. *Getting to Know Your Customers: Why Brands Must Bridge the Gap Between What They Think They Know And What Customers Really Want*. August 2019

CHAPTER 21

MULTI-FUNCTION PLATFORMS AND SUITES

Most of the MarTech platforms discussed in the chapters so far have had more than a single function, yet there is a class of platforms that expands well beyond a set of complementary features. They can be referred to as suites or multi-function platforms, and they often enable companies to buy one or more products together in a bundle. This often accelerates a company's ability to integrate core functions, become productive more quickly, and (in yet other cases) measure the results of different types of marketing more effectively.

Yet, many companies have not adopted this approach, with a 2022 study by Dun and Bradstreet showing that nearly 60% of B2B marketers are using multiple "best-of-breed" vendors rather than single platforms[1].

It should be noted that several of the platforms listed in this chapter were mentioned in Chapter 17 in the discussion of digital experience platforms (DXPs). While not always the case, a key driver of the growth and expansion of some of these platforms has been having a DXP as a central area of focus.

While the list that follows is not exhaustive, it includes many of the top multi-function platforms and MarTech suites available.

ADOBE EXPERIENCE PLATFORM (AEP)

AEP remains focused on the enterprise market, partly because of its cost of entry and implementation costs, and in part because of the breadth of its features and platforms that have robust features that may not apply to smaller organizations with simpler MarTech needs.

Path to Growth

Adobe's journey to establishing the AEP began with strategic acquisitions aimed at strengthening its digital experience offerings, including its 2009 acquisition of Utah-based Omniture for $1.8 billion, which brought Web analytics and optimization capabilities to its portfolio of products, and which would later be renamed Adobe Analytics[2].

A key acquisition took place in 2010 when Adobe acquired Day Software for $240 million, a Swiss enterprise content management company known for its Web content management system (CMS), CQ5[3]. This acquisition formed the foundation of what would later evolve into Adobe Experience Manager (AEM), a key component of AEP.

Adobe continued to expand its MarTech capabilities through acquisitions. Some of the most significant acquisitions are as follows:

- *Neolane (2013)*: Advanced cross-channel campaign management and conversational marketing platform was acquired for $600 million, and later became Adobe Campaign[4].
- *Magento (2018)*: This integrated e-commerce functionalities to offer a more holistic digital experience solution was acquired for $1.68 billion[5].
- *Marketo (2018)*: Adobe acquired this B2B marketing automation and customer engagement capabilities for $4.75 billion[6].
- *Workfront (2020)*: This acquisition ($1.5 billion) strengthened work management and collaboration functionalities within Adobe's suite of products, streamlining marketing operations capabilities[7].

Current State of Adobe Experience Platform

Formerly called Adobe Marketing Cloud, AEP was officially launched in 2019 as a unified platform designed to enable real-time customer profiles, AI-driven insights, and cross-channel personalization and content management across desktop, mobile, and other channels[8].

Throughout the course of its many acquisitions and several renamings and consolidations, some products have evolved and others have been deprecated or scheduled for deprecation. The current AEP products and features include the following:

- *Adobe Experience Manager Sites and Assets*: Provides a cross-channel CMS and digital asset management (DAM) platform.

- *Customer Journey Analytics*: Provides the ability to measure and track customer behavior across multiple channels.
- *Real-Time Customer Data Platform (CDP)*: Enables brands to unify data across various sources to create comprehensive customer profiles.
- *Journey Optimizer*: Helps businesses orchestrate and deliver personalized customer journeys. This is also available in an edition focused on B2B marketers.
- *Mix Modeler*: This component enables marketers to better understand the effectiveness of their marketing and advertising channels across channels and tactics.
- *Workfront*: This component of the platform enables workflow management and focuses on the content marketing and marketing operations components of the enterprise.

AI and machine learning have been increasingly incorporated into products in the Adobe suite. Additionally, Adobe has consistently invested in compliance with regulations such as the GDPR and CCPA to ensure robust data governance or in HIPAA-compliance for US-based healthcare companies.

OPTIMIZELY ONE

What is now Optimizely ONE has its roots in a combination of a few key platforms. The first step towards this was the merger of the Nashua, New Hampshire-based Ektron, and Stockholm-based EPiServer CMS platforms in 2015[9], which proceeded under the EPiServer name and was backed by private equity firm Accel-KKR.

Then, in 2018, Accel-KKR sold the company to Insight Partners, followed by a 2020 acquisition of San Francisco-based Optimizely, a pioneer in experimentation and A/B testing, by EPiServer[10]. This combined company rebranded as Optimizely in 2021 to reflect its broader focus on digital experience optimization.

Current Market Focus

Optimizely ONE currently sits in the large and enterprise tier in the MarTech market, offering capabilities for mid-market companies, while many of its recent acquisitions and new features target the enterprise company, with more complex personalization, measurement, and content management workflow needs.

Recent developments such as a HIPAA-ready version that complies with US healthcare patient data regulations, have broadened the platform's relevance while increasing the sophistication of the way that it handles customer data[11].

Key Acquisitions and Feature Expansions

In addition to the foundational acquisitions that built the company from the combination of EPiServer, Ektron, and Optimizely ONE has grown through a series of acquisitions and internal developments that have strengthened its position as a leader in DXPs:

- *InSite (2019)*: This Minneapolis-based B2B e-commerce software platform was acquired by then EPiServer, and has been a core e-commerce component of the Optimizely ONE platform[12].
- *Welcome (2021)*: A content marketing and collaboration platform that enhanced Optimizely's content lifecycle management capabilities[13].
- *Zaius (2021)*: A CDP acquisition that brought real-time customer insights and analytics into Optimizely ONE[14].
- *NetSpring (2024)*: This acquisition brought "data warehouse-native analytics" to the Optimizely ONE platform, enabling marketers using the platform direct access to data sitting within an enterprise data cloud[15].

Current State of Optimizely ONE

Optimizely ONE has evolved into a comprehensive DXP that integrates content, commerce, and optimization capabilities.

As of 2025, Optimizely was ranked as the leading DXP in the prestigious Gartner Magic Quadrant for DXPs, with Adobe's offering as a close second[16].

Mirroring other players in the space, Optimizely also appears to be increasingly focusing on cloud-based or Software as a Service (SaaS) products, including its CMS component.

Key features include the following:

- *Content Cloud*: A powerful CMS for omnichannel digital experiences
- *DAM*: Enables cloud-based access to files used across products and teams

- *Web Experimentation*: AI-powered A/B and multivariate testing capabilities to optimize digital experiences
- *Feature Experimentation*: Enables product teams to experiment with feature rollouts and user experiences
- *Personalization*: Allows content to be tailored to an individual or audience segment
- *E-commerce*: Provides a seamless e-commerce experience with AI-driven product recommendations and personalization
- *CDP*: Empowers businesses to unify and activate customer data for personalized marketing campaigns and support first-party marketing activities
- *Data Warehouse-Native Analytics*: Enables marketers to access business-critical data not traditionally stored or shared with marketing data platforms

SITECORE DXP

Sitecore was founded by five former University of Copenhagen students in 2001 in Denmark as a CMS provider, after their initial foray into professional services related to Web publishing[17]. Over the years, it has expanded into a full DXP, enabling brands to manage and optimize customer interactions across multiple channels.

As seen with other platforms, Sitecore's platform has expanded through a combination of product development and acquisitions, including an Ireland-based CDP, Boxever[18], and Four51, an e-commerce platform, in 2021[19].

Current Market Focus

Sitecore's focus is on the middle market to enterprise companies, with a robust set of offerings, as well as pricing and typical implementation costs that best fit companies in that profile.

Recent announcements in areas such as data privacy include a HIPAA-compliant version that meets US patient data privacy standards[20], and its CDP supports overall first-party data strategies.

Evolution of Sitecore DXP

Mirroring overall MarTech trends, Sitecore's offerings are moving almost exclusively to cloud-based, SaaS offerings, with its CMS offering being the last offering with a self-hosted option.

Sitecore's platform now includes the following:

- *XM Cloud or XP*: Offers enterprise-level content management and publishing options. The former is the cloud-based CMS offering, with the latter being the self-hosted version.
- *OrderCloud*: An e-commerce platform focused on AI-driven shopping experiences.
- *Content Hub*: A centralized DAM system for managing content creation and marketing workflows.
- *CDP*: A CDP enabling real-time personalization.
- *Personalize*: Enables personalization and testing capabilities.
- *Send*: Email marketing platform with integrations to the rest of the Sitecore platform.

ACQUIA DXP

Acquia was founded in 2007 by Dries Buytaert, the creator of the open-source Drupal CMS following in the path of Red Hat's commercialization of the open-source Linux operating system[21]. Acquia positioned itself as a leading provider of enterprise-grade Drupal solutions, offering hosting, support, and development tools tailored to businesses.

Acquia was acquired by Vista Equity Partners in 2019 for $1 billion, and following that acquisition, a series of product expansions ensued[22].

Current Market Focus

While Acquia's focus is on the mid-market and enterprise organization, it is important to make the distinction between what Acquia does as a platform, and the underlying, open-source Drupal CMS, which, as of early 2025, ranks as the sixth most-used CMS platform in the world, and more than any of the other platforms mentioned in this chapter[23]. The latter is widely used by many types of organizations, both large and small, while the former is a subscription-based, enterprise-focused set of MarTech tools that integrate with the Drupal CMS.

Key Acquisitions and Feature Expansions

Similar to the other platforms explored in this chapter, Acquia has expanded its capabilities through acquisitions and product innovations, evolving into a full-scale multi-function platform. Some of these acquisitions are as follows:

- *Mautic (2019)*: Acquia acquired this open-source marketing automation platform, integrating it into its digital experience offerings[24].
- *AgilOne (2019)*: This CDP acquisition brought AI-powered customer analytics and personalization capabilities[25].
- *Widen (2021)*: This acquisition strengthened Acquia's DAM capabilities, allowing businesses to better manage and distribute digital content[26].

Current State of the Acquia DXP

Acquia's DXP now includes a comprehensive suite of tools designed to empower digital marketers and developers, built around the Drupal CMS:

- *Drupal and Acquia Cloud*: A fully managed CMS on a cloud platform optimized for Drupal. Acquia Cloud features an enterprise-grade cloud hosting environment.
- *Site Factory*: A low-code site-building tool for rapid Web site development.
- *CDP*: A CDP that enables data-driven marketing strategies.
- *DAM*: A DAM solution for centralized content storage and distribution.
- *Campaign Studio*: AI-driven tools for personalized customer experiences.

HUBSPOT MARKETING CLOUD

Origins and Growth of HubSpot Marketing Hub

HubSpot Marketing Hub is part of the broader HubSpot ecosystem, which was founded in 2006 by Brian Halligan and Dharmesh Shah, who met at the Massachusetts Institute of Technology (MIT)[27]. The company pioneered the concept of *inbound marketing*, focusing on attracting customers through content marketing, SEO, and social media rather than traditional outbound advertising.

Current Market Focus

Hubspot includes other components, such as Sales Hub (focused on sales teams) and Service Hub (focused on customer service teams), though its Marketing Hub includes many features built for small and medium businesses in either a B2B or B2C context.

While some larger organizations use Hubspot, it is not specifically geared towards the enterprise market, and some of its features have limitations that the largest companies may find limiting.

For smaller and mid-market companies, however, the company has made several acquisitions and focused on making features easy to use with minimal technical support needed.

Current State of HubSpot Marketing Hub

HubSpot Marketing Hub provides an integrated set of tools, including the following:

- *Customer Relationship Management (CRM)*: This component is also offered in a free version, which allows companies to try this central component of the platform out.
- *Email Marketing and Automation*: Easy-to-use automation workflows for sending emails and creating drip campaigns.
- *Content Management System (CMS Hub)*: Built-in blogging, SEO, and landing pages that enable organizations to either complement or replace their existing Web site CMS.
- *Organic and paid social media management*: Enables marketing teams to manage several accounts from a centralized location, including post scheduling and advertising management.
- *Reporting*: Enables marketing teams to leverage existing reports or create their own customized ones that pull data from across Hubspot entities as well as any connected platforms.

SALESFORCE MARKETING CLOUD

The Salesforce platform, originally founded in 1999[28], serves many teams beyond marketers, including features for sales teams, customer services teams, and others within an organization. While some marketers may utilize some of those features, this description will focus on those most relevant to the marketing function.

Salesforce Marketing Cloud (SFMC) originated from Salesforce's acquisition of ExactTarget, a leading email marketing platform, in 2013. This $2.5 billion acquisition was a critical move by Salesforce to expand beyond CRM into marketing automation. ExactTarget's technology, augmented by its own recent acquisition of Pardot, served as the backbone for what would evolve into SFMC[29].

More recently, Salesforce acquired reporting platform Tableau in 2019 for $15.7 billion, giving the platform's users access to more robust enterprise analytics and insights capabilities[30].

Current Market Focus

Salesforce continues to invest in AI-driven capabilities through *Einstein AI*, improving predictive analytics, automation, and personalization across SFMC. With increasing integration into the broader Salesforce ecosystem, SFMC remains a dominant force in enterprise MarTech, focusing on data-driven customer engagement and personalization.

Current State of Salesforce Marketing Cloud

Over the years, Salesforce has expanded SFMC to include a broad range of marketing automation and customer engagement tools:

- *Journey Builder:* Enables marketers to create personalized, multi-channel customer journeys
- *Email Studio:* Powers email marketing campaigns with advanced segmentation and automation
- *Advertising Studio:* Helps marketers connect CRM data with social and digital advertising platforms
- *Interaction Studio:* Provides real-time personalization and engagement tracking
- *CDP:* Introduced to unify customer data and enable real-time personalization

STRENGTHS OF MULTI-FUNCTION PLATFORMS

There are a variety of benefits for organizations that invest in and utilize multi-function marketing platforms or suites. We'll discuss them in the next few sections.

Operational Efficiency and Lower Learning Curve

When a suite of products shares a common interface, login mechanism, and other familiar characteristics, it can often speed up the time to productivity because there is less time spent learning terminology, interface anachronisms, and other elements that may interfere with work getting done.

Data-Sharing Capabilities

While composable approaches and increasing ease of connecting via Application Programming Interfaces (APIs) are becoming more common for many MarTech platforms, many suites make accessing data and insights simpler than integrating with external, third-party platforms.

This can speed up the time to value (TTV) for the investment in the system.

Ease of Integration

Similar to data integrations, functional integrations in general are often made easier when a marketing team is utilizing products within a single suite.

The caveat here is that, because of the rapid pace of acquisition in the MarTech space, in some cases, a product will be advertised as part of an existing suite immediately upon acquisition yet may take several months to achieve full integration with the entire suite of products.

Access to New Features

As has been seen recently with many platforms' integration of generative AI functionality, a benefit of investing in a suite of products is broader access to new features as they are rolled out. While in some cases brand new classes of features may require additional subscriptions, in many cases, new features are made available to subscribers as part of an ongoing release schedule.

WEAKNESSES OF MULTI-FUNCTION PLATFORMS

Vendor Lock-in

The best risk to an organization is to be committed to an expensive MarTech platform (ranging anywhere from tens of thousands of dollars to millions per year) that does not meet the ongoing needs of the marketing teams, yet to be locked into a multi-year contract.

Many organizations can find themselves signing agreements that last several years and fail to achieve a significant return on investment (ROI) for many months, if ever.

Lack of Robust Features in Specific Areas

Another risk that a marketing suite may pose to an organization is that in some areas it has very strong functionality that is well-suited to the company's needs, yet in other areas, its features fall short of what is needed.

This is why many organizations favor a "best in breed" approach to buying an entire multi-function platform or suite. In most cases, an organization will probably augment a suite with some additional specialized tools, or harness some of the increasingly composable and API-driven functionalities of some of these larger suites to bring in targeted functionality in specific cases.

For instance, an enterprise's Web site may be run on the Sitecore DXP, yet integrated with the Algolia search and discovery platform to enhance the DXP's functionality in an e-commerce setting.

Lack of Interoperability

While the quality of their APIs is often touted, and more recently their approach to a more composable approach, broad-reaching marketing suites are not uniformly simple to integrate with other platforms.

This lack of interoperability poses a challenge to organizations that are, more than ever, being asked to take agile, nimble approaches to creating a MarTech infrastructure.

HOW TO EVALUATE

Your organization will likely have many specific requirements, there are several areas that you can use to begin an initial evaluation of marketing suites.

Existing Features and Functionality

First, ensure that the existing feature set generally maps to your organization's needs. While this sounds straightforward, many companies are caught waiting for promised features that are mentioned on a platform's marketing Web site yet months away from full rollout to the customer base.

While the future product roadmap is important (and will be discussed shortly), it is just important that marketing teams are able to gain immediate value from their investments.

Customer Base

Most marketing suites will position themselves as applicable to just about any industry and many sizes of organizations. While this is probably true for the most part, marketing teams should take a close look at the case studies and use cases posed by the products and features they find most valuable to ensure they are directly applicable to their unique situation.

Product Roadmap

Just as it is important to ensure that current features and functionality are immediately beneficial to the marketing team, it is important to understand the future focus of the organization, including the product roadmap and even the key leadership within the organization.

There are several organizations that put resources into evaluating these areas, including analyst firms such as Gartner's Magic Quadrant and Forrester's Wave. It is advisable to consult some of these as well as to do independent research to ensure the analysts' findings mirror the needs of your own organization.

CONCLUSION

While marketing suites may not be a fit for every organization, and in some cases they may sit alongside other, best-of-breed platforms that are integrated via APIs (or not integrated at all), it is a worthy consideration to determine if the broad range of features, common interface, and shared data and functionality will improve a marketing team's ability to perform.

NOTES

1. Dun & Bradstreet. (n.d.). *8th annual B2B sales and marketing data report*. Dun & Bradstreet. https://www.dnb.com/perspectives/marketing-sales/8th-annual-b2b-sales-and-marketing-data-report.html

2. Lardinois, F. (2019, September 16). *Ten years after Adobe bought Omniture, the deal comes into clearer focus*. TechCrunch. https://techcrunch.com/2019/09/16/ten-years-after-adobe-bought-omniture-the-deal-comes-into-clearer-focus/

3. Whitney, L. (2010, July 28). *Adobe buys Day Software for $240 million*. CNET. https://www.cnet.com/tech/tech-industry/adobe-buys-day-software-for-240-million/

4. Ha, A. (2013, July 22). *Adobe Systems completes $600M acquisition of conversational marketing startup Neolane*. TechCrunch. https://techcrunch.com/2013/07/22/adobe-systems-completes-600m-acquisition-of-conversational-marketing-startup-neolane/

5. Lunden, I. (2018, May 21). *Adobe to acquire Magento for $1.6B*. TechCrunch. https://techcrunch.com/2018/05/21/adobe-to-acquire-magento-for-1-6-b/

6. Rodriguez, S. (2018, September 20). *Adobe confirms it's buying Marketo for $4.75 billion*. CNBC. https://www.cnbc.com/2018/09/20/adobe-confirms-its-buying-marketo-for-4point75-billion.html

7. Koksal, I. (2020, November 10). *Adobe acquires Workfront for $1.5 billion*. Forbes. https://www.forbes.com/sites/ilkerkoksal/2020/11/10/adobe-acquires-workfront-for-15-billion/

8. Softcrylic. (n.d.). *Adobe Experience Platform (AEP): What we know and what we don't*. Softcrylic. https://www.softcrylic.com/blogs/adobe-experience-platform-aep-what-we-know-and-what-we-dont/

9. Yahoo Finance. (2015, January 27). *Episerver and Ektron combine to create the most complete digital experience cloud*. Yahoo Finance. https://finance.yahoo.com/news/episerver-ektron-combine-create-most-100000855.html

10. Lunden, I. (2020, September 3). *Episerver acquires Optimizely*. TechCrunch. https://techcrunch.com/2020/09/03/episerver-acquires-optimizely/

11. Optimizely. (2024, May 7). *Optimizely announces HIPAA-ready solutions for healthcare & life sciences organizations.* PR Newswire. https://www.prnewswire.com/news-releases/optimizely-announces-hipaa-ready-solutions-for-healthcare--life-sciences-organizations-302310526.html

12. Star Tribune. (2019, December 17). *Insite Software of Minneapolis sold to Episerver. Star Tribune.* https://www.startribune.com/insite-software-of-minneapolis-sold-to-episerver/566245222

13. Optimizely. (2021, December 1). *Optimizely to acquire Welcome to help marketers drive customer experience outcomes. Business Wire.* https://www.businesswire.com/news/home/20211201005111/en/Optimizely-to-Acquire-Welcome-to-Help-Marketers-Drive-Customer-Experience-Outcomes

14. Ring, K. (2021, March 17). *Optimizely shores up DX platform with Zaius CDP acquisition. TechTarget.* https://www.techtarget.com/searchcustomerexperience/news/252498293/Optimizely-shore-up-DX-platform-with-Zaius-CDP-acquisition

15. Optimizely. (2024, April 25). *Optimizely enters definitive agreement to acquire warehouse-native analytics leader Netspring.* PR Newswire. https://www.prnewswire.com/news-releases/optimizely-enters-definitive-agreement-to-acquire-warehouse-native-analytics-leader-netspring-302261900.html

16. Gartner. (28 January 2025). *Magic Quadrant for Digital Experience Platforms.* Gartner. https://www.gartner.com/doc/reprints?id=1-2K0LH2I7&ct=250122&st=sb

17. Sitecore. (n.d.). *The Sitecore story.* Sitecore. https://www.sitecore.com/company/sitecore-story

18. O'Hear, S. (2021, March 3). *Boxever bought by US firm Sitecore in multimillion-euro deal. The Irish Times.* https://www.irishtimes.com/business/technology/boxever-bought-by-us-firm-sitecore-in-multimillion-euro-deal-1.4500068

19. van Rijmenam, M. (2021, March 4). *Sitecore acquires Boxever and Four51 to give marketers a better view of every customer. VentureBeat.* https://venturebeat.com/business/sitecore-acquires-boxever-and-four51-to-give-marketers-a-better-view-of-every-customer/

20. Sitecore. (2024, May 21). *Sitecore announces HIPAA readiness for digital experience platform's content and experience solutions*. PR Newswire. https://www.prnewswire.com/news-releases/sitecore-announces-hipaa-readiness-for-digital-experience-platforms-content-and-experience-solutions-302279215.html

21. Acquia. (n.d.). *The story behind Acquia*. Acquia. https://www.acquia.com/blog/story-behind-acquia

22. Lunden, I. (2019, September 24). *Vista Equity Partners buys Acquia for $1B*. TechCrunch. https://techcrunch.com/2019/09/24/vista-equity-partners-buys-acquia-for-1b/

23. W3Techs. (n.d.). *Usage statistics and market share of content management systems*. W3Techs. https://w3techs.com/technologies/overview/content_management

24. Sliwa, C. (2019, May 21). *Acquia Mautic acquisition underscores open source model*. TechTarget. https://www.techtarget.com/searchcontentmanagement/news/252463113/Acquia-Mautic-acquisition-underscores-open-source-model

25. Lunden, I. (2019, December 11). *Acquia nabs CDP startup AgilOne, which raised $41M*. TechCrunch. https://techcrunch.com/2019/12/11/acquia-nabs-cdp-startup-agilone-which-raised-41m/

26. Sliwa, C. (2021, September 8). *Acquia acquires Widen digital asset management*. TechTarget. https://www.techtarget.com/searchcontentmanagement/news/252506421/Acquia-acquires-Widen-digital-asset-management

27. Demodia. (n.d.). *The history of HubSpot*. Demodia. https://demodia.com/articles/data-processes/the-history-of-hubspot

28. Salesforce. (n.d.). *The history of Salesforce*. Salesforce. https://www.salesforce.com/news/stories/the-history-of-salesforce/

29. Arnold, C. (2013, July 12). *It's official: Salesforce completes ExactTarget acquisition*. CMSWire. https://www.cmswire.com/cms/customer-experience/its-official-salesforce-completes-exacttarget-acquisiton-021723.php

30. King, R. (2019, June 10). *Salesforce's $15.7 billion Tableau acquisition: Everything you need to know*. ZDNet. https://www.zdnet.com/article/salesforces-15-7-billion-tableau-acquisition-everything-you-need-to-know/

PART 5

MEASUREMENT AND REPORTING

4 Categories of MarTech Platforms

- **Customer Data** — Helping brands understand their customers
- **Content, Campaign, and Multichannel Delivery** — Serving customers with content, offers, and experiences across the journey
- **Measurement, Reporting, and Analysis** — Enabling tracking and insights for marketing efforts
- **Creation, Workflow, and Operations** — Empowering teams to be more efficient in content and campaign creation

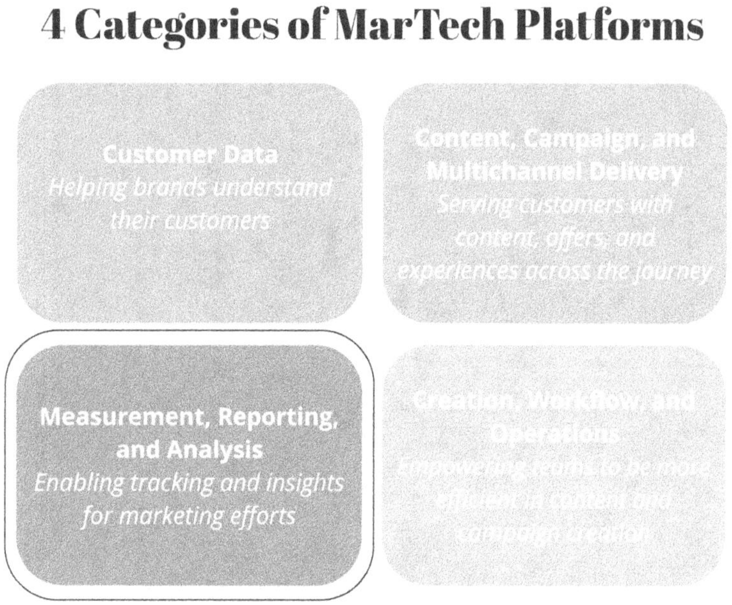

FIGURE P5.1 Measurement and reporting platforms

This last category of MarTech platforms enables marketing teams to track, measure, and analyze the effectiveness of their marketing and advertising efforts across the multitude of platforms that the company reaches its customers through.

The category includes analytics platforms such as Google Analytics and Adobe Customer Journey Analytics, as well as attribution platforms, personalization, A/B/n testing tools, and reporting and dashboarding platforms.

Without this key functionality, marketers would not be able to determine the success of their efforts, nor would they be able to discover insights that help them continuously improve, and share those insights with others in their organization using charts, graphs, and dashboards that can be updated in real time or near real time.

CHAPTER 22

KEY CONSIDERATIONS FOR MEASUREMENT, REPORTING, AND ANALYSIS

Effective marketing decision-making depends fundamentally on precise measurement techniques. Data is an essential tool in environments where each interaction provides valuable insights by affirming strategies and identifying areas needing enhancement. Marketers who use exact real-time metrics enhance their ability to refine campaigns while judiciously allocating resources to ensure prolonged business growth. A lack of strong measurement capabilities can undermine even groundbreaking marketing strategies, which forces companies to make decisions based on partial information.

This chapter will explore single-channel, multi-channel, and in-app measurement tools, with each serving a unique purpose, and with all having the potential to be used together.

DEFINING OBJECTIVES AND KPIS

A clear understanding of core business objectives is essential to initiate effective measurement practices. Marketers need to make sure that all KPIs and metrics support strategic goals, including revenue growth, brand awareness, and customer retention. A thorough examination of business priorities establishes alignment that ensures measurement activities provide direct input for strategic decision-making. When designing a sales campaign, marketers should measure conversion rate, cost per acquisition (CPA), and customer lifetime value (LTV), but for campaigns that aim to boost brand visibility, click-through rates (CTR) and social engagement metrics become essential. This essential step makes certain that data gathering and examination activities align directly with achieving targeted results.

The selection of appropriate KPIs becomes essential when objectives are established. Marketers need to create a catalog of standard performance indicators such as CTR, conversion rate, CPA, LTV, and engagement metrics, then determine the most applicable ones for their specific campaign or channel. Effective KPI prioritization requires organizations to maintain an equilibrium between immediate performance measures and extended business outcomes so that both swift assessments and enduring insights can be obtained. Stakeholder interviews combined with historical performance analysis and scenario planning enable marketers to identify the most actionable KPIs. The deliberate process establishes measurement priorities, which enables better reporting and decision-making within the marketing department.

DATA COLLECTION AND DATA INTEGRITY

The initial step towards accurate measurement lies in the selection of appropriate data sources. Data for marketing purposes originates from multiple sources such as Web traffic analytics, mobile app usage patterns, social media user interactions, and customer relationship management (CRM) systems. The insights drawn from each data source together help build a full picture of customer behavior. The variety of data sources generates challenges because different systems store information with varying formats and time intervals, which requires meticulous data consolidation to achieve a unified dataset.

Analysis quality has the potential to suffer due to problems such as duplicate records and inconsistent tagging combined with incomplete information. Therefore, organizations must establish strong standardization and tagging processes to overcome these difficulties. The use of consistent naming conventions along with UTM parameters and uniform tagging for digital assets produces reliable data that analysts can easily review. Organizations that establish these best practices are able to maintain data integrity which leads to better reporting accuracy and more informed decision-making.

PRIVACY AND COMPLIANCE CONSIDERATIONS

The General Data Protection Regulation (GDPR) and the California Consumer Privacy Act (CCPA) establish stringent rules concerning the collection, storage, and processing of personal data. The legislation mandates that organizations must receive explicit user consent before collecting their personal data and supply straightforward opt-in or opt-out choices. Marketers need to adjust their measurement practices and tools so they meet legal requirements and protect consumer privacy. Companies functioning across different regions need to understand how regional or industry-specific regulations might influence their data management processes.

Analytics tools need to be properly setup to respect user privacy preferences through consent management protocols and necessary data anonymization. Platforms need to implement functionalities that enable detailed oversight of data handling procedures to protect personal data security and limit its use to specified applications. Organizations that embed privacy and compliance considerations into their operations will build customer trust and protect their data-driven marketing strategies while fulfilling regulatory obligations.

BUDGET AND RESOURCE CONSTRAINTS

Organizations evaluating measurement and analytics tools need to thoroughly examine the total cost of ownership (TCO), which encompasses licensing fees and subscription costs, along with extra expenses for integrations and customizations. Organizations must evaluate both the initial licensing fees and the continuing costs of maintenance support along with software updates. The complete cost analysis confirms platform selection stays within budget limits while maintaining necessary features.

Businesses must assess both financial considerations and internal expertise along with their staffing needs. An analytics tool becomes highly effective when staff members with appropriate skills, such as data analysts, data engineers, and marketing specialists, handle data management and interpretation. Organizations need to evaluate their workforce's skills and decide whether further training or new team members are needed. An analysis that weighs budgetary constraints against functional needs leads to selecting an analytics tool that supports current aims while growing with organizational future ambitions.

ILLUSTRATING THE CONSIDERATIONS: VOLTAGE MOTORS EXAMPLE

The electric vehicle company VoltAge Motors faces challenges due to its fragmented marketing tech infrastructure, which obstructs its understanding of customer demographics. The company collects data from various channels, including Web site analytics, mobile app usage, social media signals, and CRM records, yet faces difficulties in merging all this data to form a unified customer profile. Inconsistent tagging alongside duplicate records and incomplete metadata intensifies data fragmentation while generating unreliable performance metrics and masking actionable insights.

VoltAge's marketing strategies face significant obstacles due to the problems associated with integrating data. Marketing teams face difficulties in customizing messages and creating targeted campaigns because they lack unified customer profiles. Marketing campaigns that attempt to personalize based on customer behavior fail to deliver specificity and consequently produce generalized tactics that do not connect with varied audience segments.

The lack of proper alignment leads to decreased customer interaction while simultaneously hindering the firm's potential to create improved customer paths.

At VoltAge Motors additional challenges emerge from privacy and compliance concerns. The current measurement systems fail to provide sufficient support for detailed consent handling and data anonymization, which creates compliance risks under the GDPR and CCPA for the company. Without strong privacy protections, VoltAge Motors faces legal non-compliance risks that could lead to financial penalties and harm its standing in a market that values privacy.

VoltAge must also cope with additional difficulties because budget restrictions and resource limitations create further complications for the company. Expensive licenses for advanced analytics along with a lack of knowledgeable data professionals force the organization to settle for second-best options. The limitations lead to slow decision-making processes while hindering the organization from utilizing its current technological capabilities, which results in lost opportunities for personalization, customer involvement, and diminished campaign results.

CONCLUSION

When choosing measurement reporting tools and analysis systems organizations should ensure their selections match their specific marketing objectives. The first step of this process requires setting specific goals and determining key performance indicators that represent essential business results. Organizations must ensure data feeding into these tools maintains accuracy and consistency because poor data quality can obscure insights and undermine decision-making. To preserve customer confidence and evade legal penalties businesses need to fulfill compliance standards such as those established by the GDPR and CCPA.

Organizations face resource limitations that drive their choice of tools because they require solutions that meet their needs without surpassing financial boundaries. Long-term success depends on performing detailed assessments of platforms for their usability, scalability potential, integration features, and analytical capabilities. An organization can establish a powerful measurement system to enable data-driven marketing strategies by balancing marketing objectives with data integrity requirements, regulatory compliance standards, budget constraints, and rigorous evaluation criteria.

CHAPTER 23

Single Platform, In-App Measurement, and Multi-Channel Measurement

MarTech measurement platforms serve as crucial elements for developing strategy and enhancing performance optimization. Marketers must determine whether to prioritize single-channel measurement or in-app analytics or adopt a multi-channel approach to obtain precise customer behavior insights. Every method presents distinct benefits and difficulties, which means choosing the right platform requires assessing how effectively a tool collects needed data for the planned analytical range. Of course, the tools and methods are not mutually exclusive either, and using multiple tools and methods can involve a strategic approach.

Single-channel measurement tools deliver a profound understanding of specific touchpoints and suit campaigns that target a narrow audience. In-app measurement platforms provide detailed user interaction data for mobile applications that brands with proprietary apps find essential to boost engagement and retention. Multi-channel measurement systems combine data from multiple digital origins to develop a complete view of the customer journey. This chapter examines how marketers can evaluate different measurement platforms to identify the strategy that fits their business goals and technological resources.

Single-Channel Measurement Tools

Single-channel measurement tools deliver precise information about specific marketing channels, including email campaigns, social media analytics, and

Web site performance. Marketers benefit from these platforms because they deliver precise metrics and track user interactions inside their dedicated channels to provide detailed insights. Marketers use Mailchimp or ActiveCampaign for email campaigns while leveraging Hootsuite or Sprout Social for social media insights and Google Analytics for tracking Web site visitor behavior. These performance tools specialize in collecting comprehensive data, which enables marketers to assess user behavior and evaluate campaign effectiveness and content engagement in precise contexts.

Through their focused specialization, single-channel measurement platforms deliver unparalleled analytical depth, which enables marketers to thoroughly examine performance across specific channels. Detailed metrics, including open rates, click-through rates, and unsubscribe rates, provided by email analytics tools enable precise optimization. Social media analytics platforms help reveal user engagement patterns and follower development while measuring how content performs across different platforms to support precise strategic enhancements. Marketers who concentrate on one or two channels benefit from this focused approach because it provides both specialized knowledge and actionable insights, which drive targeted performance improvements.

Single-channel analytics solutions possess notable strengths but they demonstrate intrinsic limitations when it comes to providing cross-channel insights. Single-channel analytics solutions usually cannot provide complete visibility into customer interactions across multiple channels, which creates blind spots during overall customer journey assessments. Marketers face difficulties in determining their campaigns' wide-ranging effects because single-channel analytics prevent them from seeing cross-channel insights that could enhance marketing tactics. Marketers must use single-channel tools for specific insights and multi-channel solutions to achieve full visibility throughout their audience's journey across different touchpoints.

Evaluating Single-Channel Analytics Platforms

Choosing the proper single-channel analytics platform requires meticulous evaluation of multiple critical factors. The initial step for marketers involves assessing both the accuracy and granularity of their data. The best analytics platform delivers accurate, timely, and comprehensive metrics tailored to each selected communication channel. Email analytics tools need to deliver trustworthy measurements of open rates and click-through rates along with deliverability, whereas social media analytics solutions must provide precise reports of engagement metrics alongside audience growth and post-performance. Marketers need to test platform data reliability with pilot tests and peer reviews to ensure the data provides actionable insights for campaign decisions.

Platform effectiveness depends heavily on the quality of the user interface and its usability. Analytics solutions need to support both expert analysts and marketing professionals with limited technical skills to rapidly obtain data insights without needing comprehensive training. Marketing teams achieve effective result interpretation and quick response through intuitive dashboards and customizable visual data representations. Marketing teams need to evaluate training resources together with customer support availability to ensure complete utilization of platform functions.

Third, consider the platform's integration capabilities. Analytics tools must be capable of seamless integration with the wider MarTech infrastructure even when operating in single-channel environments. When marketing automation platforms or content management systems (CMSs) work together with customer relationship management (CRM) systems, marketers are able to enhance data from individual channels with additional context from external sources. Platforms that provide strong API support or built-in connectors make integration straightforward, which allows data to move effortlessly into wider measurement systems.

Evaluating how well the platform can scale and support customization remains essential. Single-channel solutions need to support expanding campaign volume and manage greater data complexity while meeting detailed segmentation demands. Marketing platforms require users to be able to personalize metrics and reports while maintaining operational efficiency and user-friendliness.

With that said, it should be acknowledged that the analytics and reporting components of most of these platforms are only one component of many. Thus, some compromises will probably need to be made when striving for optimal analytics and reporting capabilities.

In-App Analytics

Specialized solutions known as in-app analytics platforms track and analyze user interactions across mobile and Web-based applications. These platforms concentrate on capturing detailed event-based user behavior, unlike broader digital analytics tools, which cover general user tracking metrics. The primary functions of in-app analytics include real-time user tracking, detailed funnel analyses, event triggers, and session duration metrics. Through the detailed observation of all taps and swipes alongside user interactions inside an app, marketers obtain a deep understanding of user behavior patterns and how their products are used. Marketers and product teams receive insights that help refine user experiences while improving navigation paths and strengthening both user engagement and retention.

Marketers utilize event-based analytics to identify important user actions such as signing up and purchasing, which they designate as conversion events for in-app measurement. Real-time tracking of events generates instant insights, enabling marketers to quickly detect problems and enhance performance. Real-time data analysis helps teams identify obstacles during crucial onboarding steps when significant user drop-offs occur so they can swiftly iterate to improve the experience. Funnel tracking provides marketers with a visualization of user journeys across designated events to identify friction points and prioritize improvements.

Popular In-App Analytics Platforms

A number of top-tier platforms have risen to prominence in the in-app analytics space and each delivers unique advantages. Mixpanel earns high praise due to its advanced event-tracking features and its simple integration process combined with powerful user cohort analysis tools. The platform stands out through its detailed user behavior tracking, which is essential for businesses focused on gaining comprehensive user engagement insights. As a Google-powered solution, Firebase Analytics has become popular due to its smooth integration with Android and iOS apps and its powerful real-time tracking, while also offering effortless connections with other Google products, such as Google Ads. Marketers already working within Google's extensive ecosystem find Firebase Analytics to be an ideal solution. Amplitude stands out as a leading analytics provider by offering advanced user-behavior insights and retention analysis, which enables marketers to study user pathways and detect crucial points in app usage that boost long-term engagement.

Designing Effective In-App Funnels

Funnel design stands as a primary consideration when developing in-app analytics solutions. A properly designed funnel provides a clear visualization of the user path to specific goals while allowing marketers to track conversion points and drop-off locations precisely. Successful funnel construction starts with marketers identifying their primary conversion events. Key conversion events for marketers to analyze include actions such as app installation and account creation, along with completing purchases, subscribing to premium features, and other critical user actions. Tracking micro-conversions becomes crucial when analyzing user behavior alongside macro-conversions because these smaller actions, such as profile completion and notification enablement demonstrate user engagement patterns. By monitoring minor user actions marketers gain insights into customer intentions and engagement quality while identifying obstacles throughout their journey.

The precise definition of funnel steps depends on a detailed analysis of user behavior patterns inside the application. Identifying the screens or features that lead up to a purchase can reveal essential touchpoints that affect conversion rates. Marketers should explore potential issues such as usability problems and navigation confusion when data shows users frequently drop off before finishing checkout. Funnel progression analysis supports ongoing refinement while enabling precise enhancements, which result in improved conversion rates and better user satisfaction.

Funnel analyses need to include retention metrics for a better understanding of long-term engagement patterns. Marketers who evaluate both initial conversions together with continuous user engagement and activity levels can pinpoint important user segments and discover both high-performing features and indicators of potential user churn. The analysis of behavior exhibited by users who maintain activity beyond their first month helps shape product development choices, as well as marketing strategies and user engagement programs. By combining data from conversion funnels with retention insights, marketers gain a complete understanding of user engagement, which helps deliver ongoing value and maintain growth momentum.

Evaluating In-App Analytics Tools

The selection of an in-app analytics platform requires marketers to evaluate multiple essential aspects, starting with the level of detail and quality of the collected data. A robust platform should capture extensive user interaction data, including clicks and swipes while tracking session duration and feature usage to help marketing teams perform detailed analysis of user behavior. Marketers need to validate how user events and conversion milestones can be defined and measured through straightforward analysis processes. With platforms such as Mixpanel and Amplitude, marketers obtain the ability to monitor intricate behaviors from onboarding completion rates to the frequency of key user actions, which facilitates actionable insights to enhance user experience and increase retention.

Integration capabilities represent another essential consideration. In-app analytics tools need seamless integration capabilities with wider marketing infrastructure components, including customer data platforms (CDPs), CRMs, marketing automation tools, and advertising platforms. By ensuring data consistency through integration we gain a single customer perspective to create precise targeting and personalized marketing strategies. Marketers need to assess whether the platform provides APIs alongside pre-built connectors and export functionalities to enable flexible data exchange between systems, which minimizes manual intervention and improves analytics accuracy.

The platform must be user-friendly while also being easily accessible. Marketers, product managers, and UX designers should be able to build funnels and segment user groups through an intuitive interface that simplifies the visualization of behavioral patterns without requiring extensive technical expertise. Marketers can use Mixpanel or Amplitude to quickly identify problems in user flows, which allows teams to make strategic and design adjustments before issues arise.

Marketers need to assess their platforms based on scalability and customization capabilities. The platform must accommodate growing user data volumes as the app expands while maintaining optimal performance. The platform stays relevant for evolving company goals and app features through customizable dashboards and event-tracking abilities combined with reporting metrics. Marketers should select platforms that adjust to shifting business needs because these platforms enable ongoing optimization alongside a more profound analysis of user behavior when products expand and customer expectations evolve.

Multichannel Measurement Tools

Multichannel measurement platforms combine customer data from multiple sources including email, social media, search engines, Web sites, apps, and offline encounters into one unified perspective. Marketers can use platforms such as Adobe Customer Journey Analytics (CJA), Funnel, and HubSpot to merge complex data sources that are often scattered across different systems. Adobe CJA delivers sophisticated journey mapping tools along with advanced attribution models, but Funnel focuses on bringing together marketing data from multiple digital platforms and ensuring its cleanliness. HubSpot excels by bringing together campaign and sales data into one ecosystem, then enabling cross-channel analysis for small teams.

Marketers face growing challenges in tracking and understanding customer interactions with brands across multiple digital and physical touchpoints as these interactions increase. The lack of integrated measurement results in insights staying confined to single channels, which contributes to fragmented and inefficient marketing initiatives. Marketers gain a more comprehensive view of customer behavior when multichannel measurement platforms aggregate data to provide a unified perspective.

The key strength of multichannel measurement platforms lies in their ability to merge customer interactions across different channels, which helps marketers visualize the complete customer journey. Marketers obtain actionable insights about consumer movement across touchpoints and conversion-driving channels while identifying friction points through comprehensive data analysis. The comprehensive perspective enables marketers to discover better engagement opportunities and precise targeting methods while simultaneously enhancing customer experiences and marketing efficiency.

Multichannel measurement tools deliver improved attribution capabilities as their primary advantage. Through the tracing of customer interactions across multiple channels, marketers obtain insights into how each touchpoint performs in generating conversions. By employing advanced attribution models, Adobe CJA reveals how brand awareness activities through social media platforms and email nurturing sequences work together with search campaigns to define each channel's contribution to customer acquisition and retention.

These platforms pose several implementation challenges. Technical difficulties arise when integrating disparate data sources since it demands significant configuration efforts and specialized knowledge. Data integrity management becomes complex when working with multiple channels while integrating offline events or older systems. Marketers should maintain vigilance towards privacy laws while making certain that customer data collection and usage meet all required compliance standards. The strategic benefits of attaining a unified customer view justify the efforts needed to overcome these challenges.

Evaluating Multichannel Measurement Platforms

Selecting a multichannel measurement platform requires the evaluation of advanced features beyond fundamental analytics functions. Marketers should first select platforms that provide precise interaction tracking across customer touchpoints such as Web, email, mobile apps, social media, paid advertising, and offline events. Cross-channel visibility capabilities allow marketers to conduct complete evaluations of customer behaviors and their entire journey throughout different touchpoints. The chosen tool should aggregate data while simultaneously reconciling user identities across multiple channels and devices through dependable identity resolution methods.

Another critical area is integration. Existing marketing and operational tools, including CDPs, CRM systems, advertising networks, and Web analytics solutions need to integrate without any issues with multichannel measurement solutions. Data consistency and completeness are maintained through effective integrations that eliminate the fragmented views created by disconnected systems. Marketers need to evaluate both APIs and pre-built connectors to assess how effectively they enable setting up and maintaining integrations.

Attribution modeling functionality represents another significant criterion. Marketers gain deeper insights into channel contributions toward campaign success through multiple attribution models, such as first-touch, last-click, linear, and custom multi-touch options offered by a robust platform. Marketers need to assess platform attribution capabilities to identify valuable channels and optimize budget allocation and marketing strategies in a world with diminishing cookie usage.

The ability to drive actionable insights depends heavily on both usability and reporting capabilities. Marketers need to assess the data visualization capabilities of platforms and their dashboard customization options to address unique business requirements. Access to real-time data and features for automated reporting and performance alerts enable marketers to quickly respond to fluctuating trends. Self-service analytics tools allow teams to conduct experiments and make iterations rapidly without needing continuous support from technical experts.

Enterprises with complex operations across multiple brands or global markets need scalable and flexible systems. Multichannel measurement solutions must sustain performance levels and deliver precise results despite growing data demands across different business units and market segments. Marketers need to evaluate whether the platform can adjust to upcoming requirements such as new measurement methodologies and advanced analytics, including AI-driven predictive analytics and changes in regulatory compliance. An advanced evaluation method guarantees long-term efficiency while business requirements and marketing environments keep changing.

CASE EXAMPLE: VOLTAGE MOTORS AND ITS MULTILAYERED ANALYTICS STRATEGY

VoltAge Motors utilizes a complex measurement strategy that combines single-channel analysis with in-app analytics, alongside multichannel platforms. Every customer interaction point receives accurate tracking of marketing effectiveness, customer engagement, and sales conversion through their approach.

VoltAge uses Mailchimp as its specialized tool for single-channel email marketing analytics. Each targeted email campaign receives close monitoring from the marketing team, which tracks open rates together with click-through rates, subscriber engagement metrics, and unsubscribe rates. The company employs A/B testing on email subject lines and content to find out which messages connect best with prospects and customers. Although VoltAge gains a significant understanding of email performance metrics through its analytics tools, it experiences problems linking this data to overall customer interactions, thereby demonstrating the inherent limitations of single-channel measurement systems.

VoltAge employs Mixpanel to conduct in-app analytics because management understands that mobile apps play a crucial role in customer experience. Through this platform, companies can closely monitor how users interact with their app while tracking vehicle configurator usage alongside virtual test-drive bookings and customer support interactions. Funnel tracking exposes user journey bottlenecks such as drop-offs during financing inquiries, which helps in creating targeted UX improvements. The in-app analytics continue to operate in isolation from other marketing channels, which makes it difficult to analyze customer behavior as a whole.

VoltAge decided to implement Adobe CJA to bridge measurement gaps with its advanced multichannel platform. Adobe CJA combines data from email campaigns with app interactions, Web site visits, social media engagements, paid search activities, and offline dealership visits. VoltAge's marketing team can follow customer journeys through various channels accurately while discovering optimal conversion routes and attributing revenue to specific marketing actions through their comprehensive view. CJA provides significant improvements for tracking cross-channel interactions, but VoltAge deals with occasional data discrepancy challenges and issues related to identity resolution.

VoltAge uses a combination of analytics tools to implement a multilayered strategy for analyzing and enhancing its marketing activities. The current fragmented state of their tools, which fails to integrate single-channel and in-app solutions with the multichannel platform, creates continuous operational difficulties. Future success hinges on VoltAge's ability to integrate its disparate platforms into a unified analytics infrastructure as it expands worldwide.

CONCLUSION

A marketer's choice between a single-channel, in-app, or multichannel measurement platform should be guided by their marketing goals as well as the intricacy of customer journeys they need to analyze and the level of analytical detail they require. Marketers can gain detailed insights into specific channels through single-channel tools, which prove optimal for enhancing performance in email and social media segments. In-app analytics platforms focus on detailed event-based user data, which helps developers enhance app experiences and increase user engagement. Single-channel and in-app analytics platforms provide narrow perspectives that fail to capture comprehensive marketing effectiveness through their isolated data views.

The use of multichannel measurement platforms resolves these shortcomings by combining information from various customer interactions into a single comprehensive perspective. The primary functionality of these analytics tools is to track customer movements through multiple channels, which allows marketers to gain deeper insights into complex customer interactions and develop comprehensive marketing strategies. Multichannel solutions frequently involve increased expenses along with technical challenges and integration difficulties.

Marketers need to thoroughly assess each analytics platform to meet their particular requirements through channel-specific insights combined with cross-channel perspectives in order to develop a scalable data-driven marketing framework.

The next chapter will discuss data visualization and reporting tools.

CHAPTER 24

DATA VISUALIZATION AND ANALYSIS TOOLS

A strong marketing strategy depends on interpreting data with precision and speed. Marketers frequently analyze large datasets to assess how marketing campaigns perform and how customers engage through different channels. Data visualization and analysis tools enable the conversion of complicated datasets into understandable and practical business insights. The platforms enable marketers to create detailed visualizations alongside dynamic dashboards and user-friendly reports that reveal essential performance metrics.

Analytics tools offer a visual representation of data that allows both technical and non-technical stakeholders to easily identify trends and opportunities compared to traditional spreadsheet methods. Tableau and its competitors, such as Google Looker and Microsoft Power BI, along with newer market entries such as Domo, enable organizations to quickly identify top-performing strategies while pinpointing operational bottlenecks and changes in customer behavior resulting in enhanced decision-making capabilities. New solutions improve user experience while deploying AI for better insights and expanding integration capabilities, which allows marketers to quickly modify their strategies and show real business results. The chapter examines these platforms by discussing their main features and the various user roles they support while showing how to select the tools that align with organizational requirements.

KEY FEATURES OF DATA VISUALIZATION PLATFORMS

While each platform varies in the way that it delivers features for data visualization and reporting on marketing analytics and effectiveness, there are several key functionalities that these platforms generally offer.

Interactive Dashboards and Workflow Management

Marketers utilize interactive dashboards as their main platform for real-time data visualization to track both campaign effectiveness and customer interaction through dynamic interfaces. Users can quickly access detailed insights through dashboards that provide intuitive navigation and drill-down capabilities without requiring specialized analytics training. By starting with an overall campaign performance overview a marketing manager can drill down to specific channels and customer segments to quickly identify successful strategies and areas for improvement.

Sophisticated interactive dashboards help users obtain actionable insights much faster than traditional methods. Marketers possess the ability to modify data views on-demand through straightforward clicks that permit parameter adjustments without needing complex query creation. A social media manager employs interactive dashboards to swiftly modify targeting criteria in a live campaign by utilizing real-time performance feedback, which then results in immediate optimization and better results.

Advanced Visualization Options

Data visualization tools provide multiple methods to present data through visual representations, such as bar graphs, line charts, scatterplots, heatmaps, and geographic mapping. Marketers benefit from having multiple visualization options because these tools present complicated data in an understandable way, which enables stakeholders to understand important insights quickly. A heatmap visualization can display the interaction patterns of users on a Web site landing page, enabling marketers to immediately identify the most engaging areas, as well as the sections ignored by visitors.

Geographic mapping functions deliver essential information that benefits location-based advertising strategies. Visualization of geographical performance data lets marketers identify successful regions and underperforming areas to optimize their campaign targeting. An e-commerce marketer can examine regional sales data on a map to discover cities with unexpectedly low sales, which allows a focused local campaign to improve performance.

Customizable Reporting

Creating and distributing custom-branded reports makes internal communication straightforward while maintaining consistent insight delivery across the organization. Marketers can produce polished presentations rapidly by utilizing customizable layouts along with flexible visuals and report templates, which eliminate extensive manual work. Marketing teams create weekly or monthly reports designed for executives that focus on KPIs, which support business objectives while adhering to branding standards.

The automation of report timing and distribution functions removes the inefficiency from sharing up-to-date insights with stakeholders. Regular performance reports delivery to key stakeholders through email or automated export minimizes manual work and maintains information flow. The leadership teams receive monthly performance summaries automatically, which keeps everyone informed while improving communication flow and speeding up decision-making.

Integration and Automated Data Ingestion

Making effective marketing decisions requires the integration of data from multiple sources such as CRM platforms, email marketing software, Web site analytics tools, and advertising platforms. The integration capabilities establish seamless data transfer across the MarTech stack, which generates complete insights for comprehensive analysis. Marketers gain precise insights by connecting CRM data with campaign analytics to establish direct links between their marketing efforts and sales results.

Through automated data ingestion, continuous analytics processes benefit from reduced errors while maintaining both the timeliness and the precision of data. Marketers achieve both reduced manual workloads and enhanced data accuracy by using automatic data importation and synchronization across systems. Automated synchronization between Google Analytics and CRM systems keeps campaign performance metrics current, which enables accurate evaluation of marketing effectiveness.

Real-Time and Near-Real-Time Data Capabilities

Marketers can observe and react to customer interactions without delay through real-time analytics platforms that process incoming data instantly. Real-time functionalities allow marketers to dynamically optimize campaigns by modifying social media advertising budgets according to live performance data. Marketers can use the real-time tracking of Web site traffic for product launches to detect unexpected interest spikes and then quickly adjust ad budgets to boost conversions.

Marketers can swiftly adapt to evolving market trends and customer actions by using platforms that deliver near-real-time analytics. This feature plays an essential role in campaigns that require immediate attention, such as product launches and promotional events. Retailers who use near-real-time sales tracking are able to swiftly detect ineffective promotions and adjust their messaging in a matter of hours or days leading to better campaign results.

Resource Allocation and Scalability

The processing of substantial data volumes becomes necessary for marketing operations within rapidly expanding or large organizations. Efficient scaling of platforms to manage growing data volumes and complexity without affecting performance remains vital. Global business expansion requires data visualization tools that maintain quick rendering and responsiveness to enable marketing teams to make decisions without experiencing delays.

High-performance infrastructure and flexible resource allocation minimize downtime while preventing bottlenecks during periods of increased data demand. Marketers benefit from systems designed to handle growing volumes of data streams along with increasing user queries and simultaneous dashboard access. A scalable infrastructure allows all teams to access real-time sales and Web analytics data during intensive events such as Black Friday without experiencing interruptions or system slowdowns.

Security and Compliance

Protecting sensitive customer and organizational information requires strict adherence to data security standards and privacy compliance regulations. Data visualization software implements robust security protocols including encrypted data storage along with user verification systems and full access control measures. Role-based permissions limit data access by displaying sensitive information only to authorized personnel, which protects confidential marketing insights from unauthorized users.

Following GDPR, CCPA, or HIPAA regulations enables organizations to reduce legal and financial exposure from data privacy breaches. Marketers can utilize data-driven insights with assurance when platforms comply with privacy protection standards. Marketers can obtain audience insights from GDPR-compliant analytics platforms because these systems automatically remove personally identifiable information thus preventing privacy violations and regulatory breaches.

OPTIONAL FEATURES

Additionally, there are several features that a data visualization platform may or may not have. Some of these may be valuable to an organization while others may be less important.

AI-Driven Insights

Marketers can swiftly identify hidden data patterns and market trends through AI technology while detecting anomalies that allow them to respond proactively to changes. Automated alerts enable marketing teams to receive instant notifications about unexpected declines in Web site conversions so they can investigate and respond without delay. Natural language querying makes analysis easier for marketers who can ask straightforward data-driven questions that produce fast and clear insights without needing complex analytical skills.

Mobile Access

Marketers with mobile access can monitor real-time campaign performance together with customer interactions and analytics dashboards no matter their location. The flexibility offered by mobile access enables marketers to make agile decisions by allowing them to respond rapidly to new opportunities and challenges without being constrained by desktop tools. A campaign manager at an event can observe real-time social media engagement data through a mobile dashboard and adjust campaign tactics immediately to boost overall effectiveness.

Embedded Analytics

Marketers can deploy interactive dashboards and visualizations on external sites and internal portals or proprietary applications through embedded analytics technology. Streamlined access to data-driven insights promotes wider usage while speeding up decision-making processes. When sales performance dashboards are embedded in CRM systems, they enable sales and marketing teams to access shared data effortlessly, which results in better alignment and deeper client discussions.

Storytelling and Presentation Tools

Marketers can use storytelling functions and integrated presentation features to transform intricate data into understandable visual narratives that capture the audience's attention. These platforms enable stakeholders without extensive data experience to communicate more effectively. A marketer might utilize storytelling tools to demonstrate the effects of recent marketing strategies at quarterly meetings through visual representation that focuses on major successes or difficulties for senior leaders to understand.

Natural Language Processing (NLP)

Data visualization platforms equipped with NLP capabilities enable marketers to conduct analytics through natural language queries that simplify data access for non-technical users. Marketers can get instant visual responses by typing simple questions such as "Which channel had the highest conversion rate last month?" instead of constructing complex queries. The enhanced speed of insight creation enables marketers to dedicate their efforts toward data utilization instead of data queries.

Advanced Data Modeling and Preparation

Data modeling and preparation tools integrated into visualization platforms simplify the process of cleaning, structuring, and organizing data for analysis directly within the platform. Marketers can quickly merge data from multiple sources into useful formats without extensive dependence on data teams. By combining data from social media platforms with Web analytics and CRM systems, marketing analysts can transform this information into unified datasets that provide enhanced customer insights.

Community and Marketplace Resources

Marketers can utilize pre-built templates along with extensions and plug-ins from community and marketplace resources to enhance visualization tool capabilities. By utilizing these shared resources, marketing teams can speed up their setup process and improve visual analytics effectiveness. A marketing team can download a template created by the community to report on social media campaigns, which allows them to adapt the tool to their needs instead of creating reports from nothing.

HOW TO EVALUATE DATA VISUALIZATION TOOLS

In addition to any specific considerations that are central to an organization's decision-making processes, the following are some of the key areas that marketers should use to evaluate data visualization tools.

Flexibility with Visualization

When selecting data visualization platforms, it is imperative to have flexibility. Marketing professionals should evaluate data visualization tools that provide multiple visual representation options, including scatterplots, heatmaps, funnel charts, and geographic maps. Deep customization options for visual components such as colors, labels, tooltips, axes, and interactive features should be available to facilitate clear storytelling that matches the needs of different target audiences. Platforms that enable marketers to create interactive dashboards with brand-specific color schemes and personalized metrics significantly improve user engagement and understanding.

The system's flexibility includes interactive capabilities to drill down into or aggregate data. Marketers gain deeper insights through platforms that enable them to explore data dynamically by filtering information in real time and switching views between broad trends and specific details. Tableau and Power BI stand out as top tools that enable marketing teams to quickly and intuitively study patterns in customer behavior as well as measure campaign effectiveness.

Compatibility with Data Sources

Data visualization tools need to integrate effortlessly with existing marketing platforms, including CRM systems (Salesforce, HubSpot), Web analytics platforms (Google Analytics, Adobe Analytics), ad networks, and customer data platforms (Segment, Tealium). The ability to effortlessly connect to data sources and maintain updated synchronization enhances both data dependability and promptness. Marketers receive continuous customer insights updates from CRMs, which enables them to run responsive campaigns and make real-time adjustments.

Marketers must choose platforms that provide strong APIs and ready-made connectors because they reduce setup time and the technical demands on internal staff while streamlining integration processes. Visualization tools that can process extensive datasets from various platforms without extensive data transformation help streamline analytics workflows while minimizing errors and saving time.

User-Friendliness vs. Advanced Features

Maintaining a proper balance between user-friendly interfaces and advanced features is crucial. The combination of drag-and-drop capabilities and easy-to-navigate dashboards allows marketing teams without extensive analytics experience to adopt new platforms more easily. Marketing teams can swiftly become proficient with Google Looker because its simple user experience facilitates broad adoption and continuous utilization.

Sophisticated analytics capabilities such as predictive analytics and custom calculations become necessary for advanced marketing teams that require in-depth statistical modeling. Tableau and Power BI platforms represent superior options for teams with advanced analytics requirements. The ability of internal teams and the extent of data exploration they seek informs marketers about how to achieve optimal usability and complexity in their organizational tools.

Pricing Structures

Data visualization platforms provide a range of licensing models including seat-based subscriptions alongside usage-based fees and hybrid structures. Tableau and Power BI use seat-based pricing where each user incurs a cost, which benefits fixed budgets but becomes expensive for expanding teams. Cloud-based visualization solutions use consumption-based pricing models that provide flexible capacity for variable usage needs yet create uncertainty in monthly financial planning.

When determining pricing models marketers need to evaluate how well they support long-term scalability through user expansion and increasing data needs and functionality requirements. Organizations make better investment decisions by understanding total ownership costs, which include initial integration expenses as well as training, support requirements, and future expansion possibilities.

VOLTAGE MOTORS' TRANSFORMATION WITH TABLEAU

The electric vehicle maker VoltAge Motors faced ongoing issues with ineffective visualization of their marketing data. Excel spreadsheets served as the primary tool for the marketing team to gather performance data from CRM systems, Web site analytics, and advertising platforms. The marketing team converted static reports into PowerPoint presentations, which were delivered to executives every week by email. The process demanded extensive time while being vulnerable to mistakes, which resulted in inconsistent reports and postponed decision-making. Leadership received outdated reports, which hindered effective campaign adjustments and budget reallocations.

VoltAge Motors turned to Tableau to enhance its marketing data visualization capabilities because it needed a more dynamic solution. Real-time access to campaign performance metrics became available to the marketing team through direct integration of Tableau with Salesforce, Google Analytics, and social media ad platforms. The transition from manual data collection from various platforms allowed the team to establish automatic live dashboards that ensured reports stayed current and matched data accurately. With Tableau's interactive capabilities, marketers can quickly access and filter insights by campaign details, regional data, or customer segments through simple clicks.

Decision-making speed and accuracy experienced dramatic improvements due to this change. Before, the Tableau leadership had to wait for weekly reports to assess marketing performance, but now they can monitor real-time data through constant access to Tableau dashboards. Immediate adjustments to an underperforming ad campaign became possible without having to wait for its weekly report. Tableau's engaging dashboard visuals enabled non-technical stakeholders to understand complex data trends, which led to better-informed strategic discussions and changes.

VoltAge Motors gained a significant marketing agility advantage through their implementation of Tableau. The marketing team achieved rapid campaign optimization and efficient budget allocation while freeing themselves from manual reporting tasks. VoltAge Motors maintained its strong market presence because real-time visualization enabled rapid decision-making that matched market dynamics.

CONCLUSION

In marketing decision-making processes data visualization serves as an essential element beyond simply its visual appeal and ability to provide branded reports. Marketing teams face the danger of missing essential trends and making budget errors when they cannot interpret complex data swiftly enough to respond to market changes. Visualization platforms enable teams to confidently act by transforming raw data into interactive and understandable insights. An effective data visualization strategy allows teams to transform raw insights from campaign performance tracking and audience behavior identification into practical applications that optimize marketing funnels.

The choice of visualization tool needs to be compatible with the organization's overall MarTech ecosystem. The best platforms function as part of a larger system by integrating seamlessly with tools for measurement, reporting, and analytics to build a unified data ecosystem. Marketing teams that commit to scalable and AI-enhanced visualization platforms gain strategic benefits by transforming their data into ongoing sources of competitive insight and optimization.

PART 6

EVALUATION OF YOUR MARTECH STACK AND PLATFORMS

MarTech can help marketers achieve amazing things, such as increasing brand awareness, generating leads, converting customers, retaining loyalty, and enhancing customer experience. But it is not a one-size-fits-all solution. Different businesses have different needs, and when combined with the continual evolution of the industry as well as with competitive pressure and customer expectations, it is crucial for marketers to evaluate their MarTech infrastructure regularly and ensure that it aligns with their current and future marketing strategies.

Doing this well can help you and your teams do the following:

- Identify gaps and opportunities in your marketing capabilities as an organization
- Optimize your marketing performance while increasing efficiency
- Enhance your customers' experience and loyalty
- Stay ahead of the competition and customers' expectations

As you can see, the stakes are high, so there needs to be a methodical approach to evaluating the marketing technology infrastructure as a whole.

According to Gartner's *The State of Marketing Budget and Strategy* 2022 report, nearly 27% of marketing budgets go toward technology spend[1]. Your marketing technology stack is a critical part of your infrastructure, enabling you to acquire and retain customers while providing your teams with the tools they need to successfully create, manage, and measure content, campaigns, and customer journeys. So, how do you strategically evaluate your MarTech stack in a way that takes all of the competing priorities into account?

In this section, we will explore five steps that should provide any organization with a complete understanding of their marketing technology needs, where their current MarTech stack fits within those needs, as well as what they need to be more successful:

- Step 1: Define your goals and approach
- Step 2: Visualize it
- Step 3: Evaluate your infrastructure
- Step 4: Evaluate individual components
- Step 5: Create a roadmap

CHAPTER 25

DEFINE YOUR GOALS

Success is unclear without clear goals and objectives, and planning, implementing, and optimizing a MarTech stack is no different. Even embarking on an effort to build or improve one should begin by establishing clear goals. This chapter will explore a framework to establish these goals and the considerations that should be made during the process.

MARTECH INFRASTRUCTURE GOALS FRAMEWORK

The following goals framework can provide a foundation for any organization undertaking a goal-setting exercise for their organization. While the strategic goals (Figure 25.1) outlined can be applied to any organization, the measurement of key performance indicators (KPIs) and the steps to achieve the goals will vary greatly from one organization to another.

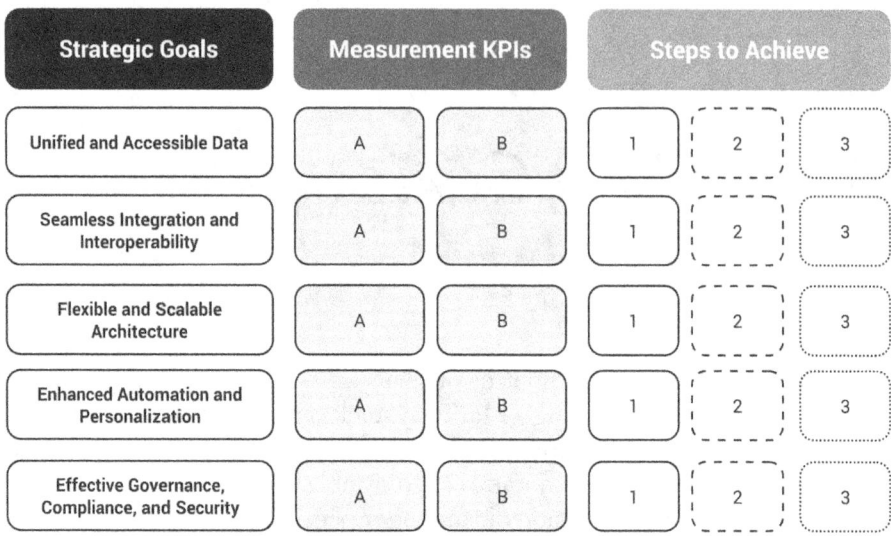

FIGURE 25.1 MarTech infrastructure goals framework

This book will provide samples of an entire framework that is completed to give an example, but make sure to evaluate this based on the unique needs, constraints, and opportunities that exist within your own organization.

Recommended Strategic Goals

The strategic goals outlined within the framework (Figure 25.1) provide any organization with the tools needed, though a valuable step to make them most valuable would be to revisit the content included in the "Objective" column to ensure it most closely aligns with your organization's objectives and the way you describe them.

TABLE 25.1 Recommended strategic goals for Martech infrastructure

Strategic Goal	Objective
Unified Data Management and Accessibility	Create a centralized, consistent data layer across all marketing tools, ensuring data cleanliness, easy access, and actionable insights.
Seamless System Integration and Interoperability	Ensure smooth data flow and collaboration among diverse tools (CRM, marketing automation, analytics, CMS, etc.), minimizing manual work and silos.
Flexible and Scalable Platform Architecture	Build an infrastructure that can grow alongside the organization, handle increasing data volumes, and adapt to new technologies or channels.
Enhanced Automation and Personalization	Leverage advanced automation capabilities to drive personalized interactions across channels and reduce repetitive tasks.
Effective Governance, Compliance, and Security	Safeguard customer data, comply with relevant regulations (e.g., GDPR, CCPA), and implement robust security protocols that protect sensitive information.

To help with the process of customizing these specifically for an organization, each of the strategic goals will be reviewed in more depth.

Unified Data Management and Accessibility

Unified Data Management and Accessibility consolidates different data sources into one dependable repository that marketing, sales, and service teams can depend upon. Organizations achieve better data integrity by establishing a "single source of truth" that resolves the errors and redundancies that occur with isolated system data storage and spreadsheet management. The unified data management framework enhances data consistency while providing all users who need insights—including campaign managers, analysts, and data scientists—with streamlined access.

Data quality must be the top priority for a reliable unified data environment that needs continuous verification and validation to maintain clean information. The system demands a role-based access structure that ensures sensitive information remains accessible only to authorized personnel. These combined practices deliver enhanced reporting capabilities and precise analytics while enabling faster decisions that lead to superior marketing performance.

Seamless System Integration and Interoperability

The connection of key platforms, such as CRM, marketing automation, analytics, and CMS, enables seamless data sharing and collaboration without any operational interruptions. Real-time communication between tools in a well-integrated MarTech ecosystem decreases manual data transfer requirements and minimizes the risk of human error. Teams gain the ability to create efficient workflows, which automatically send leads from marketing campaigns to sales departments and synchronize Web site interactions directly with analytics platforms.

Businesses gain faster access to deeper organizational insights when their systems communicate with one another effectively. Organizations gain the capability to track customer behavior to marketing activities with greater precision while detecting channel trends before they affect revenue. Through interoperability, businesses can respond to market changes more dynamically while aligning internal teams around common metrics and objectives.

Flexible and Scalable Platform Architecture

A MarTech architecture that adapts to change and scales smoothly allows your platforms and databases to adjust to new business needs and data demands. Your infrastructure should be effective enough to take in new marketing channels, regional growth, or unexpected spikes in Web traffic seamlessly while maintaining performance standards. Modular or API-first systems enable teams to integrate new tools and phase out obsolete ones while shifting their strategies all with minimal disruption.

Scalability requires systems to manage larger data volumes while preserving speed and reliability at a reasonable cost as the organization expands. Well-architected systems use load balancing together with containerization and auto-scaling features to keep resource usage aligned with demand. This approach helps maintain system speed while delivering reliable user experiences and prevents costly unplanned outages and rushed infrastructure improvements.

Enhanced Automation and Personalization

An efficient MarTech stack centers around automation, which eliminates repetitive tasks to allow teams to concentrate on strategic initiatives. Marketing leaders can utilize data triggers and machine-learning models to create automated workflows that deliver timely messages to targeted audiences without ongoing manual management. Automation enhances operational efficiency while also creating a unified brand experience throughout all communication channels.

Personalization works together with automation to use unified data to create customized content and offers based on user preferences and context.

Real-time personalization engines dynamically modify emails and Web content through the analysis of user profiles and behavioral patterns or geographical locations. The personalization results in customers experiencing more relevant interactions, which boosts engagement rates as well as conversion rates and improves brand satisfaction overall.

Effective Governance, Compliance, and Security

The main role of governance alongside compliance and security measures in a MarTech ecosystem serves to protect customer data from unauthorized access while ensuring responsible data handling. The process requires establishing explicit access rules for data while managing user consent and controlling data flow across systems. Meeting GDPR or CCPA requirements goes beyond legal obligations because it plays a critical role in building consumer trust while protecting your organization from financial loss and damage to its reputation.

A well-governed MarTech stack requires the use of foundational security protocols including encryption, multi-factor authentication, and regular vulnerability assessments. Companies reduce the risk of data leaks and unauthorized use by implementing security measures across every tool and platform they operate. Strong governance frameworks enable easier audits and reporting while promoting a transparent data culture, which ensures ethical and effective MarTech operations.

Example KPIs to Measure Success

The next component in the framework is to set measurable goals—KPIs—to ensure that the strategic goals are achieved, and to provide an objective and quantifiable set of measures of success so that subjectivity and anecdotal evidence are avoided. Because technology infrastructure projects often span months—if not years—maintaining consistent measurements of success is critical because short-term bumps in the road on a long-term project can create misperceptions of overall success, and as team members may change throughout the lifecycle, it is important not to modify the definition of done of a successful project without a broader consensus.

Additionally, as stated earlier, while the strategic goals in the framework have broad applicability to organizations, the sample KPIs provided below are directional and while some may be directly applicable, others may either need refinement or may need to be replaced altogether.

TABLE 25.2 Sample KPIs matched with strategic goals

Strategic Goal	Sample KPIs to Measure Success
Unified Data Management and Accessibility	• Data completeness: Percentage of required fields populated in the central database • Data accuracy: Rate of data errors or duplications identified via routine audits • Data latency: Time lag between data generation and availability in the unified repository
Seamless System Integration and Interoperability	• API utilization and throughput: Number of successful API calls and average response time • Integration error rate: Frequency of system failures or errors in data handoffs between tools • Time to onboard newtools: Speed at which a new platform can be integrated into the existing stack
Flexible and Scalable Platform Architecture	• System uptime: Percentage of time the MarTech environment is fully operational • Load handling: Ability to sustain peak traffic or usage without performance degradation (e.g., metrics around concurrency or peak loads) • Time to scale: How quickly additional resources (servers, licenses, storage) can be added without disruption
Enhanced Automation and Personalization	• Automation coverage: Percentage of marketing tasks or workflows automated (e.g., lead routing, email triggers) • Personalization rate: Proportion of outbound marketing communications that use dynamic content or real-time data • Reduced manual effort: Decrease in total hours spent on routine marketing tasks due to automation
Effective Governance, Compliance, and Security	• Compliance audits: Number of successful compliance checks or certifications (ISO, SOC 2, etc.) • Security incidents: Frequency of system breaches or near misses • Data retention and governance score: Internal metrics tracking adherence to data storage, deletion, and governance policies

Example Steps to Achieve Each Strategic Goal (Technology-Focused)

Table 25.3 features some hypothetical action steps that address each infrastructure goal at a platform level. They can be adapted to fit specific company sizes, industries, or existing technology stacks. It is highly recommended to more closely tailor these steps to the individual organization, as many of these factors can vary dramatically, from the necessary scope (e.g., is this a full-scale evaluation of all platforms?) to the data requirements (e.g., a healthcare company may have very different or additional patient data needs that an e-commerce platform may not).

TABLE 25.3 Example action items for each of the strategic goals

Strategic Goal	Example Action Steps

Unified Data Management and Accessibility	1. *Conduct a platform audit*
	    • Inventory all data sources and platforms (CRM, analytics, CDP, etc.).
	    • Identify redundant systems, conflicting data schemas, or missing integrations.
	2. *Establish a central data repository*
	    • Implement a CDP or data warehouse to unify data silos.
	    • Standardize data formatting and naming conventions across systems.
	3. *Implement data quality and validation processes*
	    • Deploy automated scripts that check for duplicates, incomplete fields, or invalid entries.
	    • Set up alerts or dashboards that notify teams of data discrepancies in real time.
	4. *Ensure role-based accessibility*
	    • Define permission levels for different teams (marketing, sales, data science, etc.).
	    • Maintain a clear governance policy that outlines data ownership and usage rights.

Seamless System Integration and Interoperability	1. *Adopt an integration framework*
	    • Use APIs or middleware platforms (e.g., iPaaS solutions) to link marketing, sales, and customer service tools.
	    • Focus on solutions that support real-time and batch data syncing for maximum flexibility.
	2. *Standardize data exchange protocols*
	    • Use consistent data formats (JSON, XML) and naming conventions for internal APIs.
	    • Document all endpoints and maintain an internal wiki or knowledge base.
	3. *Create a testing and monitoring environment*
	    • Set up a staging or sandbox environment to test integrations before rolling them out to production.
	    • Implement monitoring tools that provide real-time alerts when data transfer fails or throughput slows.
	4. *Establish Clear Integration SLAs*
	    • Define acceptable response times, maximum downtime, and support procedures for each connected tool.
	    • Collaborate with vendors to ensure integration performance meets your organization's requirements.

(Continued)

Strategic Goal	Example Action Steps
Flexible and Scalable Platform Architecture	1. *Adopt a modular, API-first approach* • Favor platforms and tools with open APIs that facilitate customization and expansion. • Avoid over-customization that might lock you into rigid systems or high switching costs. 2. *Leverage cloud infrastructure and services* • Use scalable hosting providers (IaaS, PaaS) to handle variable loads. • Employ containerization (e.g., Docker, Kubernetes) for portable, resilient deployment. 3. *Implement load testing and capacity planning* • Regularly run stress tests to identify performance bottlenecks. • Adjust your resource allocation (e.g., auto-scaling in the cloud) based on usage patterns. 4. *Maintain a roadmap for future expansion* • Align platform upgrades and new tool integrations with product or growth milestones. • Forecast likely technology needs 12–24 months out and plan accordingly.
Enhanced Automation and Personalization	1. *Map the automation ecosystem* • Identify repetitive marketing tasks (campaign workflows, data updates, notifications). • Implement marketing automation tools that easily integrate with your CRM and CMS. 2. *Use real-time data for personalization* • Leverage behavioral data (site visits, clicks, opens) and context (location, device) to tailor messages. • Integrate machine-learning models for next-best-action or product recommendations. 3. *Set up trigger-based workflows* • Automate key journeys (e.g., post-signup welcome series, lead nurturing sequences). • Incorporate lead scoring or customer health scoring to drive timely, relevant engagement. 4. *Monitor automation effectiveness* • Track completion rates, open rates, and usage stats for automated workflows. • Continuously optimize based on performance metrics and A/B testing results.
Effective Governance, Compliance and Security	1. *Create a formal data governance committee* • Include representatives from IT, security, marketing, and legal. • Define policies for data usage, privacy compliance, and record retention. 2. *Deploy robust security protocols* • Implement single sign-on (SSO) and multi-factor authentication (MFA). • Use encryption in transit (SSL/TLS) and at rest for sensitive data. 3. *Ensure regulatory compliance* • Align data storage and consent management with GDPR, CCPA, or relevant industry-specific regulations. • Maintain audit logs of all data access and changes for accountability. 4. *Conduct regular security and compliance audits* • Perform penetration tests and vulnerability scans. • Update policies and controls as new regulations emerge or as business operations evolve.

Key Takeaways

Additionally, there are some considerations that may help to craft both KPIs and action steps that align with the strategic goals. These include the following:

Tie MarTech Infrastructure to Business Needs

While the focus is on technology, ensure your stack remains aligned with high-level organizational goals (e.g., growth, efficiency, customer satisfaction).

Measure What Matters Technically

Metrics such as system uptime, data latency, API throughput, and error rates will show how effectively the infrastructure supports marketing teams.

Adopt an Agile, Iterative Mindset

New marketing channels and technologies emerge rapidly. Build an architecture that supports quick changes and prevents lengthy redevelopment cycles.

Promote Cross-Functional Collaboration

Successful implementations require input from IT, security, analytics, marketing operations, and beyond—fostering a holistic approach to MarTech.

Prioritize Governance and Security

As data volumes grow, regulatory requirements and customer trust hinge on secure, compliant management of customer information.

Ensure that any unique industry, regulatory, or market factors are addressed as well, to ensure that these requirements are part of the evaluation, and not overlooked at key parts in the process.

REVISITING THE MARTECH MATURITY MODEL

As first explored in Chapter 6, the MarTech Maturity Model (Figure 25.2) can provide a key component of measuring the effectiveness of your infrastructure over time.

MarTech Maturity Model

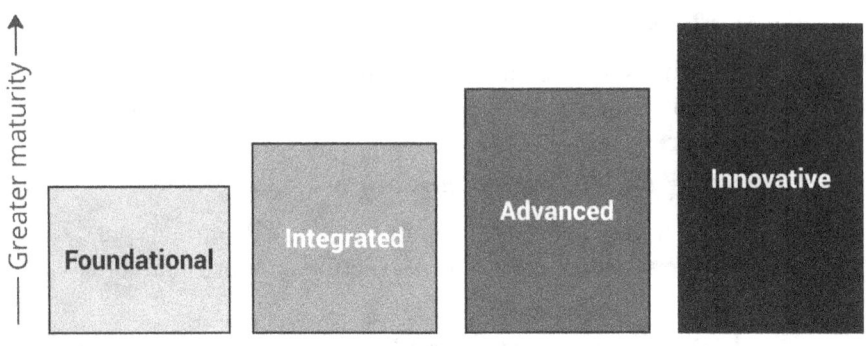

FIGURE 25.2 Martech Maturity Model

This model can be used to evaluate the current sophistication of your MarTech infrastructure and identify the next steps to advance it. Each stage outlines the typical technology ecosystem, the challenges to expect, and the opportunities for improvement.

TABLE 25.4 A MarTech maturity model

Technology Ecosystem	Primary Challenges	Main Opportunities
1. Foundational		
Minimal toolset (often single-channel)	Data inconsistencies and duplicates	Establish a stable platform
Siloed or spreadsheet-based data	High manual workload	Automate simple tasks
Limited automation and integrations	Poor visibility into performance	Basic integrations (e.g., CRM and email)
2. Integrated		
Core platforms connected (CRM, marketing automation, analytics)	Growing complexity of integrations	Streamline cross-platform data flow
Basic centralized data layer	Potential skill gaps in data/automation	Deeper data-driven insights
Introductory personalization and workflow automations	Governance policies may lag behind adoption of new tools	Set the stage for more robust automation and analytics

(Continued)

Technology Ecosystem	Primary Challenges	Main Opportunities
3. Advanced		
Sophisticated data infrastructure (CDP, data warehouse)	Technical overhead (data engineering, ML expertise)	AI-driven personalization and lead scoring
Highly automated marketing, sales, and service workflows	Complex compliance requirements	Holistic, cross-department collaboration
Cross-channel personalization and targeting	Risk of vendor or architectural lock-in	Scalable, enterprise-grade campaigns
4. Innovative		
Modular, API-first architecture (microservices, containerization)	Ongoing innovation pressure	Omnichannel, context-aware engagement
Real-time orchestration of customer journeys	High technical complexity and resource demands	Continuous competitive differentiation
Built-in experimentation frameworks (feature flags, A/B testing)	Maintaining advanced data pipelines and real-time analytics	Rapid adaptation to emerging tools (AR, IoT, etc.)

Technology Ecosystem	Primary Challenges	Main Opportunities
1. Foundational		
Minimal toolset (often single-channel)	Data inconsistencies and duplicates	Establish a stable platform
Siloed or spreadsheet-based data	High manual workload	Automate simple tasks
Limited automation and integrations	Poor visibility into performance	Basic integrations (e.g., CRM and email)
2. Integrated		
Core platforms connected (CRM, marketing automation, analytics)	Growing complexity of integrations	Streamline cross-platform data flow
Basic centralized data layer	Potential skill gaps in data/automation	Deeper data-driven insights
Introductory personalization and workflow automations	Governance policies may lag behind adoption of new tools	Set the stage for more robust automation and analytics
3. Advanced		
Sophisticated data infrastructure (CDP, data warehouse)	Technical overhead (data engineering, ML expertise)	AI-driven personalization and lead scoring
Highly automated marketing, sales, and service workflows	Complex compliance requirements	Holistic, cross-department collaboration
Cross-channel personalization and targeting	Risk of vendor or architectural lock-in	Scalable, enterprise-grade campaigns

(Continued)

Technology Ecosystem	Primary Challenges	Main Opportunities
4. Innovative		
Modular, API-first architecture (microservices, containerization)	Ongoing innovation pressure	Omnichannel, context-aware engagement
Real-time orchestration of customer journeys	High technical complexity and resource demands	Continuous competitive differentiation
Built-in experimentation frameworks (feature flags, A/B testing)	Maintaining advanced data pipelines and real-time analytics	Rapid adaptation to emerging tools (AR, IoT, etc.)

An Overview of Each Stage in the MarTech Maturity Model

Stage 1: Foundational

At the Foundational stage, organizations are just beginning to build their MarTech capabilities. The technology ecosystem typically consists of a minimal set of tools, such as basic email platforms or social media schedulers, often operating in isolation. Data is fragmented across spreadsheets and individual systems, such as a CRM. Automation is limited or nonexistent, resulting in manual processes for campaign management and reporting. Teams experience inefficiency due to siloed data and minimal integrations, but this stage offers an opportunity to establish a strong base for future growth. At this stage, the following is true:

- Marketing relies on manual processes with little to no automation in place.
- Data is inconsistent and spread across disconnected tools and spreadsheets.
- There is minimal integration between systems, limiting real-time visibility and reporting.

Stage 2: Integrated

In the Integrated stage, organizations begin connecting core platforms, such as their CRM, marketing automation, and analytics tools, through APIs or native integrations. A centralized data layer starts to form, although real-time synchronization may not be fully realized. Personalization is basic but emerging, with trigger-based campaigns and simple automated workflows. While integration improves efficiency and enables better data-driven insights, this stage introduces complexity around governance, data quality, and maintaining consistent workflows across systems. The following is true at this stage:

- Core systems such as CRM and marketing automation are connected, reducing data silos.

- Basic personalization and automated workflows (e.g., triggered emails) are in place.
- Teams face growing pains as governance and data management processes try to keep pace with expanding tools.

Stage 3: Advanced

At the Advanced stage, organizations have developed a sophisticated data infrastructure, typically leveraging a CDP, a data warehouse, or a lakehouse with near real-time pipelines. Marketing, sales, and service workflows are highly automated, and advanced segmentation supports cross-channel personalization that spans digital and offline experiences. While these capabilities provide predictive insights and scale, they also demand high technical expertise, careful governance, and an awareness of potential vendor lock-in due to deep integrations. The following is true at this stage:

- Real-time data pipelines feed advanced personalization and cross-channel campaigns.
- Workflows across marketing, sales, and service are largely automated for scale and efficiency.
- Governance and compliance become more complex due to increased data flow and privacy concerns.

Stage 4: Innovative/Next-Level

At the Innovative stage, the MarTech stack is modular, API-first, and designed for agility. Organizations leverage microservices, real-time orchestration, and advanced experimentation frameworks to deliver hyper-personalized, omnichannel experiences. Teams can adapt quickly to emerging technologies, such as AR/VR, IoT, and voice, positioning themselves as industry leaders. The technical complexity is significant, however, requiring continuous innovation and dedicated expertise to maintain and evolve these environments. The following is true at this stage:

- Customer experiences are orchestrated in real time, with dynamic personalization across all channels.
- The architecture is modular and highly adaptable, supporting rapid integration of new tools and platforms.
- Continuous testing and experimentation are embedded in both product and marketing processes.

INVOLVING THE RIGHT TEAMS AND UNIQUE ROLES

A robust MarTech environment is not solely an IT or marketing effort. It demands *cross-functional collaboration* to ensure every stakeholder's requirements are properly captured, aligned, and implemented. Every organization is structured, with teams playing often unique roles or overlapping responsibilities. That said, Table 25.5 outlines the ideal roles that typical teams should play both in the initial evaluation of a MarTech stack and in its ongoing improvement:

TABLE 25.5 Teams and roles in relation to a MarTech stack evaluation

Department and Role	Key Responsibilities
IT / Engineering: Leads on system architecture, integrations, and ensuring scalability/security	API development
	Data infrastructure
	Vendor evaluations (technical fit)
	Ongoing maintenance
Marketing Operations (MOPs): Owns day-to-day platform usage, ensures that marketing requirements (automation logic, campaign needs) are met	Configuring marketing automation
	Building workflows
	Identifying integration pain points
	Championing new platform features
Data and Analytics: Drives data governance, analytics, and predictive modeling	Data quality assurance
	Building dashboards
	Machine learning initiatives
	Measuring platform effectiveness
Information Security and Compliance: Manages data protection, privacy regulation compliance, and security protocols	Implementing encryption
	Auditing data usage
	Creating policies for secure data storage/transfer
Finance/Procurement: Ensures MarTech investments align with budget constraints and ROI expectations	Negotiating with vendors
	Overseeing licensing structures
	Validating ROI or total cost of ownership (TCO)
Sales / Customer Success: Provides feedback on customer-facing activities and uses the data/ insights to close deals or enhance retention	Providing real-world user feedback
	Assisting in designing lead management processes
	Verifying integrated systems (e.g., CRM) function as required

(Continued)

Department and Role	Key Responsibilities
Legal / Compliance Officers: Advises on data privacy laws, consent management, and contractual obligations with vendors	Monitoring data-sharing policies
	Ensuring third-party agreements align with privacy standards
	Verifying compliance with regional regulations

CASE STUDY: VOLTAGE MOTORS' MATURITY JOURNEY

In recent years, VoltAge Motors has expanded its marketing operations to support its ambitious growth targets. VoltAge began with a simple email marketing platform alongside a CRM system for their sales team but soon understood the need for advanced tools to handle customer relationships and campaign management. During their company-wide marketing transformation initiative, leadership determined it was time to evaluate their position on the MarTech maturity scale.

VoltAge Motors completed a comprehensive review of its MarTech stack and operational processes by collaborating with internal teams and external consultants. The company soon discovered that it had outgrown Stage 1 operations, which featured isolated data and manual processes. VoltAge Motors integrated its marketing automation platform with its CRM system to allow trigger-based email campaigns and automate portions of its lead nurturing workflows. Its analytics solutions started to deliver better insights into customer engagement even though real-time data synchronization across platforms had not yet been achieved.

The evaluation revealed that VoltAge was functioning at Stage 2: Integrated based on multiple indicators. Its central systems, including CRM, marketing automation, and basic analytics, were linked, but they faced restricted data exchange between platforms due to inconsistent definitions of data and missing governance protocols. Lead scoring and personalization were implemented but basic because segmentation primarily relied on geography or vehicle model preferences. The marketing team encountered new operational challenges, including data mapping problems and insufficient staff automation skills that exposed the gap between their tools' potential and their team's expertise.

The company recognized the common challenges for its current stage while celebrating its progress. The company required a formal data governance structure and needed to fully integrate systems to close customer insight gaps. Nevertheless, being at Stage 2 offered clear opportunities: VoltAge Motors planned to achieve Stage 3 maturity through investments in a CDP and improvements to their data pipeline strategy. The leadership team prioritized developing beyond basic automation capabilities while creating an integrated customer view to enable sophisticated personalization and journey management soon.

CONCLUSION

Determining your MarTech maturity level helps you understand which infrastructure investments to prioritize next and reveals whether you need to unify data or create seamless integrations and scale advanced automation while exploring cutting-edge personalization. Each phase demands the involvement of appropriate teams, which includes IT and legal professionals, to manage functionality while balancing compliance and cost-effectiveness.

Approach solution development strategically by choosing between building or acquiring platforms and selecting a composable or all-in-one suite while deciding between open-source and proprietary software. How you make choices in these domains determines the effectiveness of your MarTech infrastructure in achieving organizational objectives by affecting scalability and security.

You can build and optimize your MarTech infrastructure with confidence after understanding your current position and planning your evolution to address today's needs while preparing for future possibilities.

NOTE

1. Gartner. "How CMOs Are Spending Their Marketing Biudget—and What It Means For You." Retrieved April 26, 2023 from: https://www.gartner.com/en/marketing/research/annual-cmo-spend-survey-research

CHAPTER 26

A MARTECH STACK EVALUATION FRAMEWORK

First, it is important to understand where your MarTech stack currently stands. After all, how can you truly know where your current opportunities and challenges lie until you look deeply at your current platforms and the work they support? Additionally, it pays to look at the processes that your teams use to create, manage, and optimize their work.

When assessing your MarTech stack, it's important to understand the tools each team uses and how they work. A few things to keep in mind are as follows:

- The impact that a change will have—either positive or negative—on the teams.

- The current adoption of existing platforms and the projected adoption of new ones.

- How extensive the integrations will be, and how many integrations will be necessary.

- The competitive landscape from your direct competitors and potential disruptors or best-in-category brands that are innovating in your work areas.

- Customers' expectations based on either the competition or what they might be experiencing in other interactions with leading brands. Think about how Amazon has set an expectation for next-day delivery with even the smallest of retailers.

- The initial investment, as well as the projected total cost of ownership, and how this might change based on key platform decisions.

A strong MarTech assessment can save time and money and create a greater understanding of how marketing teams work with each other and with other teams within the organization. It can also be part of a strong foundation for the future customer experiences that the brand can build upon.

START BY VISUALIZING

They say that a picture is worth a thousand words, and when it comes to doing something as complex and nuanced as evaluating your MarTech infrastructure, having a visual to refer to can be incredibly valuable indeed.

Successfully doing an assessment means that you create a visualization of your MarTech infrastructure.

Suffice it to say that you won't get very far in your internal conversations unless you have a compelling (or at least accurate) set of visuals that you can refer to as you are having strategic conversations about your MarTech landscape.

The inner workings of your MarTech stack are certainly a complex thing to explain with words or text. Sometimes, relationships of systems and the flow of data from one platform to another with boxes and arrows can convey what many other media simply cannot.

For example, refer to Figure 26.1 to see an example of a MarTech map for an organization with several Web sites, a customer portal, a mobile app, and a partner portal. This diagram is meant to show the data and the basic integration relationships between each touchpoint.

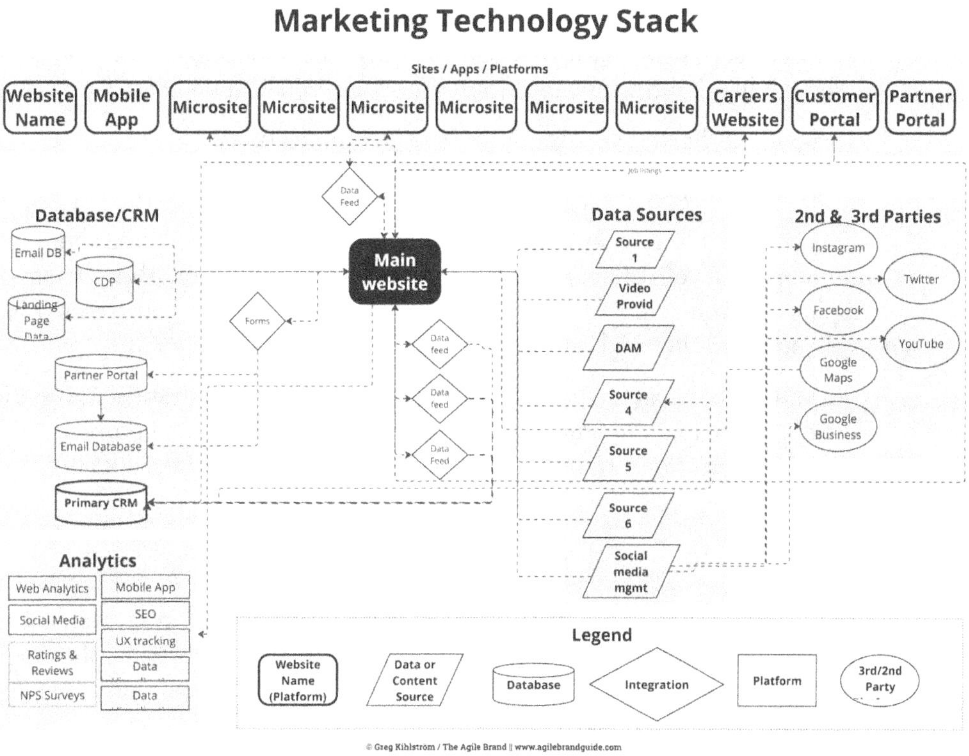

FIGURE 26.1 MarTech infrastructure map example

If you don't have a visual of your MarTech infrastructure that is recent, it is highly recommended that you create one. Even if it is recent, these things can change frequently, it is best to closely examine any existing maps you might have. You will probably need to refer to this several times (or even continuously) as you create your time-based roadmap.

REQUIREMENTS FOR A MARTECH INFRASTRUCTURE MAP

While needs may vary based on the complexity of your marketing infrastructure and the teams involved, here are some things recommended for mapping and displaying:

- The individual platforms and functions (e.g., if you have an all-in-one platform, I still recommend distinguishing between its many functions, such as email marketing, Web analytics, etc.).

- Integration points are often connected via data flow, so you can probably connect your platforms according to how data flows between them, though highlighting any other integrations could also be helpful.
- SaaS/cloud-hosted versus any self-hosted applications.
- Internally managed versus managed by an external agency or team.

You and your team may find other items useful as well. While you shouldn't over complicate your static map with unnecessary information, the very act of mapping out your current state of things will give you and your team what they need down the road.

Additionally, because you may have many platforms that comprise this static infrastructure map, it is highly recommended categorizing them in terms of the *House of the Customer* framework we discussed in Part 1 of this book.

WHO TO INVOLVE

Additionally, ensure you get enough eyeballs from the different teams and functions within your organization to look at the map to ensure it is thorough and takes all of the areas into account. Each organization varies, but some key parties to involve are as follows:

- All the marketing teams and functions within the overall marketing umbrella
- Data and data science teams
- IT and technology teams
- Sales teams
- Customer experience and/or customer service teams

By involving the right people early and at this critical stage, you will be able to ensure you are taking all of the proper considerations as you might have to make some tough decisions later on in the evaluation.

EVALUATING YOUR MARTECH STACK

The assessment of a MarTech stack requires continuous evaluation beyond just major updates as it maintains marketing tools in sync with organizational changes and customer demands. The MarTech stack's performance is crucial in enabling marketing teams to provide personalized and seamless customer experiences across multiple channels and touchpoints in today's fragmented digital ecosystem. Organizations without structured evaluation frameworks face risks, including bloated technology portfolios, data silos, missed automation opportunities, and security vulnerabilities. The evaluation process works as a health monitoring system and strategic framework to make sure technology investments support business goals.

Evaluating a MarTech stack comprehensively involves assessing factors beyond just platform functionality. Table 26.1 illustrates the steps in the process and key activities and considerations to take at each stage.

This evaluation process investigates platform architecture and data management systems alongside integration capabilities and governance frameworks to ensure proper alignment with organizational readiness and business strategy. Organizations can measure technical and business performance through this systematic approach while recognizing capability gaps and determining which improvements to prioritize. Marketing operations stay agile and scalable while meeting compliance demands as regulations and competition grow. Organizations in any development phase can obtain clear insights into how to enhance their technology performance and customer engagement while boosting ROI through a structured evaluation process.

TABLE 26.1 MarTech stack evaluation framework overview

Step	Key Activities	Key Considerations
1. Define Purpose and Scope	Clarify business objectives and tech requirements	Alignment between technology goals and broader organizational strategies
	Identify key stakeholders and use cases	Stakeholder commitment to the evaluation process
	Set timeframe and budget constraints	
2. Assess Current Maturity	Map existing platforms and integrations	Degree of data silos and manual processes
	Determine maturity stage (Foundational, Integrated, Advanced, Innovative)	Immediate vs. long-term system requirements
	Document known gaps	
3. Evaluate Core Infrastructure and Architecture	Review hosting (cloud vs. on-premises vs. hybrid)	Future-proofing capabilities (microservices, containers)
	Assess system scalability, load handling, disaster recovery	Cost and complexity of scaling
	Examine API-first vs. monolithic architectures	

(Continued)

Step	Key Activities	Key Considerations
4. Review Data Management and Accessibility	Audit data repositories (CDP, warehouse, lake)	Data latency and real-time sync needs
	Check data quality, cleansing, governance	Regulatory requirements for data retention and consent
	Evaluate role-based access and permissions	
5. Examine System Integration and Interoperability	Analyze integration methods (APIs, middleware, iPaaS)	Potential integration bottlenecks
	Assess monitoring and logging for data flows	Complexity of maintaining multiple vendor solutions
	Confirm CRM, analytics, CMS connectivity	
6. Analyze Automation and Personalization	Identify workflow automation (marketing, sales, service)	ROI from reducing manual tasks
	Check personalization frameworks (segmentation, dynamic content)	Feasibility of real-time personalization based on data maturity
	Explore AI/ML	
7. Evaluate Governance, Compliance, and Security	Verify data privacy measures (GDPR, CCPA)	Handling of sensitive customer data
	Assess encryption, access controls, and compliance checks	Penalties for non-compliance and breach risks
	Conduct security and vulnerability audits	
8. Assess Vendor and Platform Roadmaps	Review vendors' product vision and release cycles	Vendor lock-in risk
	Examine support quality and SLAs	Alignment with your long-term tech direction
	Explore open source vs. proprietary ecosystems	
9. Examine Organizational Readiness and Skills	Gauge internal team capabilities (IT, Marketing Ops, Data, Security)	Training and onboarding for new tools
	Evaluate cross-functional collaboration	Resource allocation and budget for specialized talent
	Plan for change management	
10. Quantify Performance and ROI	Track system uptime, latency, error rates	Balancing high performance with budget constraints
	Calculate total cost of ownership (TCO)	Justifying ROI for MarTech investments
	Relate tech metrics to broader business impact	
11. Identify Gaps and Prioritize Next Steps	Map findings to maturity goals	Budget and resource availability
	Recommend remediation (integration upgrades, data governance, security)	Timing for upgrades vs. organizational objectives
	Develop short-/long-term roadmap	

(Continued)

Step	Key Activities	Key Considerations
12. Evaluation Summary	Summarize insights for leadership	Ensuring stakeholder buy-in for future initiatives
	Highlight major strengths and weaknesses	Regular audits to update stack as technology and business needs evolve
	Plan ongoing reviews to maintain continuous improvement	

Now, we will review each of the steps in more depth, and for each step, an example from VoltAge Motors will illustrate how the step can be carried out.

Step 1: Define the Scope and Purpose

The first step is to establish both the purpose and the scope parameters of the evaluation process. The foundation of any successful MarTech stack evaluation rests on establishing a clear understanding of its intended purpose and its defined scope. The process serves not only as a technical audit but as a strategic initiative that aims to synchronize technology spending with business results. Determining the evaluation's purpose and desired outcomes will maintain focus and prevent unnecessary expansion of the project scope. Organizations must establish clear objectives from the beginning, whether they aim to streamline workflows, improve customer data integration, or expand into new channels. Defining the evaluation boundaries is crucial because it determines which systems will be assessed and identifies who will participate while also explaining how the results will guide decisions.

Clarify Business Objectives and Tech Requirements

The primary focus of MarTech evaluations must always be their alignment with the organization's wider business objectives. It is essential to evaluate the technology stack based on its capacity to aid vital business initiatives such as customer acquisition growth, personalization enhancements, operational efficiency boosts, and market expansion efforts. Early clarification of business objectives guarantees that the evaluation process centers on the most critical elements. The documentation phase requires setting marketing objectives followed by KPI identification and specifying technical needs essential for reaching these objectives including real-time data processing and advanced segmentation capabilities.

As VoltAge Motors planned to enter European markets they discovered the need to implement stricter data privacy controls and customer engagement processes that were compliant with GDPR regulations. Their business goal focused on expanding market share across new regions while ensuring full compliance with privacy regulations. Their assessment process revealed insufficient consent management practices and data portability issues. The team established their objectives at the beginning, which allowed them to concentrate on locating platforms capable of dynamic consent frameworks and localized personalization, which became essential criteria for platform evaluation.

Identify Stakeholders and Primary Use Cases

A successful evaluation of a MarTech stack requires input from every stakeholder who interacts with or experiences effects from the system. The primary groups involved in evaluation processes typically encompass marketing teams alongside sales representatives as well as IT staff and data analytics experts, with legal/compliance officers and customer experience specialists also taking part. Involvement from these groups makes the evaluation inclusive of various departmental needs, which helps to establish interdepartmental alignment. The stakeholders need to determine primary use cases such as marketing automation workflows and customer journey orchestration and clarify how the technology should assist these functions.

The marketing operations team at VoltAge Motors concentrated on campaign automation and lead scoring while the legal team handled consent management and data-sharing policies. They compiled an extensive list of use cases for the evaluation by bringing together representatives from both stakeholder groups. The team's use cases included setting up automated event-triggered emails and maintaining opt-out compliance throughout every channel. Early stakeholder collaboration prevented future process conflicts and made sure platform choices satisfied all requirements.

Determine Timeframe, Budget, and Resource Constraints

The evaluation scoping process demands an accurate evaluation of available time and budget resources for the task. The evaluation process requires knowledge of available internal resources that will support platform assessments and RFP creation as well as implementation efforts. Defining clear budgetary and temporal boundaries establishes evaluation priorities for stack components while clarifying the necessary compromises between cost and functionality.

VoltAge Motors established a three-month evaluation period because they had to introduce new customer experiences before launching their upcoming product. Due to budget constraints, they chose to evaluate tools that provided pre-built integrations instead of those demanding costly custom development efforts. VoltAge kept its evaluation process within practical limits by clearly defining its time and resource limits to prevent overextension.

The definition of purpose and scope for a MarTech evaluation establishes the foundation for a targeted and efficient assessment process. A process aligned with organizational priorities and realistic execution results from clear business objectives and stakeholder engagement alongside an understanding of constraints. Clear upfront parameters enhance efficiency in future evaluation steps, which results in more implementable findings.

Step 2: Assess the Current Maturity Level

Successful technology implementation within your marketing stack requires a clear understanding of your existing system position. All future decisions will be anchored by the baseline provided in this step. Organizations face the danger of adding complexity to delicate systems without proper evaluation of their platforms and integration maturity levels. The evaluation of your MarTech ecosystem's maturity level shows how effectively it supports your strategic marketing goals and identifies fundamental weaknesses that could jeopardize these efforts. This evaluation determines whether your team needs to build foundational abilities before moving to complex solutions.

Map Existing Platforms and Integrations

A complete list of all MarTech tools allows you to understand how each component fits into the overall ecosystem. The process involves documenting core platforms including CRM, CMS, CDP, and marketing automation together with point solutions such as SEO tools and chatbots and understanding their integration connections. The analysis of data flows and system dependencies helps to identify how information progresses through systems and pinpoints potential failure points. The exercise needs to monitor all licensing fees while evaluating user adoption rates and determining if each technology fulfills its planned objectives.

The marketing operations team at VoltAge Motors completed a comprehensive audit of their entire technology stack. The team documented all their tools including their Salesforce CRM system together with their email platform and social media management tool. Their CRM system worked well with their email system but did not connect with their Web site CMS, which resulted in missed lead capture opportunities. The audit revealed redundant tools because two teams operated separate webinar platforms without awareness of each other's usage. The mapping process revealed where system integration needed completion and where workflow efficiency could benefit from consolidation.

Organizations must determine their current position in the maturity model between the Foundational, Integrated, Advanced, and Innovative stages.

Organizations must assess their MarTech stack through a maturity model after mapping their platforms and integrations. The evaluation process examines key capabilities in data centralization, automation, personalization, integration, and governance to find which maturity stage accurately represents the existing state. Identifying which stage of maturity (Foundational, Integrated, Advanced, or Innovative) you are in enables better prioritization of improvements and helps establish achievable transformation goals.

Upon evaluation, VoltAge Motors found its MarTech stack to be at the Integrated stage according to the maturity model. It had a connected CRM system along with a marketing automation platform and analytics tools, but the personalization remained restricted to basic segmentation, while its data governance methods continued to function on an ad hoc basis. It did not possess AI-driven personalization capabilities or real-time decision-making features that characterize higher maturity levels. Its understanding of its current limitations directed its attention toward enhancing cross-channel orchestration and personalization instead of investing too early in complex machine learning initiatives that they were not prepared to implement.

Identify and Record Existing Operational Challenges Such as Data Silos and Security Issues

Marketing teams can address performance issues and improve efficiency by conducting truthful assessments of operational pain points. Organizations frequently encounter barriers such as platform-specific data silos, manual reporting systems, data entry procedures, and poor personalization resulting in inconsistent customer experiences, along with security and compliance vulnerabilities. Teams need to record both the pain points they encounter and their underlying causes together with how these issues affect business results. The information acts as both a guide for enhancements and a standard for the assessment of future technological solutions.

The documentation exercise at VoltAge Motors uncovered multiple critical gaps that need addressing. Their lead scoring process depended on manual updates between connected CRM and email systems, which resulted in campaign targeting delays. The company's customer data resided with multiple regional teams, which caused the creation of duplicate records and inconsistent communication messages. The process of GDPR compliance reviews became difficult due to the lack of centralized consent data. The identified issues demonstrated how crucial a customer data platform (CDP) and improved automation would be for enhancing data accuracy as well as operational speed and data governance.

Evaluating your current maturity level means obtaining a direct assessment of your MarTech stack's strengths and weaknesses. A realistic and actionable improvement plan begins with a comprehensive analysis of current platforms and integrations against the maturity curve along with documentation of both gaps and pain points. Organizations that skip this step face the danger of making misaligned investments alongside wasted resources. Knowing your baseline position allows you to make decisions that reflect your operational situation while directing attention toward measurable advancements.

Step 3: Evaluate Core Infrastructure and Architecture

The growing complexity of MarTech ecosystems makes the underlying infrastructure essential for success. Your selected infrastructure model—be it on-premises or cloud-based with possible hybrid elements—and its scalability and reliability features impact how well you can provide personalized customer experiences at a large scale. Efficient infrastructure allows systems to remain operational during high-stress situations while enabling growth and reducing expensive downtime risks. Understanding your core architecture serves as a vital exercise beyond IT management because it determines how well your MarTech investments can adjust to new requirements while maintaining effective integration and performance standards.

Platform Architecture

Platform architecture serves as the fundamental groundwork for any MarTech stack. Companies need to select whether their operations will be managed through on-premises solutions or cloud-based platforms or they will use a combination of them in a hybrid environment. Cloud platforms gain popularity for their flexible infrastructure and scalable resources with minimal maintenance needs, but on-premises solutions provide superior control and security at the expense of agility. The hybrid strategy offers a compromise by storing sensitive data on-premises and using cloud services to achieve better speed and scalability. The type of architecture—either API-first, microservices-based, or monolithic—defines the level of system integration and evolutionary capability. API-first and microservices architectures provide superior scalability and flexibility but monolithic solutions remain easier to manage for small deployments despite modification difficulties.

The marketing and IT teams at VoltAge Motors examined their existing setup and discovered they had outdated on-premises systems combined with modern cloud-based applications. The CRM operated in the cloud while the CMS remained on-premises, which complicated integrations and delayed Web site updates. The team found that their existing platforms were built upon a mostly monolithic architecture, which restricted their capacity for rapid innovation. The company opted for API-first and microservices-based solutions as their future investment direction to enable swifter integration of new tools while enhancing their omnichannel strategy support.

Scalability and Load Handling

The traffic and data volume generated by marketing campaigns produce unpredictable spikes. Your systems must be capable of scaling efficiently without experiencing any performance degradation, which remains a critical requirement. Scalability measures a system's potential to expand to meet increased demand levels while load handling defines how systems manage unexpected spikes in traffic and activity. Capacity planning, alongside auto-scaling and stress testing, serve essential functions to ensure peak loads do not cause delays or downtime in infrastructure operations. The absence of proper safeguards leads to poor customer experience and campaign performance degradation during critical events such as product launches and seasonal promotions.

VoltAge Motors encountered significant operational delays during their recent electric SUV product introduction. The announcement event caused their Web site and lead capture forms to slow down considerably, which resulted in potential customers getting frustrated and conversion rates falling measurably. IT specialists helped them implement auto-scaling cloud services alongside load-balancing tools after the event. The organization started regular peak traffic stress testing to make sure their systems could sustain a five-fold increase in visitors and transactions during new product launches.

Redundancy and Reliability

A sophisticated MarTech stack becomes completely useless if it fails to maintain its online status. Planning for redundancy and reliability prevents system failures and reduces downtime and potential data loss. Redundancy means setting up backup systems and failover processes while reliability refers to providing uptime guarantees together with disaster recovery protocols. Mission-critical marketing systems including campaign management, analytics, and personalization engines need reliable failover processes to maintain customer trust and protect revenue streams.

A regional server outage that took VoltAge Motors' CMS and personalization tools down for four hours served as a wake-up call for the company. Their Web site visitors faced broken pages and outdated content while the outage lasted. They put their resources into a cloud-based redundancy solution and developed a comprehensive disaster recovery plan that involved data replication across different regions. The marketing systems at VoltAge Motors now transition without disruption to backup systems in case of localized outages, which sustains continuous customer experiences.

The evaluation of your core infrastructure and architecture ensures that your MarTech stack operates on a stable, scalable, and resilient foundation. Organizations that evaluate their platform architecture choices alongside scalability capabilities and redundancy measures can protect their systems from growth demands and operational risks. The stability and reliability of integrated platforms are essential for IT teams and marketers who depend on them to maintain consistent customer experiences. The refined architecture supports marketing operations by delivering agility together with innovation and reliability.

Step 4: Review Data Management and Accessibility

Any MarTech ecosystem depends on data to function effectively. Sophisticated platforms cannot provide meaningful insights or effective personalization without access to clean and timely data that users can retrieve. A review of data management practices requires an assessment of data collection methods alongside storage systems and governance practices before examining how data moves through the MarTech stack. This step focuses on establishing secure and efficient data flow between systems while making data accessible to necessary teams and trustworthy for marketing decision-making. Organizations aiming to compete on customer experience must prioritize their data practices because poor management leads to fragmented customer profiles, ineffective campaigns, and compliance risks.

The concept of a centralized data repository includes solutions such as a CDP, data warehouses, or data lakes.

A centralized data repository allows marketing teams to access a unified source of truth instead of multiple fragmented data sets. The target is to unify customer and campaign data into a structured and accessible format using either a CDP, a data warehouse, or a data lake. Data models and schemas establish information organization methods while synchronization choices between real-time and batch processing determine the speed of insights and actions. Real-time synchronization powers prompt personalization and customer engagement but batch processing remains adequate for data reports and trend evaluation.

VoltAge Motors learned that its customer data resided separately within its CRM system, email platform, and Web site analytics tools because it did not have a centralized repository. The scattered data created duplicate marketing campaigns and provided uneven customer experiences. VoltAge Motors chose to deploy a CDP to consolidate their customer profiles after reviewing their alternatives. The company developed real-time behavioral data synchronization, which allowed immediate personalization across their Web platform and email marketing efforts. The centralization process enabled VoltAge to eliminate duplicate communications while establishing unified customer experiences.

Data Quality and Governance

No matter how sophisticated analytics tools are, high-quality data remains essential for their effective operation. Data quality requires deduplication together with cleansing and validation processes to achieve accurate and consistent relevant information. Data stewardship governance defines policies and responsibilities such as identifying dataset owners and maintaining data hygiene while ensuring compliance standards. Strong data governance serves as protective guardrails for data integrity while ensuring that all decisions derived from data maintain trustworthiness and actionable value.

VoltAge Motors faced segmentation and targeting challenges because their customer records contained inconsistencies. The same customers showed up multiple times across various systems, which resulted in excessive communication and incorrect reporting. The company tackled its data issues by introducing deduplication and cleansing tools into its CDP. Their marketing operations team now includes a data steward who oversees data quality standards as part of their governance structure. The implementation of clear policies along with regular audits maintains customer data reliability and ensures compliance with privacy regulations.

Data Accessibility and Usability

Data delivers value when teams needing this information can access and use it. Role-based access control (RBAC) serves to protect sensitive data while giving marketers and analysts the ability to access necessary insights. Organizations need to create simple data extraction methods for analytics and business intelligence (BI) through dashboards or API access so teams can make quick, informed decisions without extensive IT support.

The marketing team at VoltAge Motors experienced delays in obtaining reports from their data warehouse because they had to depend on IT for each data request before implementing standardized data access protocols. The bottleneck created delays in both campaign optimizations and strategic decision-making processes. The company enabled marketers to directly access segmented customer lists and real-time engagement metrics with predictive scoring through role-based permissions in their CDP and self-service BI tool integration. The new system improved campaign planning speed while enabling marketers to respond more flexibly to changes in customer behavior.

A successful MarTech stack depends fundamentally on effective data management. A centralized data system maintains consistency across the organization while governance practices alongside quality control measures safeguard data reliability. Ensuring secure and efficient data access for proper teams holds equal importance. Organizations that establish effective data management will enhance their personalization capabilities while gaining sharper insights and maintaining a competitive advantage. Organizations that neglect data management will face challenges with disjointed customer experiences along with compliance issues and ineffective marketing campaigns. Step 4 delivers clean data that remains accessible and actionable for use in modern marketing.

Step 5: Examine System Integration and Interoperability

The true strength of a MarTech stack comes from its capacity to work as a cohesive system with full interoperability. The real benefit of a MarTech stack emerges when seamless data transfer between systems occurs alongside automated processes and real-time sharing of insights. Step 5 examines the performance of platform connections and their scalability along with their support for complete marketing workflow processes. When integration fails systems operate in isolation, which results in fragmented customer experiences and operational inefficiency. An advanced integration approach makes certain your technology infrastructure functions together seamlessly while growing alongside your business demands.

Integration Architecture (Direct APIs, Middleware, iPaaS)

The integration architecture encompasses the structural design that enables systems to share data and initiate business processes. Direct APIs establish direct links between different platforms that provide maximum control but need extensive maintenance because of system evolution. Middleware and integration platform as a service solutions bridge systems by providing centralized management alongside scalable integration with monitoring capabilities. An integration architecture that is well designed will minimize complexity while maximizing system flexibility and allowing system evolution without affecting the entire stack's stability.

VoltAge Motors started with direct API connections to link its CRM system with its email platform and Web analytics tool. Their expanded stack comprising a customer journey orchestration platform and a data lake made managing point-to-point integrations difficult and prone to mistakes. The adoption of an iPaaS solution allowed them to centralize integration management, which facilitated new platform onboarding and relieved their engineering team from extensive workloads. The modification led to faster data synchronization, which enabled real-time adjustments for personalized content across digital channels.

The primary platform integrations that support the customer lifecycle include CRM systems, marketing automation platforms (MAPs), analytics tools, content management systems (CMSs), and e-commerce platforms.

The assessment of these integrations requires validation of bi-directional data movement alongside system-driven actions based on shared data and maintaining uniform customer profiles across platforms. Ensuring high-quality integration between systems creates both seamless customer experiences and efficient internal processes.

The separation of CMS and CRM at VoltAge Motors resulted in both inconsistent lead information and content personalization errors. Site visitors who submitted forms typically experienced a delay in response due to the CRM system not instantly capturing Web engagement data. VoltAge combined its CMS with its CRM and marketing automation platform to ensure Web site content was dynamically updated from CRM data and started automated lead nurturing workflows immediately after form submissions. The new system decreased lead response times and enhanced Web site conversion rates.

Monitoring and Maintenance

The implementation of strong monitoring and maintenance procedures proves essential for maintaining the health of integration systems. The absence of automated logging and alerting mechanisms means integration failures remain hidden until they cause data gaps and dissatisfied customers. Continuous monitoring guarantees proper data flow, but documentation, alongside version control practices, preserves integration stability and makes troubleshooting and upgrades easier. Follow these practices to decrease risk while building trustworthiness in your marketing information.

Integration problems between VoltAge Motors' e-commerce platform and CRM system produced lost sales data and disorderly customer records. Its iPaaS provider enabled automated logging and alerting, which allowed it to receive real-time failure notifications and fix problems in minutes rather than days. Its IT team achieved better coordination with vendors during platform upgrades and reduced downtime by documenting API integrations and applying version control methods.

Effective MarTech platform integration enables organizations to deliver unified customer experiences while achieving operational efficiency. Organizations that establish a strategic integration architecture supported by strong core platform connections and ongoing monitoring capabilities can achieve system and workflow unification. Businesses such as VoltAge Motors can build flexible and resilient technology systems that are prepared for future demands through scalable integration solutions and strong governance processes. The connection between separate data systems and a unified intelligent marketing platform depends entirely on seamless integration.

Step 6: Analyze Automation and Personalization Capabilities

A MarTech stack's level of maturity depends heavily on its ability to automate operations while creating personalized user interactions on a massive scale. This step requires an assessment of your platforms' workflow automation effectiveness as well as the advancement of your personalization techniques and AI and machine learning's contribution to these functions. Through automation, organizations experience better efficiency while minimizing manual mistakes, which enables teams to dedicate their efforts to strategic and creative tasks. Personalization boosts customer engagement along with conversion rates and customer retention through the provision of relevant experiences at the right moments. Evaluating these capabilities comprehensively enables your MarTech stack to maintain operational excellence while delivering superior customer experiences.

Workflow Automation

Workflow automation enables organizations to establish repeatable procedures, which minimize human input within marketing operations, along with sales and customer service departments. Marketing operations implement automation systems to streamline lead distribution while also managing email nurture sequences and campaign workflows. Sales departments use automation tools to identify leads with strong purchase intent while service groups deploy chatbots and self-service portals to respond to fundamental customer questions. Efficient workflow automation improves operational flow while maintaining standardization and supporting growth.

VoltAge Motors discovered that its lead management process was inefficient because sales reps received leads after losing the opportunity to take timely action. The integration of its marketing automation platform with CRM allowed VoltAge Motors to score leads based on engagement data and direct them straight to the relevant sales rep. The implemented automation achieved reduced response times together with better conversion rates for inbound leads.

Personalization Framework

A personalization framework becomes effective when it utilizes real-time customer behavior and contextual data collection to drive actions. Audience segmentation uses demographics and behavior-based criteria to create groups while dynamic content utilizes real-time adaptation for personalized user experiences. Advanced personalization systems implement these tools to develop customized customer paths that enhance relationships and deliver better campaign results.

Before implementing their real-time personalization systems, customers who browsed VoltAge Motors' electric vehicle models online received generic promotional offers. The CDP enabled real-time personalization, which allowed visitors to receive promotions customized according to their browsing history, geolocation data, and established preferences. VoltAge delivered sustainability content to eco-conscious customers while offering urban dwellers promotions for city-friendly compact models. The use of personalized marketing strategies led to more Web site visitors booking test drives and generating more leads.

AI/ML Models

Predictive analytics and intelligent automation rely heavily on AI and machine learning models. The predictive scoring system identifies valuable leads while detecting potential customer churn risks and prioritizes necessary actions for sales and service teams. Recommendation engines operating in real-time deliver personalized product suggestions while using both historical and behavioral data to determine the next best actions or content for users. The combination of these functionalities enhances operational efficiency and delivers hyper-personalized customer interactions resulting in increased revenue and stronger customer loyalty.

VoltAge Motors employed AI-based lead scoring to determine which B2B fleet sales leads deserved an immediate follow-up. Through analysis of engagement metrics such as email opens and Web site visits, along with social media interactions, its machine learning model determined which leads held the highest conversion potential. Their real-time recommendation engine enabled the mobile app to provide service packages and accessory suggestions tailored to customer vehicle usage patterns. AI-driven solutions boosted sales team productivity while also growing cross-sell revenue.

Marketing scalability and superior customer experiences depend on the implementation of automation and personalization. Organizations achieve significant improvements in efficiency and effectiveness through workflow automation in combination with real-time data personalization and AI/machine learning-driven predictive insights and recommendations.

Step 7: Evaluate Governance, Compliance, and Security

The expanding amounts of personal and behavioral data processed by MarTech platforms make governance and security along with compliance essential considerations. Organizations that fail to follow data privacy laws or safeguard sensitive data face financial penalties along with damage to their reputation and diminished customer trust. Organizations verify their MarTech stack's capability to meet regulatory standards while maintaining user access controls and customer data security during Step 7. Marketing leaders should partner with IT, legal, and security departments to implement strong policies and technical controls that safeguard data while showing accountability.

Privacy and Data Protection

The introduction of regulations such as the GDPR and CCPA has established higher standards for data management practices in organizations regarding customer data collection, storage, and usage. Marketing platforms should provide organizations with transparent consent management capabilities alongside user-friendly tools that allow customers to exercise their data rights. Businesses require detailed auditing systems to document compliance activities and generate proof for regulatory investigations or audits.

VoltAge Motors encountered compliance challenges with GDPR and CCPA regulations during its expansion into European and California markets through its marketing channels. The company deployed a consent management platform (CMP) together with their CDP to unify consent tracking throughout Web site forms, mobile app interactions, and email subscriptions. The auditing functionality of the system automatically recorded consent events and customer data requests to streamline compliance reporting and lower legal risks.

Identity and Access Management

IAM frameworks control which users gain access to particular systems and data sources. Single sign-on (SSO) and multi-factor authentication (MFA) mechanisms secure login processes while mitigating the risks of weak password usage through password reuse. RBAC establishes user permissions by allowing employees access only to resources essential for their roles, which reduces data breach risks and internal abuse.

The marketing team at VoltAge Motors achieved greater customer data access through new integrations between their CDP and personalization platforms. VoltAge Motors implemented SSO and MFA across all marketing tools to protect sensitive data by requiring user authentication through their secure identity provider. Access was limited through RBAC policies, which designated that only data analysts could export customer records while campaign managers were granted restricted access to segmentation tools and reporting dashboards.

Security Protocols and Certifications

Effective security protocols must protect data at every stage of its existence. The practice of encryption for data in transit and storage protects sensitive information from unauthorized access by making it unreadable. The use of continuous penetration testing and vulnerability assessments allows organizations to detect and repair security weaknesses before attackers have a chance to exploit them. SOC 2, ISO 27001, and PCI DSS certifications deliver external validation for how platforms manage their security methods.

VoltAge Motors set strict security standards requiring MarTech vendors to follow encryption protocols for data stored on their CDP and marketing automation platforms. The IT security team worked with them to run annual penetration tests, which helped to identify and fix vulnerabilities. VoltAge required vendors processing personal data to hold an SOC 2 certification under their procurement terms to ensure stakeholders trust their security measures.

Modern marketing operations require unwavering commitment to governance standards as well as compliance and security measures. With the advancement of data privacy laws and growing cyber threats, organizations need to appraise if their MarTech stack implements privacy-by-design principles along with secure access management and strict security protocols. Through centralized consent management together with RBAC and proactive security testing VoltAge Motors demonstrates a complete strategy to protect customer data and fulfill global compliance standards. Such practices establish customer confidence, which creates a durable base for enduring marketing achievements.

Step 8: Assess Vendor and Platform Roadmaps

Assessing your MarTech vendors' current abilities must be complemented by examining their future development paths. The current technology platform may fulfill your immediate requirements, but if it fails to match your company's growth plans through its innovation pipeline and product vision, you could exceed its capabilities sooner than expected. Step 8 examines vendor product roadmaps together with their support infrastructure capabilities and community ecosystem depth. Your MarTech stack's scalability and adaptability and its competitive advantage depend on these factors, which affect its performance in an ever-changing digital environment.

Product Vision and Update Cycles

The product vision of a vendor demonstrates their dedication to innovation and staying relevant in the market. The platform stays ahead by regularly introducing new features that meet user requirements and new technological trends, while compatibility with your business growth strategy makes it a scalable and adaptable investment choice. Suppliers who demonstrate a transparent path for AI integration or omnichannel personalization together with data privacy solutions stand to better fulfill your extended strategic goals.

During its MarTech expansion, VoltAge Motors conducted an evaluation of two digital experience platforms (DXPs). The vendor displayed a product roadmap that revealed quarterly updates to feature sets including AI-powered personalization and real-time analytics that matched VoltAge's objectives for dynamic customer-focused experiences. The second vendor demonstrated irregular update cycles and minimal innovation during their evaluation, which questioned their ability to meet long-term requirements. VoltAge chose the first vendor due to how its product vision backed its digital transformation efforts and expansion into new market territories.

Vendor Support and SLAs

A MarTech platform's success depends heavily on technical support, which becomes crucial during important campaigns or system failures. Technology vendors need to provide several support options, including live chat functions and dedicated account managers along with round-the-clock helplines and well-defined escalation procedures. Service-level agreements (SLAs) establish standards for response times while setting resolution benchmarks and ensuring uptime expectations to maintain reliable service delivery and accountability.

The legacy email system at VoltAge Motors caused delayed issue resolution, which disrupted their campaign schedules. The evaluation process for their new platform led them to choose vendors with strong SLAs that promised a 99.9% uptime rate and dedicated support teams for enterprise-level clients. The reliable support VoltAge received enabled them to scale campaigns securely while eliminating concerns about troubleshooting delays or extended system downtime.

Community and Ecosystem (Open Source vs. Proprietary)

A dynamic vendor network that features both a third-party marketplace and an engaged user base enhances both the capabilities and worth of a platform. The availability of pre-built plugins and integrations along with developer assistance substantially reduces the time required to launch new initiatives. The open-source model allows customizable solutions while proprietary platforms deliver validated enterprise solutions with guaranteed support.

VoltAge Motors opted for an open-source CDP that features both an engaging developer community and a comprehensive marketplace available for pre-built connectors. The platform enabled them to rapidly integrate additional systems including CRM and personalization engines while accessing extensions built by the community. By utilizing an open-source model, VoltAge's internal developers were able to modify features as needed while remaining independent of vendor release schedules, which resulted in more agile data strategy execution.

Evaluating vendor and platform roadmaps maintains your MarTech stack's future-readiness while keeping it in line with business objectives. Through regular innovation and responsive support along with building a strong ecosystem, vendors provide lasting value combined with adaptability. VoltAge Motors demonstrates risk mitigation and competitive scalability in MarTech investments through its selection of platforms that feature distinct product visions and strong SLAs alongside active communities.

Step 9: Examine Organizational Readiness and Skills Alignment

The performance of a sophisticated MarTech stack depends entirely on the effectiveness of the teams who operate and utilize it. Step 9 evaluates whether your organization is prepared to both implement and sustain technology investments to realize their full value. The assessment process examines internal capabilities while evaluating cross-functional collaboration alongside change management readiness. New platform adoption demands technical expertise together with organizational alignment and requires a training and onboarding plan.

The organization's IT, marketing operations, data science, and security teams make up its core team capabilities.

Organizations must assess their team's technical skills and strategic understanding before they decide to use new MarTech platforms. Data security management and integration responsibilities fall under IT teams; marketing operations professionals oversee workflow management; data science teams handle analytics and modeling functions; and security teams maintain compliance standards. Companies should consider hiring new personnel or engaging consultants and managed service providers when skill and knowledge gaps exist. Staff should have ongoing learning opportunities to learn about new technologies and best practices.

The DXP evaluation revealed VoltAge Motors' internal teams had insufficient API integration skills and faced challenges managing data governance at a large scale. The organization employed an external consultancy to manage platform integration and data pipeline configuration while they enrolled their MOPs team in advanced customer journey orchestration and personalization best practices training. The dual approach offered an immediate boost in capabilities while simultaneously building the organization's long-term knowledge base.

Cross-Functional Collaboration

For MarTech stack implementation and optimization teams must work together because each department contributes specialized priorities and expertise. Marketing develops campaign strategies and engages customers while IT manages system integration and security measures. The creation of joint targets and management systems such as steering committees or cross-functional teams enables departments to overcome functional barriers and synchronize their MarTech strategy with broad company objectives.

VoltAge Motors prioritized cross-functional collaboration during its expansion into markets with more stringent privacy regulations. Legal teams focused on compliance adherence while IT departments handled security protocols and the marketing department concentrated on delivering personalized customer experiences. A management council composed of representatives from each department was created to supervise the deployment of their CDP. Through consistent meetings and common KPIs, the organization made platform choices that maintained a balance between marketing goals and security compliance standards.

Change Management and User Adoption

The introduction of new MarTech platforms demands extensive change management when they impact existing workflows or create new responsibilities. Plans for stakeholder engagement need to counteract resistance through clear benefit communication and leadership support. Structured onboarding processes and detailed documentation together with accessible training materials help speed up user adoption and enhance platform usage.

The introduction of a new marketing automation platform at VoltAge Motors encountered numerous adoption difficulties. The team failed to account for the steep learning curve faced by non-technical users, which resulted in slow adoption rates and inconsistent utilization of automation features. They developed an official onboarding program featuring live training sessions along with platform certifications and internal super user support to address these challenges. The company created custom documentation in collaboration with the vendor to match their specific workflows, which resulted in increased platform usage and user confidence.

The success of any MarTech evaluation and implementation depends heavily on organizational readiness. The highest quality technology will fail to deliver expected results without expert teams working together across functions and an effective change management strategy. VoltAge Motors demonstrated that addressing capability gaps and fostering cross-departmental collaboration while managing user adoption emphasizes the crucial need to align organizational processes and people with technology investments.

Step 10: Quantify Performance and ROI

Organizations need to measure their MarTech investments' value during the final evaluation stage of their MarTech stack. Organizations should track technical performance indicators while analyzing costs to measure business outcomes that support strategic objectives. Accurate investment justification and future forecasting depend on performance measurement through clear metrics and disciplined evaluations to enable data-driven stack optimization decisions. The main goals include validating operational efficiency while proving business results and maintaining ongoing investment returns.

Technical KPIs

Through technical KPIs, organizations can assess whether their MarTech platforms fulfill operational performance requirements. The measurement of uptime and latency metrics safeguards the availability and responsiveness of platforms. API error rates and data synchronization times help evaluate system health and integration performance. Assessing automation coverage reveals the extent of manual process conversion to automation, thereby offering clear insights into operational efficiency gains.

VoltAge Motors deployed an advanced customer journey orchestration platform that automates lead nurturing across various channels. The company observed that manual data uploads and campaign setups fell by 40% after implementing their new system for three months. The API error rates decreased from 5% to less than 1% and data synchronization between the CDP and marketing automation platform progressed from 24 hours to near real time. Performance data demonstrated that the new system operated as expected and warranted further investment to expand automation initiatives into additional business units.

Cost Analysis

Cost analysis delivers an all-encompassing picture of MarTech investment that goes past the initial licensing fees. The Total Cost of Ownership (TCO) consists of direct costs, including platform subscriptions together with indirect expenses such as maintenance and staff training. To determine ROI businesses must compare these expenses with quantifiable benefits, including better operational efficiency, increased lead volume, higher conversion rates, and time savings for teams to undertake more valuable tasks.

During their MarTech stack expansion VoltAge Motors executed a TCO assessment to evaluate their new digital asset management (DAM) platform. The analysis showed that the implementation of the new system resulted in a 25% reduction in creative production time and a 15% decrease in duplicate marketing activities despite hefty licensing fees and implementation expenses. The improved efficiency resulted in lower costs and quicker campaign starts that demonstrated a measurable return on investment after one year.

Business Impact Metrics

The primary benefit of MarTech investments emerges from their potential to produce significant business results. Measuring essential business impact metrics, including lead conversion rates and customer retention, helps to explain whether technology investments propel organizational objectives. The effectiveness of a MarTech stack can be measured through a shorter time to market and enhanced customer lifecycle engagement.

VoltAge Motors experienced a 20% boost in lead conversion from their Web site following the integration of their DXP and personalization engine. Improved workflow automation led to a 35% reduction in their average campaign launch time. The measured metrics displayed both enhanced operational effectiveness alongside increased customer interaction and prospective revenue expansion.

Performance measurement and ROI analysis verify that MarTech investments remain consistent with strategic objectives while demonstrating tangible value. VoltAge Motors utilizes evaluations of technical KPIs combined with cost analysis and business impact measurement to make educated choices about MarTech stack expansion, optimization, or replacement. Organizations that implement systematic performance tracking and cost analysis achieve accountability while supporting sustainable innovation and growth.

Step 11: Identify Gaps and Prioritize Next Steps

After completing the evaluation the next essential step involves transforming the data into clear actionable insights. The task requires pinpointing the differences between present MarTech capabilities and future operational goals. This stage determines the organization's position on the MarTech maturity curve while recommending improvement areas and establishing a strategic roadmap to prioritize subsequent actions. The phase turns evaluation into actionable plans by creating strategies that combine technical requirements with business goals, budget constraints, and resource availability.

Map Findings to Maturity Model Stages

Foundational → Integrated → Advanced → Innovative

Utilizing the MarTech maturity model for mapping findings enables organizations to understand their present capabilities while establishing achievable goals for progress. Determining whether an organization operates at a foundational, integrated, advanced, or innovative level enables better tracking of progress and enhances stakeholder communication. The mapping exercise reveals lagging areas while providing clear guidance on where to direct future improvement efforts.

VoltAge Motors realized in their assessment that their CRM and marketing automation platforms were integrated, which showed they were at the Integrated stage, yet their customer data stayed fragmented while personalization only reached basic segmentation levels. The evaluation results positioned their stack at Stage 2 of the maturity model while showing that optimizing data unification and personalization functions could lead to progression to the Advanced stage.

Recommend Remediation

After discovering gaps organizations must suggest specific remediation actions. Organizations can strengthen their systems by updating platform integrations and improving data governance alongside addressing security vulnerabilities. The recommendations provided must align with past evaluation data and connect clearly to the organization's maturity level and its business goals.

VoltAge Motors found a critical integration gap between its CDP and personalization engine. The organization decided to make real-time API integration upgrades its top priority while simultaneously establishing uniform data governance procedures throughout all regional operations. It determined that stronger identity and access management tools were necessary for security protocol enhancement to address GDPR compliance gaps.

Create a Roadmap

The next step involves building a structured phased plan that identifies initiatives for short-term execution as well as medium-term projects leading toward long-term goals. The roadmap requires prioritization of actions based on both their impact and feasibility while ensuring alignment with budgetary limits and staff availability as well as organizational preparedness. The document directs MarTech development by achieving immediate benefits while setting strategic objectives for the future.

VoltAge Motors' short-term roadmap featured fast-impact goals such as tighter CRM and CDP integration and a staff training program for marketing operations. The medium-term objectives included the development of advanced customer journey orchestration capabilities together with the deployment of AI-based personalization techniques. The long-term plan aimed to move towards a composable DXP framework that allows enhanced scalability and flexibility in delivering customer experiences.

The process of pinpointing missing elements while sequencing immediate actions converts MarTech assessment from theoretical planning to actionable steps. The maturity model framework helps reveal progress through mapping findings and targeted recommendations to solve critical issues. Organizations achieve long-term success through continuous improvement when their MarTech investments follow well designed roadmaps that connect initiatives to both business objectives and available resources. VoltAge Motors and similar organizations can follow this structured approach to progress systematically toward an advanced integrated MarTech ecosystem.

Step 12: Evaluation Summary

The ultimate step of a complete MarTech stack evaluation requires consolidating all findings into a clear and actionable report. The summary needs to transform intricate insights into straightforward conclusions that operational teams and executive stakeholders can easily understand. The evaluation delivers a comprehensive overview of the MarTech ecosystem status while pinpointing competitive edges, improving necessities, and detailing steps for future action. The evaluation summary needs to set a schedule for regular reviews to maintain alignment between the technology stack and changing business requirements along with market trends.

Key Takeaways

This section provides a brief summary of the current state of the MarTech infrastructure. The assessment determines whether existing platforms and processes, along with their integrations, provide adequate support for achieving strategic marketing goals. The assessment must demonstrate how the organization's technology stack stands up against established industry benchmarks and competitive standards.

The evaluation completed by VoltAge Motors showed that its MarTech infrastructure was stable and adequately scalable for current needs but found that it was missing advanced personalization and customer data unification capabilities present in mature organizations. While its technology stack served current operational needs effectively, it needed targeted investment to match industry leaders who provide seamless omnichannel experiences.

Executive Summary

The executive summary transforms the full evaluation into a concise report designed for leadership audiences. The summary identifies essential findings by pointing out strengths and weaknesses and suggests strategic actions while avoiding technical details. This document should not exceed two pages and should guide executive decision-making about future actions and investment choices.

In VoltAge Motors' case, the executive summary outlined three primary challenges: the company faced challenges with disjointed customer data systems alongside restricted personalization capabilities across different channels and outdated legacy systems that reduced organizational agility. The strategic guidance emphasized speeding up their transition to a composable DXP while improving customer data management and making investments in AI-powered personalization tools. The leadership team consulted the document to organize technology investments for the upcoming budget cycle.

Ongoing Review Cycle

The changing marketing environment requires continuous assessments and one-time evaluations cannot meet these needs. A structured review cycle implementation by organizations enables continuous enhancements in their MarTech ecosystems. The process requires establishing review intervals such as yearly or semi-annual evaluations while developing performance metrics and creating governance structures for platform performance monitoring and data quality control.

VoltAge Motors implemented yearly reviews along with quarterly assessments based on performance metrics such as customer interaction rates as well as data synchronization dependability and campaign launch speed. VoltAge Motors formed a cross-functional MarTech governance council to manage future platform choices as well as integration strategies and vendor partnerships while maintaining alignment with long-term business objectives.

The evaluation summary transforms the assessment procedure into distinct takeaways and practical insights. The process maintains executive team cohesion and enables data-driven decisions while creating a structure for ongoing MarTech enhancements. Organizations such as VoltAge Motors can maintain their MarTech systems' health while improving agility and effectiveness through high-level strategic oversight and regular review processes, which set them up for lasting competitive success.

OTHER CONSIDERATIONS

Identify Overarching Gaps

The first thing you'll want to do is similar to what we did above with the hypothetical score and situation: look for broad, overarching gaps or areas for improvement. While they won't prescribe any exact course of action, they will point you in the right direction.

You may have several areas of challenge, in which case you should document all of them but highlight only a few areas to focus on first. After all, even the largest of organizations can't focus on *everything* at once!

Identify Specific Gaps

Next, look more granularly at where problem areas or shortcomings in your assessment and experience point to a need to make changes, additions, or rethinking areas of your MarTech stack altogether. Here are some dimensions you should explore.

Performance Gaps

Then, you need to analyze how well each tool or platform contributes to your marketing objectives and KPIs. You need to collect and compare data from different sources, such as the Web, social media, email, and CRM analytics. You also need to evaluate how each tool or platform affects your customer experience and satisfaction across different touchpoints and channels.

Feature Gaps

Related to performance, but potentially including other opportunities and challenges, look for features and functionalities that are either lacking or missing altogether.

This could point to a platform that needs to be acquired and integrated from scratch or a serious shortcoming in existing components of the MarTech stack.

Data or Connectivity Gaps

Look for areas where "we could do x if we only had access to the data from…" could be solved or areas where one set of data isn't syncing with another.

Often, your customer satisfaction data or consistent complaints can point to areas where customers *feel* this data disconnect.

Additionally, your marketing teams may have a wish list, particularly regarding attribution and reporting across channels.

Competitive Gaps

It is important to understand how your organization compares to the norm, the leaders, the trends driving customer expectations, and the category leaders in your industry.

For instance, keep in mind the following items:

- Current best practices within your industry and with the types of MarTech platforms you've currently adopted should be taken into consideration. Additionally, take into account any industry or regional regulatory restrictions or requirements and ensure you are adhering to them (and that any new plans keep them in mind).

- Ensure you understand the trends within your own industry, but don't stop there. Remember that your customers aren't just customers of a single industry or type of product, so their customer experience expectations are shaped by all their experiences.

- Understand what your competitors are doing and recognize that your future competitors may be different from your current ones. Think about how companies such as Amazon (moving into healthcare) and Apple (increasingly moving into financial services) may move laterally and disrupt your industry, making your competitive set increasingly complex.

Human Resource Gaps

Do you have the right platforms to fulfill your marketing and customer experience needs, but you don't have the right talent to use them? In this case, your team either needs augmentation or you need to shift team members around to utilize your MarTech stack better.

Financial Resource Gaps

It is possible to buy too much and invest in a platform or system that you use only a small portion of and that simply costs too much for the benefits it provides.

Additionally, it is possible that not enough budget is allocated to infrastructure, and with that and the demands on the marketing team, more financial resources are needed.

These gaps cover a lot of ground, but it's possible that you may find others. Starting with these gaps, begin to uncover the areas of greatest need. They will inform some strategic planning and decisions you will be making soon in the process.

CONCLUSION

Once you have identified the gaps, take them along with the visualization, and now you have a report of your findings. This will show your approach and analysis and provide some next steps.

CHAPTER 27

EVALUATION OF INDIVIDUAL MARTECH PLATFORMS

The average enterprise has 1,295 cloud services[1], meaning there are many individual components within your MarTech stack alone. Thus, evaluating every platform will not be feasible. Instead, prioritize the areas of weakness or where there is the greatest opportunity.

Now that the big picture has been reviewed in terms of both setting goals and evaluating maturity (Chapter 25), as well as mapping and evaluating the entire infrastructure (Chapter 26), it is time to look at how the individual pieces of a MarTech infrastructure contribute to the whole, and to achieving the organizational goals.

Another reason for doing this is that if a major new component needs to be considered, it is often approached from the big picture standpoint first, and then individual platforms would be evaluated and integrated.

One caveat before we get too far along, however: because there are myriad types of MarTech platforms with just as many purposes, there will inevitably be specific factors that apply to one software tool that may not apply to another. Thus, a guide such as this meant to talk about *all* of the potential platforms will never be able to get specific enough. That is, in fact, why we are writing guides on individual types of platforms (e.g., customer data platforms (CDPs) and customer journey orchestration (CJO) are the first two, with many more on the way).

That said, the ideas in the rest of this chapter should apply to just about anything, including the following:

- CDPs and customer relationship managers (CRMs)
- Content management systems (CMSs) and digital experience platforms (DXPs)
- Analytics and reporting tools
- Email marketing and marketing automation
- CJO
- Conversational marketing
- Enterprise search

WHEN SHOULD YOU EVALUATE A SINGLE PLATFORM IN YOUR MARTECH STACK?

We start with the most fundamental question of all because let's remember that it takes time and resources to do these evaluations. So, there need to be some good reasons to do so. Here are a few that might spur some additional ideas:

- Your MarTech philosophy dictates a change in approach (e.g., from all-in-one to composable).
- There are compliance or data privacy issues with your platform (e.g., you have a new need for GDPR compliance, or a region you work in has enacted new privacy regulations).
- The platform's utilization is low, and its cost is high.
- Another platform you are already using, or planning to incorporate, has the same or similar functionality.
- The performance of your marketing efforts related to this platform is failing to meet goals.
- Finally, when your MarTech roadmap dictates you need a new approach—we'll get to this in the next section of the book.

AN EVALUATION FRAMEWORK

Similar to the framework introduced for evaluating an entire MarTech infrastructure, it is important to have a consistent method of evaluating the individual platforms that make up that ecosystem. Refer to Figure 27.1, which shows the categories and subcategories in our framework.

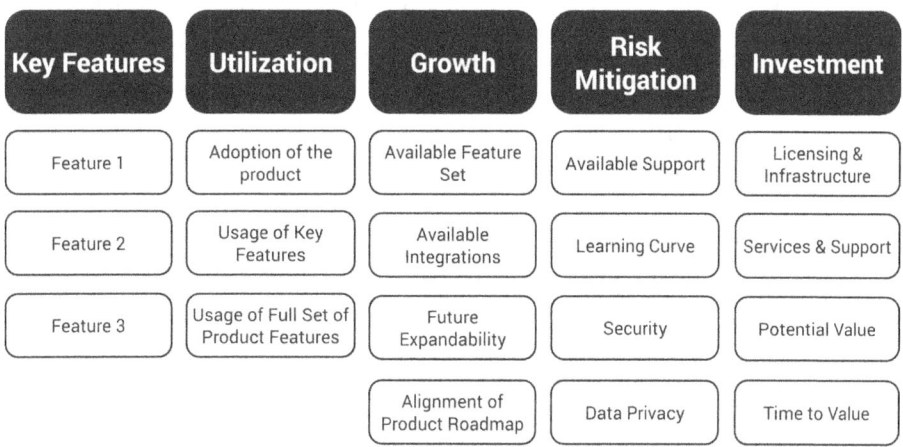

FIGURE 27.1 Evaluating an individual MarTech platform

This is where your evaluation methodology comes in. In addition to the categories and definitions below, we are including an Excel spreadsheet that will allow you to rate and evaluate a set of platforms based on these criteria in the accompanying digital resources for this book. Now, let's look at each of these categories and the definitions of each of them.

Key Features

Determining key features can sometimes be difficult, but often there are a few "must have" areas that the product must deliver on. In fact, they are usually why you are buying it in the first place or why it is being recommended as part of your strategic roadmap.

I don't recommend more than 3-4 key features at most because it can be difficult to do the due diligence necessary to fully compare so many during this process. Make sure these key features include the primary reasons you need the product, however.

Utilization

Remember, if you are evaluating a brand new product, instead of making this about the *current* utilization of the product, you can make it about *anticipated* utilization. Evaluate the following aspects of utilization:

- *Adoption of the product,* or how many members of your team (or teams within the marketing function in your enterprise) are utilizing it
- *Usage of key features,* or how many teams will utilize the identified key features, and how often those key features will be used
- *Usage of the full set of product features* or the potential for your teams to benefit from the entire feature set

Growth

Evaluate the following aspects of growth:

- *The available feature set,* or the quality of the "out-of-the-box" features and their applicability to your teams' needs, even if they are not immediately applicable
- *Available integrations,* or the platforms' ability to connect to other currently-used or potential platforms, as well as how easy it is to connect from a complexity and cost standpoint
- *Future expandability,* or the overall ability of the platform to scale to meet the future strategic needs of your organization
- *Alignment of product roadmap,* or how well the strategic focus of the platform aligns with your own industry and growth plans (e.g., if the platform's focus is on financial services companies and your brand sells athletic apparel, there may not be a good match here)

Risk Mitigation

Evaluate the following aspects of risk mitigation:

- *Available support,* or how extensive the help will be if you and your team have questions, need help with implementation, or other customer support

- *Learning curve,* or how much training and onboarding time will be needed for your team(s) to get up to speed and productive with the platform
- *Security,* or the level of cyber and other security the platform incorporates
- *Data privacy,* which includes the type(s) of regulation support, whether that is regional (GDPR or CCPA), industry-specific (HIPAA for healthcare), or other more general consumer or corporate data privacy concerns

Investment

Evaluate the following aspects of investment:

- *Licensing and infrastructure,* or the costs to access the product as well as to support it on internal or third-party clouds or other infrastructure
- *Services,* or the internal or external costs to implement and maintain the platform
- *The potential value* that the platform has to deliver to the organization
- *Time to value,* or how long it will take to deliver that potential value to the organization

PRIORITIZE THE CATEGORIES

While the above categories are important to you and your organization, they can't all be the most important. That's why, in the Excel sheet we're providing as a template to help you do evaluations, we have a "priority" ranking for each of the above main categories (Growth, Risk Mitigation, Investment, etc.) so that you can tailor it to your organization's exact needs.

By prioritization and creating weighted averages, you can get a better picture of the platform that meets your requirements best.

BENEFITS OF THE EVALUATION PROCESS

Ultimately, if you use the spreadsheet we're providing or create your own ranking/ordering methods, you will gain valuable insights and have a ranked-order list of potential platforms according to the evaluation criteria. This should provide some strong guidance on the approach you should take, though there have been cases where a platform that ranked number one on a list like this actually ended up losing out to the second-place platform. In that case, there was an issue with the agreement once the deal went to procurement, but there may also be other reasons.

The evaluation should serve as an objective source of information that can then be weighed against any other number of factors that either can't fit in the spreadsheet or—like the legal example—are mitigating factors outside the scope of an evaluation.

That said, there are several benefits of evaluating in this manner:

- You have an objective view that you can present to stakeholders that represents that due diligence performed.
- You have a record of your evaluation that can be referred to later if questions pop up or you need to answer questions like "Why didn't we choose platform x originally?".
- You have established criteria to re-evaluate the same platform when the time comes or evaluate new platforms as your needs arise.

VENDOR EVALUATION

While it's not the subject of this guide, another important consideration—sometimes as important as the platforms themselves—is *how* you will implement a new platform or make the necessary upgrades to an existing one and *who* will do it. In other words, is this an in-house job, or will you need a systems integrator to help?

Evaluating vendors is not exactly a straightforward affair either. Still, one rule of thumb is to look at their direct experience with the platform you're considering and talk with the platform vendor as well to get some recommendations. I am pretty sure we are going to write a guide about this topic in the coming months, so stay tuned!

WHEN DO YOU REPLACE A PLATFORM?

We've talked about strategy, evaluating our current MarTech stack, and evaluating individual platforms, but how do you act based on your evaluation, and when is the right time to do so? Let's explore that in this chapter.

From Big Picture to Small Picture

First, we have the context. If you have determined that there are large-scale changes needed in your MarTech stack, it's not time to think about the individual pieces but rather the big picture. If that's the case, you should start (or continue) building your roadmap to better determine your high-priority areas and focus on those platforms first.

If, however, you completed your roadmap recently, it is already up to date, or there is a particular issue that needs to be addressed immediately, you're in the right place. It is time to make a change to a specific platform in your infrastructure.

As an example, I worked with a company in the healthcare space that had a rather pressing issue come up related to patient privacy requirements (as many in that industry did around the same time due to some regulation changes), and for this reason, we needed to pivot our roadmap efforts to this area immediately. It didn't mean we abandoned our strategic plans, but it did mean that there was suddenly a priority number one that couldn't wait for other dependencies to be in place.

Of course, what you *do* about either your internal evaluations or external mitigating factors can vary widely.

WHAT TO DO WITH WEAK LINKS IN YOUR MARTECH CHAIN

So, what do you do with your MarTech stack evaluation in regard to an individual platform? Ultimately, you will need to consider your current challenges with an existing platform, opportunities to improve, the potential to consolidate functionality or costs, and many others.

There are several possibilities here, so let's discuss them all in a bit more detail:

- *Replace*, which is a very common option to switch one platform out for another when the features or functionality is necessary, but the platform is not the best fit for the task.
- *Add*, evaluate, procure, and implement a net new platform into your MarTech infrastructure. This may happen when a new technology is added to your MarTech solutions, such as a CDP, or augmenting your customer interactions with conversational AI.

- *Improve* or augment an existing platform with new and needed functionality. This may be an upgrade to a tool currently in use or an integration to expand capabilities through APIs, composability, and other means.
- *Consolidate* or deprecate some features or products that may be redundant within your stack in order to streamline workflows or save time and money.
- *Remove,* which means that the features and functionality are either no longer needed, are accounted for somewhere else (similar to consolidation), or that cuts need to be made and this tool or platform is not considered vital.

The approach you take will depend on your strategic needs, your resources available, and your current and pending infrastructure. It is advised that you to keep all options open, however, as sometimes big thinking from a strategic standpoint means rethinking individual components. What may *seem* vital as an independent platform may be causing bottlenecks in your marketing work and execution.

PROCURING A NEW PLATFORM

If you have read one of our other Agile Brand Guides on individual platform types, such as CDPs or CJO, you've most likely seen our process for buying a new platform outlined.

Because the focus of this guide is evaluation and not buying and implementing, we won't go into much detail here except to show a visual of the process.

Buying a MarTech Platform

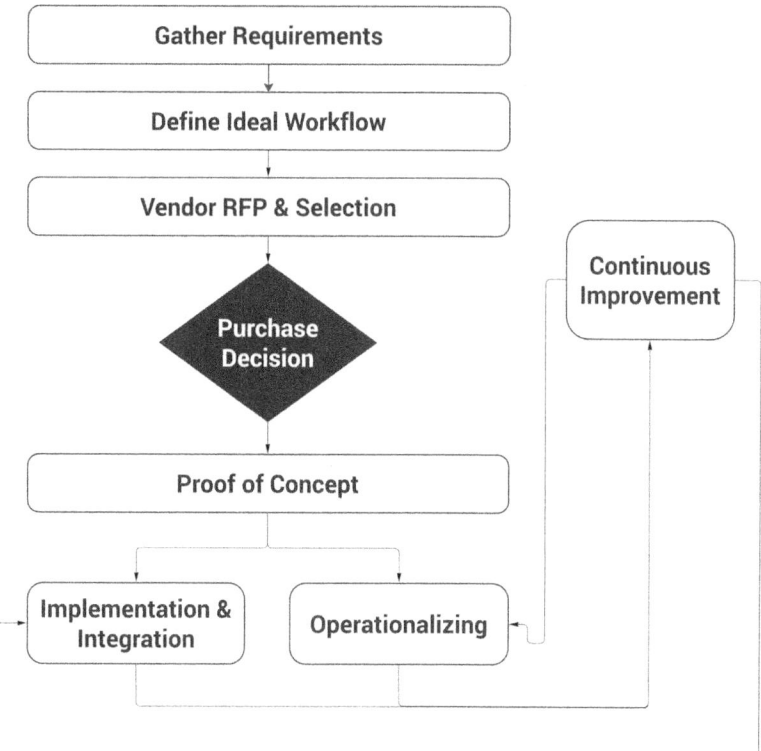

FIGURE 27.2 Buying a MarTech platform

If you're interested in learning more about the process of evaluating and buying a new platform, it is recommended that you check out one of our other guides that focuses on the platform type you're interested in.

Now that we've looked at both the MarTech infrastructure as a whole as well as individual platforms and how to determine when they need to be replaced, the next section of our book is going to explore how to create a roadmap that takes your evaluations and prioritizes and lines them up on a timeline.

VOLTAGE: EVALUATING ITS WEB SITE CMS

VoltAge Motors experienced a pivotal moment in its growth within the electric vehicle market while it extended its product lineup and expanded its customer reach globally. Their innovative and stylish vehicles attracted attention on the streets yet their online presence remained undeveloped. The marketing and IT teams at the company realized that their current CMS was inadequate because it had been established during their startup phase. The original CMS, which served as a quick solution for basic Web content management, failed to meet the brand's requirements for multichannel personalization and real-time customer engagement alongside seamless user experiences.

The catalyst for the evaluation was multifaceted. As VoltAge Motors expanded into new markets, it experienced growing digital requirements that demanded language localization, as well as region-specific promotions and personalized content experiences for both leads and current customers. The existing CMS system couldn't connect effectively with their marketing automation tools, CRM system, and CDP. The CMS system could not provide dynamic personalized interactions that adapted to each user's behavior and preferences when needed to support VoltAge's evolving marketing strategy.

The Evaluation Process

The VoltAge Motors leadership team determined through their MarTech roadmap assessment that the Web site stands out as a vital customer interaction point, while their existing CMS limitations emerged as a significant bottleneck. Using the evaluation framework outlined in their MarTech playbook, the team initiated a platform assessment process focusing on key areas. The platform evaluation process at VoltAge Motors concentrated on five main areas: feature evaluation, utilization assessment, growth potential analysis, risk mitigation strategies, and investment considerations.

Key Features

VoltAge Motors identified robust personalization capabilities as an essential feature requirement. The team needed a platform that could automatically modify Web content, product recommendations, and calls to action using real-time data from their CDP and behavioral analytics systems. A system capable of headless content delivery was essential to VoltAge Motors for generating seamless user experiences across their Web platform, mobile applications, and upcoming in-car infotainment systems.

Utilization and Growth Potential

The digital marketing team used the current CMS for simple content updates, but VoltAge Motors identified the potential to broaden its application to customer support and product teams. The DXP candidates evaluated featured scalable capabilities that could enable future initiatives including owner portal integration and brand community user-generated content management.

Risk Mitigation and Investment Considerations

Their assessment heavily relied on evaluating security standards, privacy compliance requirements, and integration simplicity. The DXP had to adhere to GDPR and CCPA requirements while providing built-in consent management features. VoltAge's procurement team completed a comprehensive total cost of ownership (TCO) analysis from an investment perspective by comparing licensing fees with infrastructure costs and implementation services to determine the expected time to value.

The Decision: Moving to a DXP

VoltAge Motors concluded its evaluation process by choosing Adobe Experience Manager (AEM) as its DXP. The company chose AEM because of its sophisticated personalization engine alongside its adaptable content delivery architecture and its ability to integrate flawlessly with existing Adobe Marketing Cloud tools. The modular design of the platform permitted VoltAge Motors to develop and control reusable content components, which accelerated campaign rollout and maintained brand uniformity throughout different markets.

The transition to AEM gave VoltAge Motors the ability to deliver hyper-personalized experiences for its customers. Web site visitors at VoltAge Motors now get personalized vehicle suggestions that use their location data and history of interactions with VoltAge email campaigns. Real-time analytics from the platform provided continuous insights into content performance, which enabled the marketing team to make ongoing optimizations to experiences.

Results and Next Steps

The move to a DXP platform by VoltAge Motors has generated substantial advantages. The brand has achieved faster page load times along with better engagement measures and higher conversion rates on important landing pages. The marketing team now wields enhanced power to orchestrate customer journeys, which results in delivering personalized experiences consistently at scale.

VoltAge intends to use AEM's headless CMS to broaden its omnichannel approach by implementing deeper connections with its mobile application and vehicle interfaces. The company uses the successful evaluation and migration to a DXP as a blueprint for managing other aspects of its MarTech stack.

CONCLUSION

There will be many specifics to work through when evaluating an individual platform, though having a method to do this in an objective and consistent manner will help to avoid potential gaps or other issues that may get in the way of achieving the best possible outcomes for the organization and the end customers who rely on the MarTech infrastructure.

NOTE

1. Chiefmartech. "The Average Enterprise Uses 1,295 Cloud Services." Retrieved April 26, 2023 from: https://chiefmartec.com/2020/02/average-enterprise-uses-1295-cloud-services/

CHAPTER 28

BUILDING A MARTECH ROADMAP

To create long-term success for your MarTech infrastructure and the critical initiatives and teams it supports, as well as to understand what that plan requires to be successful, you need an actionable and realistic plan to move your organization and its marketing efforts in the right direction. A good roadmap provides an incremental plan to transform your areas of weakness into strength and to keep your brand competitive and meeting customer expectations.

A well-done roadmap is also something teams across your organization can collaborate on and use as a reference for related workstreams and complementary initiatives. In this section of the guide, we will discuss the roadmap and share some ideas on how to approach creating it.

Your MarTech roadmap will translate your goals, evaluations, and recommended approaches and map them out over time so that you can create a feasible and realistic plan that is ambitious enough to keep you ahead of your competition and in line with your customers' rising expectations. Therefore, it shouldn't be a simple diagram that's for reference only; instead, it needs to take into account all of the details and nuances that will make you and your teams successful.

YOUR MARTECH ROADMAP

Documenting and having a shared understanding of the intricacies of your MarTech infrastructure is essential to ensure it achieves your desired outcomes. Additionally, it is important to understand that you and your teams are on a *journey* together to create the ultimate MarTech stack and that, because of that, things may need to change and shift priorities over time.

Because of this, a roadmap for your MarTech infrastructure is extremely valuable and will help you ensure that all elements are connected and working together effectively and that the timing of phased rollouts is considered, which allows greater efficiency, increased ROI from campaigns, and reduced risk when deploying new initiatives. In this chapter, we will talk about how creating a strategic roadmap for your MarTech stack can benefit your organization and some steps you can take to build one.

WHY A STRATEGIC ROADMAP IS NECESSARY FOR YOUR MARTECH INFRASTRUCTURE

With so many platform options available and potential areas of need, it can be difficult to determine where to make the best and most strategic investments in your MarTech stack. Because of this, creating a comprehensive roadmap for your MarTech infrastructure is key.

With a strategic roadmap, you can better align your technology with business objectives and ensure that resources are being efficiently allocated. Furthermore, an effective roadmap is instrumental in identifying short-term tactics and long-term goals to target the right audiences with the right messages, ensuring the technological infrastructure is equipped with all the necessary tools to bear positive results.

Ultimately, having a proactive roadmap that others within your organization (whether they sit on the marketing team or not) gives you greater control over your infrastructure while providing insights into what lies ahead so that you can make decisions accordingly. It also provides necessary transparency and visibility to the many other parts of the organization that will probably be involved in implementing and even using the tools within your MarTech stack.

STEPS TO CREATE A COMPREHENSIVE ROADMAP

Let's now discuss how best to create a MarTech roadmap that will ensure key areas are addressed and that other parts of the organization will gain value from reviewing it as well. After all, you will probably need allies in other areas of the company to make your MarTech dreams a reality!

Thoughtful consideration of the entire technology landscape should be reviewed to identify current gaps or areas of improvement, opportunities for automation, and tools that may need to be retired or replaced. It is also beneficial to ensure you're keeping up with the ever-changing technology trends, ensuring your roadmap remains future-proof. With a road map in place that aligns with objectives, it will ensure everyone on the team is working together towards successful outcomes.

Roadmaps can vary greatly in complexity, and much of that is determined by the audience. As a very simple example, the roadmap below (Figure 28.1) simply shows the stages of an initiative over time:

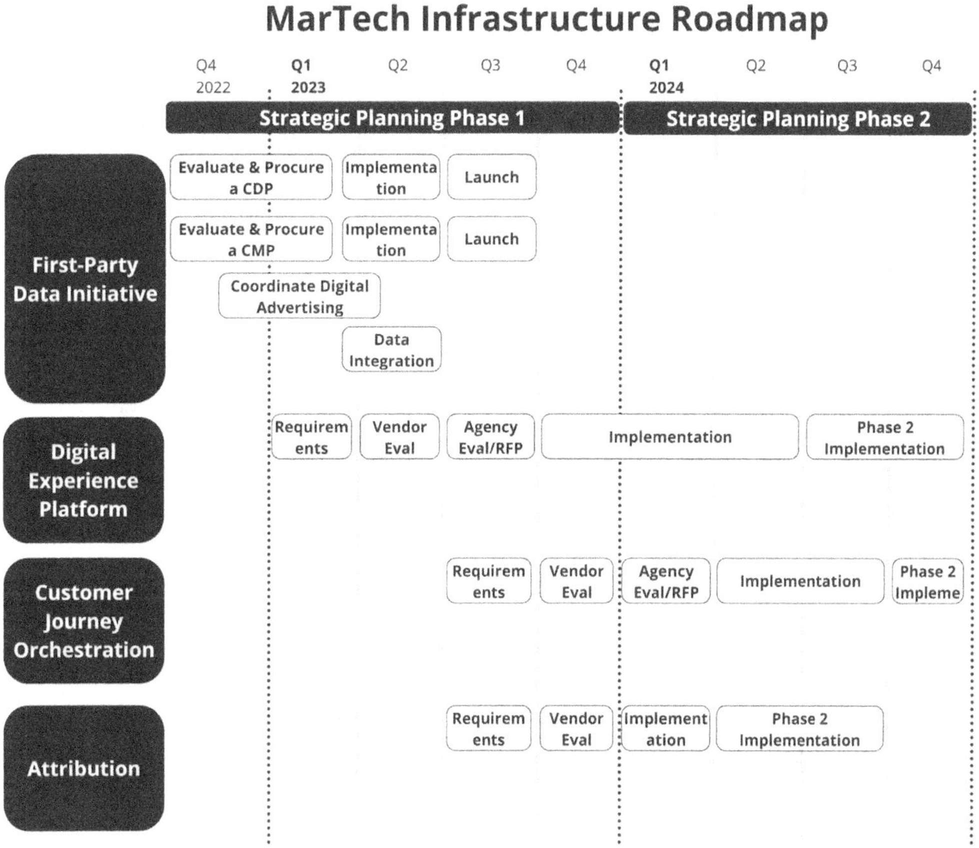

FIGURE 28.1 Simplified MarTech infrastructure roadmap

There are many other components that you may want to include in yours, including dependencies, decision trees, and roles/responsibilities. The steps below apply regardless of the complexity of your roadmap.

One: Start with a Static Infrastructure Map

To see how to create a valuable map of your MarTech infrastructure, please refer to Chapter 27. This is a well-laid-out map of your MarTech stack that can effectively illustrate how platforms relate and interact with one another, and the functions they serve will go a long way as you build your roadmap.

Two: Determine the Best Format for Your Roadmap

While this might sound like a simple decision, trust me, it's not! In my work, I've seen many tools used to create this roadmap, which includes the following:

- Microsoft Visio (one of the more popular choices) or LucidCharts (a Google Docs alternative)
- Miro (my personal favorite)
- PowerPoint
- Adobe Illustrator
- Microsoft Excel

The platform you use is going to depend on several things, such as which platforms your team can access, how many people will be making active edits, whether you need live or remote collaborations, and what format the output of the work will need to be shared and distributed with others in.

Additionally, here are some other things to consider:

- *Timescale*, which might be a matter of months in the case of many fast-moving pieces, or more likely more than a year when planning a longer-term strategy.
- *Level of detail*, or how many data points you want to include about each individual item. This will not only increase complexity but may make it harder to get a full picture when "zoomed out."
- *Audiences* or who will be reviewing this. For instance, if both marketing executives and hands-on technology integrators will need to review, you might consider making multiple views or having layers that can be shown and hidden depending on the audience.

- Items to include within the roadmap:
  - Platforms and their functions
  - Teams that both use and maintain them (internal or external)
  - Timing of implementation and integration, including major features and functionality
  - Data connectivity between platforms
  - Identifying SaaS, internally hosted, or a third party (e.g., Google Analytics) or second party (e.g., social media platforms)

As mentioned in the previous section, there may be other items that would help in your case as well. Finally, if you will be using the *House of the Customer* framework to do your evaluation, as discussed earlier, ensure you include those "parent" categories when identifying tools within the infrastructure.

Three: Socialize Your Roadmap

As well as you and your immediate team might think they know and understand the law of the land, it is always helpful to get input from other teams and groups that interact with and support your MarTech infrastructure when creating your roadmap.

There are numerous benefits from socializing roadmaps. Even if you have done a thorough job, it is still helpful for others in the organization to know and understand the current state of things. After all, they may be involved later in the process as you are getting approval on new investments or potentially need some of their teams' time during the implementation phases.

Additionally, other teams may have strategic initiatives of their own that your MarTech stack can support or from which it can benefit. Understanding this sooner rather than later can help you know, understand, and potentially leverage opportunities that may arise. Therefore, some of the teams you should consider involving are as follows:

- The broader marketing team (for instance, if you sit on a digital marketing or digital experience team, consider involving all teams tasked with marketing)
- Customer experience teams
- Data and data science teams
- Technology and IT security teams
- Customer service or other teams that regularly interact with customers

There are other groups that are increasingly being brought into at least high-level discussions, such as CFOs. Keep an open mind here. While involving *everyone* may slow things down, understanding points of view and potential partnering opportunities can go a long way down the road. Ultimately, socializing the roadmap ensures that the best insights from the full spectrum of teams supporting your MarTech infrastructure are gathered and incorporated.

Four: Analyze Gaps Between the Current and Desired State of Technology

A thorough analysis in the form of a roadmap of your current MarTech infrastructure is an essential step in improving it; identifying existing gaps between your desired and current state is the key to closing them.

With the right roadmap and direction, you can ensure your MarTech is optimally coordinated, highly effective, and strategically aligned. Understanding these gaps will help you develop a framework that defines how new technologies should be implemented into the system, creating a more well-rounded experience for both internal and external customers.

Five: Align KPIs with Your Roadmap

To ensure that your MarTech infrastructure is successful and well-aligned, prioritizing the right measurement of success through key performance indicators (KPIs) needs to be factored into the roadmap. KPIs vary depending on the type of business. They can provide metrics such as customer engagement, consumer behavior, ROI, and expenses.

Without knowing these metrics, it becomes difficult to understand if your campaigns or services are achieving the desired objectives or if there's room for improvement. By making KPIs an essential component of the roadmap, you build a clear understanding of how investments in distribution technology channels and marketing activities are driving business results.

Six: Identify and Prioritize Current Marketing Technologies

Understanding the technologies currently in place is a necessary step in developing an effective MarTech roadmap. The next step is to review each system and determine how it fits into the organizational view of marketing objectives. From there, you can devise strategies to ensure these systems are effectively integrated, providing additional value and ROI from their use.

This evaluation process should be ongoing as new technologies enter the market and as priorities related to goals change. Developing a framework for technology decisions, prioritization, and budget is essential for the successful implementation of a MarTech infrastructure.

ENSURE YOUR ROADMAP STAYS UP TO DATE

Finding and maintaining the right MarTech infrastructure is not a one-time event; it needs to be consistently optimized over time. Thus, the roadmap you use to evaluate and communicate it throughout your teams and organization needs to stay up to date as well. This will enable quicker evaluation of potential changes and enable a more proactive approach to capturing new opportunities.

Staying on top of your roadmap updates to make it as accurate as possible will ensure you maintain an efficient and effective MarTech infrastructure throughout its lifecycle. We'll be talking about this a little more in the next chapter, which is on continuous improvement.

VOLTAGE MOTORS: USING A MARTECH ROADMAP TO ALIGN DXP AND CUSTOMER DATA INITIATIVES

VoltAge Motors reached a critical turning point as a leading contender in the electric vehicle market. The company needed to expand its digital marketing capabilities to handle new customer acquisition while delivering personalized experiences and maintaining long-term customer relationships due to rapid growth in both domestic and international markets. The current MarTech infrastructure began to demonstrate its constraints as VoltAge Motors expanded its operations. Manual processes combined with platform integration gaps and fragmented systems prevented the team from acting on real-time customer insights.

The company's leaders understood that separate enhancements to individual platforms would not resolve their fundamental systemic obstacles. To achieve alignment between customer experience goals and technology investments, they required a strategic roadmap that could link their tech investments to their overall vision for customer experience. VoltAge Motors established a comprehensive MarTech roadmap that enabled them to understand the interdependencies between various initiatives such as launching

Adobe Experience Manager (AEM) as their new digital experience platform (DXP) and optimizing their customer data infrastructure and to plan these projects in a sequence that would maximize their impact.

Establishing the Roadmap: From Static Map to Dynamic Strategy

VoltAge began its roadmap initiative by performing an audit to evaluate its current MarTech stack. It examined how its CMS, CRM, marketing automation software, customer data platform (CDP), and analytics tools were integrated or operated independently. Its old CMS system was unable to meet the business requirements for omnichannel personalization and seamless user experiences. Its customer database existed as isolated silos across different systems, which created inconsistent customer profiles and fragmented communication.

The roadmap process from VoltAge enabled the organization to determine the sequence of its essential initiatives. The decision to move to AEM as its new DXP went beyond upgrading its Web site because it became the essential base for creating personalized experiences at every customer interaction point. The effectiveness of AEM's personalization features depends on access to a unified customer database of high quality. The understanding of customer data importance led VoltAge to focus on data unification, which sped up customer profile consolidation in its CDP before it could fully implement AEM.

How the Roadmap Brought Teams Together

VoltAge encouraged cross-functional collaboration between marketing, IT, customer experience, and sales by distributing its roadmap to all departments. All teams received clear visibility about the DXP project's alignment with their objectives. While the customer experience team planned new post-purchase journeys within AEM, the IT team prioritized integrations between the DXP and the CDP. The sales and customer service teams started getting ready to deliver more personalized experiences through improved customer data.

The roadmap provided clarity on dependencies, which ensured that platform migrations and integrations proceeded in a systematic phased order. The team needed to complete their GDPR compliance and data governance policies in the CDP before they could start launching localized Web sites through AEM for the European market of VoltAge. The roadmap clearly incorporated these factors to enable better coordination and prevent expensive rework.

Results and Impact: A Clear Path to Personalization and Efficiency

The MarTech roadmap from VoltAge Motors served as a strategic guide to evolve its digital ecosystem. It managed to prevent isolated technology implementations by understanding the connections between its AEM implementation and customer data unification initiatives. The phased rollout allowed AEM to be launched with access to enriched customer data, which enabled advanced personalization.

The early results demonstrated accelerated campaign launches together with better localized content and uniform customer experiences on multiple platforms, including Web, mobile, and email. Previously basic segmentation-based personalization initiatives now utilize real-time data from the CDP to provide dynamic content to users through the new Web site and mobile app.

The roadmap process provided VoltAge Motors with essential tools for continuous evaluation and prioritization. With its expansion into new markets and the exploration of emerging channels, such as connected car experiences, VoltAge Motors now utilizes a living collaborative strategy to direct its MarTech investments.

CONCLUSION

Creating a roadmap for a MarTech infrastructure should be an initiative at the top of any company's wish list. It will help ensure that technology investments are wisely managed, existing resources are used more effectively, and customer experiences are improved across all channels. When creating such a roadmap, it is important to be aware of current trend-setting technologies, have a clear vision of success, and anticipate the risk of failure.

Having an actionable plan in place will save precious time in the long run and provide the necessary assurance that you are heading in the right direction. With these steps taken and KPIs for tracking progress, companies can maximize their potential for growth and streamline operations by understanding what technology will best serve their current needs and long-term objectives.

CHAPTER 29

MEASURING SUCCESS

To maximize the impact of your investments in MarTech, you need to look at it from a number of perspectives, including your overall business objectives and your more specific customer and marketing goals. Then, because there are so many different possibilities of how a MarTech stack can be implemented, you should look at how to prioritize investments based on well-defined measures of success. Let's explore these aspects in more detail.

It is not enough to simply implement MarTech within your organization. It is something that needs continual measurement and refinement as customer expectations evolve and as platforms and technologies develop and mature. Let's define some measures of success for MarTech in your business.

OVERALL RETURN ON INVESTMENT (ROI)

The ultimate goal of most endeavors within an enterprise is to achieve a return on their investment, commonly referred to as ROI. With complex systems, teams, and processes, this can be difficult to measure. A helpful way to do this is to look at this investment in terms of four dimensions, as illustrated in Figure 29.1. In this case, the following dimensions are considered: total cost of ownership (TCO), adoption rate, level of integration, and benefit to business.

Overall Return on Investment (ROI)

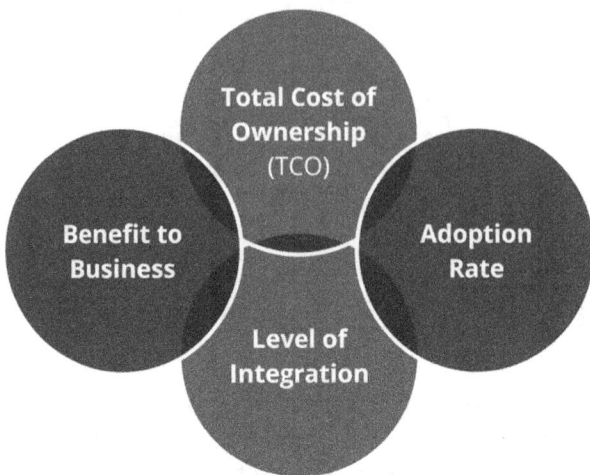

FIGURE 29.1 Measures of ROI

Each of these dimensions will be explored in more depth in the rest of this chapter.

TCO

TCO represents the full financial evaluation of expenses related to a MarTech platform or MarTech infrastructure throughout its entire lifecycle. The cost of technology ownership extends beyond the upfront licensing fees or subscription payments. The TCO encompasses implementation expenses, integration costs, customization fees, maintenance costs, staff training expenses, upgrade costs, vendor support charges, security compliance requirements, and the cost of eventually decommissioning or migrating the platform. The TCO for an individual MarTech platform reveals the financial commitments needed to maintain the tool throughout its lifecycle. TCO analysis provides insights into cost accumulation across multiple systems and offers organizations a comprehensive framework to manage budgets and investment priorities while assessing ROI with precision.

Organizations that understand TCO can decide wisely on tool investments and determine whether platform consolidation or contract renegotiation alongside staff enablement will enhance platform value.

Speed to Delivery

Marketing teams' speed to delivery measures how quickly they can execute new campaigns, customer experiences, or products upon identifying an idea or need. The entire process, which includes strategic planning, creative development, and technical execution and deployment, covers MarTech platforms.

The measurement process usually determines the average duration between developing a campaign concept and executing its launch, which is called time-to-market. Key performance indicators (KPIs) might include the following:

- Time from campaign briefing to execution
- The number of iterations required to reach approval
- Platform setup and configuration time
- SLA compliance percentages for marketing service requests

Collect baseline and benchmarking data to evaluate delivery speed against previous records or industry competitors.

To optimize speed to delivery, organizations should be focused on the following:

- Investing in platforms that feature intuitive interfaces and workflow automation will help to streamline the production and approval process for campaigns.
- Streamline organizational operations by merging redundant systems and making integrations more straightforward.
- Organizations should implement standardized procedures and utilize pre-approved modular content blocks to enable quick deployment.
- Building collaborative relationships between marketing, IT, compliance, and creative teams will help to eliminate bottlenecks.

By implementing a MarTech roadmap that emphasizes low-friction systems with automated workflows, organizations can achieve faster delivery timelines.

Speed to Insights

Speed to insights represents the rapidity with which an organization collects data analysis results to support informed decision-making. Analytics tools, dashboards, and reporting workflows demonstrate their efficiency by providing marketing teams and stakeholders with meaningful actionable intelligence.

Some essential speed to insights measures are as follows:

- Time from data capture to report generation.
- The period required to derive actionable insights from campaign launch data includes optimization recommendations.
- Reporting cycles should maintain regular intervals and timely delivery in forms such as real-time, daily, and weekly updates.
- The ratio of manual data collection and analysis methods to automated systems.

By benchmarking performance metrics, organizations can assess their analytics infrastructure effectiveness and identify areas for improvement.

Optimization strategies for speed to insights include the following approaches:

- Organizations should invest in centralized analytics platforms that combine data from multiple sources while minimizing manual data collection efforts.
- Real-time or near-real-time data pipelines enable faster reporting and decision-making processes.
- Utilizing AI-powered analytics platforms enables automated detection of trends and anomalies while generating predictive insights.
- Organizations need to develop their teams to understand insights rapidly and use them for strategic decision-making.

Data teams and marketing leaders must prioritize quick insight delivery to ensure strategic decisions are based on insights rather than solely on reports.

Customer Satisfaction (Internal and External)

Customer satisfaction regarding MarTech affects both external customers, who interact with brand marketing, and internal teams, who use MarTech platforms to execute their responsibilities.

For external customers, common measures include the following:

- Customer satisfaction scores (CSATs) are collected via surveys.

- Net promoter score (NPS) measures how likely customers are to recommend a company to others based on their loyalty.

- Customer responsiveness is measured through engagement metrics, which includes open rates, click-through rates, and conversion rates.

MarTech platforms for external customers should deliver seamless personalized experiences across multiple channels. Deploy relevant content and interactions based on real-time data while ensuring privacy protection and adherence to compliance standards.

Of course, external customers are not the only audience for MarTech platforms. In fact, the people who often use the tools the most are the internal teams that must manage them on a daily basis. Thus it is critical that their satisfaction is optimized. For internal customers, measurements may include the following:

- User satisfaction surveys evaluating measures of system usability together with its reliability and support options

- Adoption rates and platform utilization metrics

- The amount of time dedicated to manual tasks compared to automated process workflows

- Structured interviews or focus groups serve as the method for collecting feedback

Provide internal customers with training and onboarding programs to increase their familiarity with platforms. Consistently request feedback about customer difficulties and proactively focus on implementing feature enhancements or additional support measures. Make user interfaces as simple as possible while eliminating complex features that do not serve a purpose.

A customer-focused MarTech strategy makes sure tools achieve external engagement objectives while supporting internal operational efficiency.

TCO delivers a full picture of MarTech investment expenses by analyzing costs beyond just licensing fees for organizations. Through evaluation of TCO, marketing leaders achieve a balance between financial limitations and the requirements for strong capabilities and scalable solutions. The MarTech stack delivers true value and meets its mission to enable faster and smarter marketing at scale by optimizing customer satisfaction along with speed to delivery and speed to insights. A properly managed TCO strategy forms the foundation of both sustainability and competitive advantage for an organization's marketing operations.

ADOPTION RATE

The next metric you can use to measure the success of your MarTech stack is the overall adoption rate. You can look at this in terms of a few aspects.

Cross-Team Adoption

This refers to how many teams are utilizing all or at least parts of the MarTech infrastructure you've created, or if they are relying on outside vendors or siloed platforms in order to accomplish their work. There are obvious cost savings to be had when multiple teams can rely on a single set of platforms, from licensing costs to onboarding and cross-training time and costs for employees.

Level of Usage of Specific Products

Next, you can look at teams' usage of specific products or platforms within your MarTech stack and how fully they are utilizing them. For instance, has a team invested in a multi-million pound per year license, only to use one small feature within that product? Or are they fully utilizing it to the point that the license fee is fully justified?

Level of Usage of the Entire MarTech Ecosystem

Similar to the way we look at the usage of an individual product within the MarTech infrastructure, we can also look at the adoption and usage of the entire platform by teams. Keep in mind, of course, that some portions of a large, all-in-one marketing cloud platform may only be relevant to one team (e.g., an email marketing component may only apply to the team(s) that send emails), but with this measure, you are looking at how fully the ecosystem's many products and features are being utilized.

LEVEL OF INTEGRATION

While adoption looks at how fully a platform is utilized by internal teams, as well as how many of its features are being utilized, level of integration refers to how deeply the data, features, and systems themselves that make up your MarTech infrastructure are interconnected and able to operate in conjunction with one another. You should look at this in terms of the following:

- *Data integration*, or how easily information from one product, system, or platform within your MarTech infrastructure is able to be sent and received.
- *Feature conflicts*, or how many duplicate feature sets exist across the different platforms that exist within your MarTech infrastructure, if any. This can potentially cause confusion both internally and with customers.
- *Interoperability*, or how well features from one platform within your MarTech ecosystem are able to complement one another and provide an additive effect, as opposed to creating a conflict, as we mentioned in the previous point.
- *Rate of innovation*. Finally, we need to ensure our MarTech stack is keeping pace with the rate of change in the marketing industry, our specific industry vertical, and with growing customer expectations. To do this, we want to look at the rate of innovation within our MarTech infrastructure.

We can do this by looking at the following:

- *New feature addition frequency and quality*, or how often high-quality features and enhancements are being added to the platforms and products within our MarTech stack. Depending on the makeup of your ecosystem, this could be a combination of how often platform vendors are adding features, it could be how often your internal teams are building or integrating features, or it could be a combination of both.
- *Product roadmap*, or the overall vision of the component parts of your MarTech infrastructure. This evaluates the overall approach that your MarTech stack is taking and the goals that guide its future product and feature releases.

BENEFIT TO THE BUSINESS

The fourth component is one that is often most top of mind for executives across the organization from Chief Marketing Officers to Chief Financial Officers. While it is important, it shouldn't overshadow the previous three.

A complete evaluation of a MarTech stack's business impact requires linking technology investments to specific business results. A MarTech ecosystem delivers value beyond campaign support and task automation by effectively empowering organizations to generate revenue while maintaining customer loyalty and improving operational efficiency through innovation.

Organizations must set precise metrics and collect data from various sources to validate MarTech's value while ensuring alignment with business goals.

Business benefit evaluation can be directly achieved by tracking revenue growth generated from MarTech initiatives. Organizations can track this performance by measuring lead generation volume and quality together with conversion rates, average deal size, and customer lifetime value (CLV). Different marketing attribution models including first-touch, last-touch, and multi-touch enable businesses to determine which campaigns and technologies drive revenue through closed deals. When MarTech stacks are properly integrated, they support better attribution accuracy, which leads to proper credit assignment and improved budget allocation decisions. The predictive analytics tools found in the stack enable marketers to anticipate revenue outcomes based on patterns in customer behavior and campaign performance, which translates into a data-supported understanding of marketing's financial impact.

Customer retention functions as an essential metric when determining the effectiveness of MarTech solutions. Businesses spend less to maintain current customers compared to acquiring new ones and customer data platforms (CDPs), personalization engines, and customer journey orchestration tools strengthen ongoing customer relationships. Businesses need to monitor retention-related statistics, which include metrics such as churn rate, repeat purchase rate, NPS, and CSAT scores.

The integration of MarTech platforms that facilitate timely personalized communication across channels through consistent data directly influences customer loyalty and lifetime value. A successful MarTech stack delivers customer engagement strategies that utilize data analysis and adaptability to boost retention results.

The importance of MarTech in fostering innovation demands equal evaluation. The rate at which an organization tests new marketing strategies, adopts new communication channels, or introduces fresh customer experiences serves as a measurement of innovation. Experimentation platforms such as A/B testing tools and personalization frameworks serve as primary catalysts for innovation. A stack's ability to drive innovation can be measured through the volume of tests executed, the speed at which new initiatives reach launch, and how well customer experiences can be improved through iterative changes. Organizations benefit from modular and API-first MarTech platforms because they enable quicker adoption of new tools while keeping ahead of market trends without needing significant rework.

A modern MarTech stack has the potential to deliver significant operational efficiency benefits that organizations often overlook. Recent innovations and integrations with AI tools built for efficiency and speed are a good example.

Organizations can measure efficiency through decreases in manual work and campaign deployment time together with better team collaboration. Operational gains can be effectively measured through metrics such as automation coverage (percentage of workflows automated), time-to-market for campaigns, and decreased IT support dependency for marketing activities. A properly arranged MarTech stack that merges tools and centralizes data eliminates unnecessary repetition while speeding up response times and reducing operational expenses. This shift enables marketing teams to dedicate more time to strategic initiatives and creative projects instead of administrative responsibilities.

The evaluation of a MarTech stack's business value demands both quantitative and qualitative assessment methods. Quantifiable indicators such as revenue expansion and efficiency improvements give direct measures of worth, but qualitative aspects such as better teamwork and customer sentiment provide essential contextual understanding. Thorough evaluations of KPIs based on stakeholder feedback enable the MarTech stack to stay in sync with changing business objectives. Through disciplined measurement of MarTech value organizations can recognize its importance both as a marketing excellence enabler and as a contributor to business success.

CASE STUDY: VOLTAGE MOTORS ALIGNS MARTECH ADOPTION WITH BUSINESS BENEFITS

Over the last three years, VoltAge Motors invested heavily in developing its MarTech stack. Its aim was clear: the company wants to offer top-notch customer experiences while enhancing operational efficiency and protecting its market position amidst rapid automotive industry changes. Even though the leadership team observed significant improvements in customer engagement and revenue generation through MarTech platforms, they recognized that employee adoption rates were not meeting expectations.

VoltAge Motors Searched for Reasons Why Its Business Results Didn't Align with How Its Teams Used MarTech Platforms

A series of audits and feedback loops revealed that VoltAge Motors' marketing operations team found a discrepancy between business results and internal tool usage. On the surface, the data was promising. The company achieved a 15% increase in lead conversion rates, while customer retention rates grew by 10% over the last fiscal year. The implementation of a CDP led to improved personalization which increased their NPS, while their marketing automation tools reduced time-to-market for digital campaigns significantly.

The assessment of platform utilization showed an entirely different scenario. Cross-team adoption was inconsistent. The digital marketing team consistently used automation workflows and customer journey orchestration while regional sales teams and customer service departments continued to depend on manual processes and separate tools. The digital experience platform (DXP) usage reports demonstrated that many of Adobe Experience Manager's premium features remained unused. Organizations utilized only fundamental CMS functions because they did not fully implement the advanced personalization features that were key reasons for selecting the DXP.

Diagnosis: The Root Causes of Limited Adoption

Several factors contributed to this limited adoption. In its initial MarTech rollout, VoltAge Motors focused mainly on platform selection and technical implementation but failed to make sufficient provisions for user enablement. The training sessions lacked regularity while internal documentation remained out of sync with updated workflows. The complexity of new platforms presented a steep learning curve, which became an issue primarily for teams outside the digital marketing department.

Second, integration gaps created friction. Data accessibility challenges remained prevalent across other departments despite the core marketing team achieving full system connectivity. Sales reps experienced frustration because customer insights from the CDP failed to integrate properly with CRM workflows and they reverted to old methods. The absence of interoperability between systems reduced the stakeholders' trust in technology and resulted in decreased adoption rates.

The Turning Point: VoltAge Motors Developed a Practical Adoption and Alignment Roadmap to Address Its Issues

VoltAge Motors designed a detailed MarTech roadmap to address these challenges. The strategy addressed platform capabilities and user engagement. The roadmap prioritized user training through intensive workshops for key teams, which later extended to role-specific onboarding materials and self-service resources. User support received investment from the company when they established a cross-functional MarTech Center of Excellence (CoE) to deliver guidance and gather feedback.

In accordance with its roadmap, VoltAge Motors carried out an integration audit. VoltAge Motors improved its CRM and DXP data flows, so sales and service teams gained seamless access to real-time insights that marketing teams already utilized for personalization. It improved system confidence and reduced operational friction by resolving data silos and optimizing its integrations.

Results: Aligning Adoption with Business Benefits

VoltAge Motors experienced noticeable advancements within its operations after implementing its roadmap for six months. The adoption rate across teams increased by 30% because customer service teams employed the personalization modules of the DXP to offer appropriate content through service portals. Regional sales teams embraced automated lead nurturing workflows, which reduced their reliance on manual processes.

The increased use of essential platform features enabled VoltAge Motors to achieve improved customer satisfaction levels and enhanced operational performance. The number of customer complaints regarding inconsistent messaging reduced, while internal surveys demonstrated a 25% boost in user satisfaction with MarTech tools. The leadership team could now establish a strong connection between the adoption of MarTech platforms and enhanced customer experiences that led to measurable business results.

CONCLUSION

While short-term metrics and cost savings can be beneficial to an organization, the best way to approach the measurement of success of a MarTech stack, as well as the individual platforms within them, is to evaluate their long-term return on investment using a number of different metrics to do so. This holistic measurement will enable an organization to review its investments through a number of lenses and perspectives to determine the biggest gaps to address, as well as the areas of strengths that need continual support.

EPILOGUE

While it has been the intent of this book to provide as complete an understanding as possible of what a successful MarTech infrastructure consists of, there are, of course, too many details in this fast-growing area to contain in a single book.

That said, while details may change, and new features, buzzwords, and must-have features may evolve from time to time, the concepts behind the evaluation frameworks as well as the strategic approaches should provide a consistent point of reference that can be used, regardless of small ebbs and flows in the industry and the practices in marketing.

Thus, this book will conclude by looking at six key areas that marketers should incorporate into their short-term and long-term marketing strategies.

DEFINE YOUR GOALS WITH CONTINUOUS IMPROVEMENT IN MIND

When defining your MarTech infrastructure goals, it is important to keep in mind that the MarTech landscape has never stopped changing and doesn't show any signs of slowing down. Technologies and applications that were once successful can quickly become outdated, making it critical to stay on top of the latest solutions and strategies to stay competitive.

Taking time to continually evaluate what you need from your MarTech infrastructure and adjust as necessary will ensure that your business can reach its current or future goals. A few metrics and considerations here are as follows:

- Determine a measure of overall return on investment (ROI), which will probably be unique to your organization.
- Look at the adoption rate of the MarTech stack in general, individual tools within the infrastructure, and the adoption across your marketing teams.
- Assess the level of integration between platforms because this can help you evaluate the efficiency and data connectivity (or a lack thereof).
- Find a way to measure your teams' rate of innovation, as well as the platforms that you've invested in.

KEEP UP WITH THE CHANGING TECHNOLOGY LANDSCAPE

Keeping up with the ever-evolving MarTech landscape can be difficult, but an approach focusing on infrastructure goals as a moving target can make it much easier. The key is to be flexible and adaptive, redirecting your resources and approaches when needed.

For instance, many forms of AI, from generative AI to predictive analytics and agentic AI, are being quickly adopted by many MarTech platforms, and while some of this adoption brings significant improvements to features and functionality, other applications of AI may be less beneficial and potentially be a case of the product's team rushing a feature to market that is half baked.

AI is only the most recent trend that is transforming MarTech; you must also remain vigilant in staying ahead of new advancements to continue providing effective solutions for your clients. Consistent research of the current industry trends is essential, allowing you to remain competitive and properly utilize the latest technologies to maximize efficiency for all involved.

ANTICIPATE FUTURE NEEDS AND TRENDS

The continually evolving nature of the MarTech infrastructure can make building a solid infrastructure challenging. To get ahead of the game, it is important to take a forward-thinking approach and anticipate future market needs and trends.

Stay up to date on emerging technologies that may become critical, and develop strategies that are nimble enough to account for changes while also providing necessary foundations and frameworks. Utilize predictive analytics models to gain insight into potential future scenarios while also drawing data from recent activities or events as they happen. Companies can continuously prepare their marketing strategies for the ever-shifting marketplace through this approach.

INTEGRATE YOUR EXISTING SYSTEMS

Several approaches to integration have been discussed in the preceding chapters: from utilizing all-in-one marketing suites to API-driven integration or a composable approach.

Changing technologies and market needs can make it difficult to ensure your MarTech infrastructure is optimized for your business. Integrating existing systems into the new program and models available, however, can help you stay on top of your goals. By combining the capabilities of the current and emerging technology, you can create a system that is up to date and tailored specifically for your business.

This allows you to reap the benefits of new advances without having to start over completely every time a new tool becomes available. Spending some time working out the best way to use existing systems with new tools helps create a streamlined workflow that is unique to your organization and saves you money in the long run.

MAINTAIN FLEXIBILITY

While integration is key to productivity and in creating a seamless customer experience, changes in the MarTech landscape can take place quickly and, to ensure that your company is prepared for whatever comes its way, it's important to build a MarTech infrastructure that maintains flexibility. That way, you can make adjustments when needed, allowing you to stay on top of trends and keep up with changing technologies.

With an approach that focuses on being agile and flexible, companies can weather any changes that come their way without sacrificing their operational performance or efficiency.

REGULARLY REVISIT AND RECONCILE YOUR STRATEGIC GOALS

With consumer expectations, competitive demands, and industry shifts occurring on a seemingly more frequent basis, it is increasingly important to check in with strategic goals for MarTech infrastructure periodically.

Accounting for environmental changes can help ensure that progress is being maintained and customer expectations are properly addressed. Setting aside time intervals for these check-ins and analyzing approaches that have worked well (or otherwise) can provide insights into how to reconcile goals with current strategies better. A comprehensive understanding of what works best and why is essential to move forward with agility and a clear understanding of objectives.

Your MarTech infrastructure goals are dynamic, and you should plan for them as if they are a moving target. You need to pay close attention to changes in the MarTech landscape and anticipate future needs and trends. Integrating your current systems and technologies for maximum efficiency and productivity is prudent. Remember, flexibility is key for making changes when necessary.

Finally, don't forget to reconcile your strategic goals at key intervals; this will enable you to adjust as needed to achieve them efficiently. By doing all of these things, you'll be well positioned to reach your MarTech infrastructure goals to remain relevant today while helping you stay ahead of what's coming tomorrow.

About the Author

Greg Kihlström is a best-selling author, speaker, and entrepreneur, and serves as an advisor and consultant to top companies on MarTech, marketing operations, and digital transformation initiatives. He has worked with some of the world's top brands, including Adidas, Coca-Cola, FedEx, HP, Marriott, Nationwide Insurance, Victoria's Secret, and Toyota.

He is a multiple co-founder and C-level leader, leading to his digital experience agency being acquired in 2017. He successfully exited an HR technology platform provider he co-founded in 2020, and led a SaaS startup to be acquired by a leading edge computing company in 2021. He currently advises and sits on the board of a MarTech startup.

In addition to his experience as an entrepreneur and leader, he has earned an MBA, is currently a doctoral candidate for a DBA in business intelligence, and teaches several courses and workshops as a member of the School of Marketing Faculty at the Association of National Advertisers. He has served on the Virginia Tech Pamplin College of Business Marketing Mentorship Advisory Board and the University of Richmond's CX Advisory Board, and was the founding Chair of the American Advertising Federation's National Innovation Committee. Greg is Lean Six Sigma Black Belt certified, is an Agile Certified Coach (ICP-ACC), and holds a certification in Business Agility (ICP-BAF).

Greg has written over 20 books on marketing and MarTech, including several books for an imprint of De Gruyter, several books in *The Agile Brand Guide*® series on MarTech platforms and practices, and his *Agile*, *House of the Customer*, and *Priority is Action* series of books.

He executive-produces eight business and marketing-related podcasts and hosts three, including the award-winning *The Agile Brand with Greg Kihlström,* now in the top 10 of Apple's US marketing charts and in its 7th year, with over 600 episodes and millions of downloads, which discusses MarTech and its role in the customer experience with some of the world's leading experts and leaders.

Greg is a contributing writer to MarTech, CustomerThink, and CMSWire, and has been featured in publications such as Advertising Age, Business Insider, Financial Times, Forbes, and The Washington Post. Greg has been named #1 on its list of the Top Global Marketing Thought Leaders by Thinkers 360, was named one of ICMI's Top 25 CX Thought Leaders two years in a row, and a DC Inno 50 on Fire as a DC trendsetter in marketing. He's also participated as a speaker at global industry events and has guest lectured at prominent universities and colleges.

INDEX

A

A/B testing, 5, 37–41, 46, 50, 67, 85, 230, 274, 298, 340
Accel-KKR, 325
access controls and rights management, 218–219
accessibility, 371
Acquia, 269
Acquia DXP, 328–329
    acquisitions and expansions, 329
    current state of, 329
    market focus, 328–329
ActiveCampaign, 348
    automation platform, 112
Ada, 313
Adobe Analytics, 8, 23, 41, 51, 58, 62, 126, 281, 324, 363
Adobe Campaign, 38, 324
Adobe Experience Manager (AEM), 268, 324, 429, 438
    sites and assets, 324
Adobe Experience Platform (AEP), 323–325
Adobe Illustrator, 434
Adobe Marketing Cloud, 278, 324
Adobe Target, 7, 114
adoption rate, 441
AdRoll, advertisement platform, 123
advanced data modeling and preparation, 362
advanced deletion software, 181
advanced search and filters, 218, 222
advertising; *see also* digital advertising
    DMPs, 149, 172
    mobile, 22
    optimization and media buying, 121–123
    paid social, 20
    platforms, 36, 352
    programmatic, 38, 44, 123, 139, 149, 169–170, 172, 250, 305
AEM, *see* Adobe Experience Manager (AEM)
AEP, *see* Adobe Experience Platform (AEP)
AgilOne, 329
Ahrefs, 307
AI-assisted brainstorming platforms, 192
AI-based systems, 194
AI-driven automation, 315
AI-driven content automation tools, 252
AI-driven personalization, 270
AI-driven segmentation tools, 267
AI-powered chatbots, 312–313, 318
AI-powered conversational platforms, 315
AI-powered personalization, 245, 267, 406
    and customer journey orchestration, 275
AI-powered search, 218
Airtable, 195
Amazon, 25, 72, 312, 314, 320, 385, 416
Amazon Alexa, 312, 320
    for Business, 314

Amplitude, 351
analytics
    and performance teams, 203–204
    and performance tracking, 206, 209
    platforms, 247
    and reporting, 237
    teams, 51–52
    tools, 298, 357
anonymization of data points, 183
API-first and headless capabilities, 267–268
API-first systems, 372
Apple company, 21–22, 76, 171, 416, 458
application programming interfaces (APIs), 99
    access and custom integrations, 222
approval workflows and version control, 236, 240
artificial intelligence (AI), 4–5
    ad optimization and media buying, 121–123
    audience targeting and segmentation, 120–121
    brand monitoring and sentiment analysis, 123–124
    chatbots, 312, 317
    in content creation, 196, 205
    content optimization and analysis, 118–120
    cross-channel attribution, 124–126
    customer journey orchestration, 128–129
    customer support, 116–118
    engine automatically tags images, 221
    generative, 107–109
    and machine learning capabilities, 221, 224
    and machine learning models, 403
    in marketing, 25–26
    personalization, 112–114
    personalization tools, 246
    predictive analytics, 109–111
    and privacy, 182
    synthetic personas and research, 114–116
    visual recognition and image processing, 126–127
    workflow automation, 111–112
Asana, 194
asset
    analytics, 220
    management, 192, 206
    sharing, 220–221
    version control, 229
AT&T banner ad, HotWired, 18
Attentive, 315
attribution modeling functionality, 354
audit trails, 236
augmented reality (AR), 251; *see also* artificial intelligence (AI)
automated AI systems, 318
automated lead qualification systems, 315
automated scheduling, 205
automated tagging and file organization, 223
automated workflows, 229
automatic data importation, 359
automatic security patches, 278
automation, 285
    and AI-driven recommendations, 194
    and workflow efficiency, 239–240

B

B2B conversational marketing, 313
B2B software company, 252, 257
benchmarking performance metrics, 444
benefit to business, 441
big data and automation era (2010s)
    CRM and CDP, 23–24
    data explosion and advanced analytics, 22–23
    marketing automation, 23
    privacy and consent management, 24–25
Bizzabo, event management software, 47
BlackBerry, 21

blogging, 264
BlueVenn, 121
brand
   alignment, 215
   awareness, 192
   communication, 311
   management, 207, 219–220
   monitoring and sentiment analysis, 123–124
   storytelling, 214
brand-building, 248
branding designers, 229
Brandwatch, 42, 124
browser cookies, 179
Buffer, 303
built-in scalability, 278
business intelligence (BI), 399
business objectives and tech requirements, 391
Bynder, DAM platform, 39, 127

C

California Consumer Privacy Act (CCPA), 24, 176–177, 181–182, 343, 423
campaign
   and content teams, 229
   performance tracking, 230
   planning and calendar management, 234–235
   setup, 301
CCPA, *see* California Consumer Privacy Act (CCPA)
CDPs, *see* customer data platforms (CDPs)
centralized asset library, 218
change management and user adoption, 409
chatbots, 25, 65, 69, 78, 87, 116–118, 140, 251–252, 269, 272, 275, 279, 311–320, 393, 402; *see also* AI-powered chatbots; conversational marketing
ChatGPT, 25, 27, 40, 107, 196

CJO, *see* customer journey orchestration (CJO)
Clarifai, visual recognition platform, 127
click-through rates (CTR), 342
cloud and on-premises options, 224
CMPs, *see* consent management platforms (CMPs); content marketing platforms (CMPs)
collaboration
   and communication features, 239
   features, 208
   tools, 205, 219
communication tools, 225
compliance with industry regulations, 223
connected TV (CTV), 251
consent management, 184
consent management platforms (CMPs), 182–183, 185, 404
consent mechanisms, 179
consumer data privacy, 76–77
Consumer Privacy Protection Act (CPPA), 177
content
   collaboration tools, 195
   creation tools, 205, 210
   creators, 202
   curation platforms, 211
   distribution, 205–206
   duplication, 248
   opportunities, 204
   optimizers, 202
   performance, 204
      management, 201
      tracking, 201
   repositories, 194
   strategists, 201–202
content delivery systems
   marketing channels, 250–256
   marketing lifecycle and platform selection, 246–250
   training and adoption challenges, 256–260

Voltage Motors' multichannel marketing challenges, 260–262
Contentful, 265, 278
Contently, 210
content management and distribution, 267
content management and workflow automation, 208–209
content management systems (CMSs), 3, 39, 210, 220, 225, 244–246, 248, 263–264, 324, 349, 400
    Contentful, 265
    Drupal, 265
    and DXP, difference, 271
        content creators and editors, 273
        developers, engineers, and IT teams, 273–274
        integration and extensibility, 270
        marketers, 272–273
        personalization and AI capabilities, 270
        platforms, 274–282
        primary users of, 272
        scope and capabilities, 269
        target users and use cases, 271–272
        UX and design teams, 274
    evaluation, 277
    features
        plugin and extension support, 277
        SEO and performance optimization tools, 277
        user-friendly content editing interfaces, 276
    headless CMS, 279
        benefits, 279–280
        Composable CMSs *vs.* Headless CMSs, 280
        proprietary *vs.* open source, 281–282
        working, 279
    Hub, 330
    non-technical users, 264–265
    SaaS DXP/CMS and, 278
    Webflow, 266
    WordPress, 265
content marketing, 39–40
    team, 229
content marketing platforms (CMPs), 199
    evaluation, 207
        analytics and performance tracking, 209
        content management and workflow automation, 208–209
        integration capabilities, 208
        scalability and future-proofing, 209–210
        usability and user experience, 207–208
    features, 204
        analytics and performance tracking, 206
        asset management, 206
        brand management, 207
        collaboration tools, 205
        content creation tools, 205
        content distribution, 205–206
        content planning and strategy, 204
        integration capabilities, 206–207
    influencer marketing platforms, 211
    marketers and, 199–200
    optimizing content distribution, 201
    personalization efforts, 200
    quality of content, 200
    related platforms, 210
        collaboration tools, 210
        content curation platforms, 211
        content management systems (CMSs), 210
        digital experience platforms (DXPs), 210
        focused content creation, 210
        video content marketing platforms, 211
    streamlining the content process, 200

users, 201
    analytics and performance teams, 203–204
    content strategists and marketing managers, 201–202
    project managers, 203
    SEO specialists and content optimizers, 202
    social media and email teams, 203
    writers, editors, and content creators, 202
content planning
    and ideation tools, 192
    and strategy, 204
conversational AI and voice assistants, 314
conversational channels, 251
conversational marketing
    platform, 321
    solutions, 318
conversational marketing and real-time communication platforms, 311–312
    24/7 customer support and automation, 316
    data collection and behavioral insights, 316–317
    evaluation, 319
        AI and NLP capabilities, 320
        customization and personalization features, 321
        integration capabilities, 320–321
        omni-channel functionality, 320
        scalability and automation, 321
    example platforms, 312–313
        AI-powered chatbots, 313
        conversational AI and voice assistants, 314
        live chat solutions, 313
        messaging apps and social media chat, 314
        SMS marketing and business texting, 314–315
    features, 318–319
    instant engagement and lead conversion, 315
    marketers and, 315
    omni-channel communication, 316
    personalized customer experiences, 315–316
    users of, 317
        customer support and customer service teams, 317–318
        e-commerce and retail teams, 318
        marketing teams, 317
        sales teams, 317
        technology teams, 318
conversion rate, 342
cookies, 178–179
    and pixels tracking, 179
copywriters, 214–215
corporate web sites, 264
cost per acquisition (CPA), 342
CQ5, 324
Crazy Egg, heatmap tool, 41
creative and design teams, 229–230
creative efforts with marketing objectives, 192–193
creative marketing, 192
creative teams, 66–68, 214–215
CRM systems, 194, 220
cross-border data transfer rules, 184
cross-border data transfers, 181
cross-channel attribution, 124–126
cross-functional collaboration, 408
cross-functional teams, 230–231
cross-functional workflows, 185
cross-team collaboration, 229, 239
cross-team coordination, 195–196
customer acquisition
    MarTech, use of, 36
    platforms, 36–37
    primary function, 36
customer channel switching, 253
customer data, 176
    AI and privacy, 182

MarTech platforms for privacy and
compliance, 182
consent management platforms (CMPs),
182–183
consistency, 140
cross-border data transfers, 181
customer data platforms (CDPs), 183
data minimization, 181
data privacy regulations, 176
    balancing historical data insights with
privacy concerns, 180
    customer data retention periods, 180
    data collection approaches, 178
    data retention, 179–180
    industry self-regulations, 177–178
    industry-specific regulations, 177
    opt-in and consent mechanisms,
178–179
    public policy regulations, 176–177
    secure and compliant data deletion,
180–181
    tracking cookies and pixels, ethical
implications of, 179
    transparency in data collection
methods, 178
enrichment, 140–141
example platforms, 6
first-party data strategy, 137–139
helping brands, *131*
hygiene, 140
identity resolution platforms, 183
management, challenges of, 147–148
quality, 139
recency and relevance, 139–140
source reliability, 141
storage and access, 141–143
types of, 134–137
VoltAge Motors, 143–145
Voltage Motors and GDPR compliance,
183–185
customer data foundation, 190

customer data platforms (CDPs), 5, 24, 148,
183, 246, 298, 352, 395, 419, 438, 448
    analytics and insights, 162
    benefits of evaluation, 423–424
    categories, 423
    and CRM, differences, 149, 167–169
    data activation and integration, 162–163
    data collection and ingestion, 161
    data management and storage, 161–162
    data unification and identity resolution,
161
    evaluation of, 164–167, 420–421
        features, 421–422
        growth, 422
        investment, 423
        risk mitigation, 422–423
        utilization, 422
    features of, 148–149, *160,* 160–163
    history and usage by marketers, 159–160
    marketing leadership, 35
    new platform, procuring, 426–427
    real-time processing and activation, 163
    replacement, 424
        big picture to small picture, 425
    segmentation and audience
management, 162
    self-service analysis and customization,
163
    vendor evaluation, 424
    VoltAge Motors, 172–173, 428
        DXP, moving to, 429
        evaluation process, 428
        key features, 428
        risk mitigation and investment
considerations, 429
        utilization and growth potential, 429
    weak links in MarTech chain, 425–426
customer data retention periods, 180
customer engagement, 214, 270
customer experience (CX) teams, 68–69
customer journey analytics, 324

INDEX • 465

customer journey orchestration (CJO), 244,
    267, 284, 352, 419
  platforms, 7, 86–87, 284–285, 293
    evaluations, 295–296
    examples, 294
    features, 293
    strengths, 294
    weaknesses, 294–295
  tools, 245, 448
customer lifetime value (CLV), 342, 448
customer relationship
  consumer data privacy, 76–77
  digital channels, 71–72
  mobile-first behavior, 74–75
  omnichannel customer experience,
    75–76
  personalization, expectation of, 77–78
  self-serve options, 78–79
  social media mindset, 73–74
customer relationship management (CRM),
    3, 5–6, 23–24, 225, 248, 330, 342, 349,
    352
  analytics and reporting, 155, 158–159
  automation and workflow efficiency, 158
  components of, 152–156
  contact and relationship, 156–157
  contact management, 153
  customer service support, 154
  data management, 155
  evaluation, 156–159
  functions of, 149
  integration and customization, 157–158
  integration capabilities, 155
  marketers, usage by, 152
  marketing, 52–54
  marketing team support, 154
  sales and pipeline, 157
  sales team, 153–154
  VoltAge Motors, 172–173
  workflow automation and collaboration,
    156
customer responsiveness, 445

customer retention, 448
customer satisfaction scores (CSATs), 444
customer service teams, 64–66
customer support and customer service
  teams, 317–318
customization and scalability, 238–239
custom reporting, 240
Cvent, event management platform, 46–47
CX team, *see* customer experience (CX)
  teams

D

DAM, *see* digital asset management (DAM)
data; *see also* customer data
  explosion, 22–23
  siloes, 85, 142–145, 148, 158, 208, 239,
    246, 380, 394–395, 451
  storage and access, 141–143
  teams, 61–64, 216–217, 288, 362, 444
data accessibility and usability, 399
data acquisition, 181
database marketing (1960s-1980s), 14–15
data collection
  approaches, 178
  and data integrity, 342
  methods, transparency in, 178
data-driven decision-making, 91–93, 192
data-driven marketing, 176
data integration, 447
data localization, 184–185
data management platforms (DMPs), 6
  features of, 149, 170
  first-party data strategy, 171
  future of, 171–172
  using of, 170–171
  VoltAge Motors, 172
data minimization, 181
data ownership, 134–141; *see also* customer
  data
  customer data types, 134–137, 136–137
  enrichment, 140–141

first-party data strategy, 134–135, 137–139
    reliability, 141
data privacy regulations, 176
    balancing historical data insights with privacy concerns, 180
    customer data retention periods, 180
    data collection approaches, 178
    data retention, 179–180
    industry self-regulations, 177–178
    industry-specific regulations, 177
    opt-in and consent mechanisms, 178–179
    public policy regulations, 176–177
    secure and compliant data deletion, 180–181
    tracking cookies and pixels, ethical implications of, 179
    transparency in data collection methods, 178
data quality, 371
    and governance, 398–399
data retention, 179–180
    policies, 185
data synchronization, 236
data-visualization and analysis tools, 357–358, 363
    evaluation, 363
        compatibility with data sources, 363–364
        flexibility with visualization, 363
        pricing structures, 364
        user-friendliness *vs.* advanced features, 364
    features
        advanced data modeling and preparation, 362
        advanced visualization options, 358
        AI-driven insights, 361
        community and marketplace resources, 362
        customizable reporting, 359
        embedded analytics, 361
        integration and automated data ingestion, 359
        interactive dashboards and workflow management, 358
        mobile access, 361
        natural language processing (NLP), 362
        real-time and near-real-time data capabilities, 359–360
        resource allocation and scalability, 360
        security and compliance, 360
        storytelling and presentation tools, 362
    Voltage Motors' transformation with Tableau, 364–365
Day Software, 324
deep customization, 363
designers, 214–215, 229
designing effective in-app funnels, 350–351
design software, 220
design software compatibility, 222
digital advertising, 72, 148
    birth of, 17–18
    management, 304–307
    MarTech, use of, 43–44
    platforms utilized, 44
Digital Advertising Alliance (DAA), 177
digital asset management (DAM), 192, 213–214, 225, 246, 248, 410
    evaluation of, 221
        asset organization and searchability, 222
        collaboration and workflow automation, 223
        integration with existing systems, 222
        scalability, performance, and future-proofing, 224
        user access, security, and compliance, 223
    features, 217

access controls and rights
    management, 218–219
advanced search and filters, 218
AI and machine learning capabilities,
    221
asset analytics, 220
asset sharing, 220–221
brand management, 219–220
centralized asset library, 218
collaboration tools, 219
file format support and conversion,
    219
integrations, 220
metadata and tag management, 218
scalability, 221
user-friendly interface, 220
version control, 219
marketers and, 214
platforms, 39–40
users of, 214
    creative teams, 214–215
    external agencies and partners, 217
    IT and data teams, 216–217
    legal and compliance teams, 216
    marketing teams, 214
    product management teams, 215–216
    sales teams, 215
digital channels, 71–72
digital content organization, 213
digital experience platforms (DXPs), 210,
    243, 245, 263, 266, 323, 406, 438, 450
    and CMS, difference, 271
        content creators and editors, 273
        developers, engineers, and IT teams,
            273–274
        integration and extensibility, 270
        marketers, 272–273
        personalization and AI capabilities,
            270
        platforms, 274–282
        primary users of, 272
        scope and capabilities, 269
        target users and use cases, 271–272
        UX and design teams, 274
    evaluation, 275–276
    examples, 268
        Acquia, 269
        Adobe Experience Manager (AEM),
            268
        Optimizely, 268
        Sitecore, 268
    features of
        AI-powered personalization and
            customer journey orchestration,
            275
        deep integration with the MarTech
            stack, 275
        multi-channel content delivery, 275
    multi-channel focus, 267
        API-first and headless capabilities,
            267–268
        content management and
            distribution, 267
        e-commerce integration, 267
        marketing and analytics connectivity,
            268
    shift from traditional CMS to, 266–267
digital marketer, 229
digital marketing
    campaigns, 37–38
    environment, 249
digital media, 213
digital platform, 252
digital publishing, 264
digital rights management (DRM), 216, 223
digital touchpoints, 250
direct APIs, 400
direct sales conversations, 317
disjointed processes, 184
DivvyHQ, 210
DMP, *see* data management platforms
    (DMPs)
Domo, 357
Drift, 313

Drupal, 265
Dun and Bradstreet company, 3, 323
DXP, *see* digital experience platforms (DXPs)
Dynamic Yield, personalization platform, 41, 50, 114, 291

E

e-commerce, 8, 89, 264
    AI-driven cross-channel attribution, 125
    AI-powered campaigns, 120
    chatbots, 315
    company, 220
    integration, 267
    marketer, 358
    marketing, 49–51
    platforms, 400
    and retail teams, 318
    stores, 264
editorial and campaign planning, 208
editors, 202
electric vehicle (EV) startup, 9
electronic data interchange (EDI), 15
email, 15, 25, 316
    campaign creation, 108
    campaigns, 193
    chains, 225
    communication, 299
    marketing, 6, 45–46, 244
        and automation, 330
        platforms, 245, 299–302
            automation and personalization, 300
            campaign creation and management, 300
            deliverability optimization and compliance, 300
            design and customization, 300
            testing and analytics, 300
        teams, 203
    newsletter, 193

enterprise-level marketing departments, 233
EPiServer CMS platforms, 324
ethical data collection, 179
Eventbrite event management platform, 46–47
events marketing, 46–48
    MarTech, use of, 46–47
    platforms utilized, 47
Evidenza, synthetic research platform, 116
experiential and emerging channels, 251
external agencies and partners, 217
external collaboration, 239
EZ Texting, 315

F

Facebook, 19–20, 36, 38, 42, 74, 125, 135, 305–306, 312, 314–315, 320
Facebook Messenger, 312, 320
feature conflicts, 447
Feinler, E., 16
file format support and conversion, 219
Firebase Analytics, 350
first-party cookies, 179
first-party data, 26, 134–135
    benefits, 137–138
    CDPs, 148, 150–151
    DMPs and, 171
    enrichment, 140–141
    lack of, 143–144
    third-party data, role in, 139
focused content creation, 210
folder structures and asset relationships, 222
Friendster, social media platform, 19
Funnel, 352

G

GAN, *see* generative adversarial networks (GANs)
GDPR, *see* General Data Protection Regulation (GDPR)

General Data Protection Regulation (GDPR), 24–25, 176–177, 180–182, 343, 391, 423
generative adversarial networks (GANs), 107
generative AI, 107–109, 194
geographic mapping functions, 358
Google Ads, 36, 38, 43–44, 48–49, 122, 305, 350
Google AdWords, 17–18
Google Analytics, 8, 23, 35, 38–41, 44, 49–51, 62, 110, 126, 240, 309, 340, 348, 348, 359, 363, 365, 435
Google Analytics 360, 51, 110, 126
Google Assistant, 312, 320
Google Dialogflow, 314
Google Looker, 357
Google Search Engine launch, 17
Google Vision AI, 127
Google Workspace, 8, 195, 235, 239
Gramm-Leach-Bliley Act (GLBA), 177
graphical user interface (GUI), 264–265

H

Headless CMS, 279–282
Health Insurance Portability and Accountability Act (HIPAA), 177
heatmap visualization, 358
high-resolution photography, 215
historical data, 180
history of MarTech
    AI and personalization era (2020s), 25–27
    big data and automation era (2010s), 22–25
    database marketing (1960s-1980s), 14–15
    early (pre-1960s), 14
    eras of, 13
    internet (1990-2001), 16–18
    mobile emergence (2007-2011), 20–22
    social media era (2002-2009), 18–20
Hootsuite, 302
    Insights, 124
    social media management platform, 7, 38, 42, 302, 348
Hotjar, 38, 41, 50
HubSpot, 23, 36, 38, 45, 47, 352, 363
    CMS, 278
    Marketing Cloud
        current market focus, 330
        current state, 330
        origins and growth, 329
    Marketing Hub, 299–300
    workflow automation, 112
hyper-personalization, 26–27

I

IBM Watson Assistant, 314
identity and access management (IAM), 404
identity resolution platforms, 183
implementation methods
    all-in-one MarTech solutions, 98–99
    best-of-breed approach, 99–100
    composable approach, 100–101
    hybrid approach, 104–105
    open source software, 101, 103–104
    proprietary software, 101–102
in-app analytics, 349–350
    designing effective in-app funnels, 350–351
    evaluating in-app analytics tools, 351–352
    popular in-app analytics platforms, 350
    real-time user tracking, 350
in-app analytics platforms, 349
inconsistent messaging, 248
individual privacy, 180
industry self-regulations, 177–178
industry-specific regulations, 177
influencer marketing platforms, 211
InSite, 326

Instagram, 20, 38, 42, 73–74, 108, 119, 127, 129, 193, 218–219, 252, 305, 314, 320
integrations, 220
    capabilities, 206–207
    with communication tools, 239
    with other marketing tools, 236
interactive dashboards, 358
intercom, 313
internal marketing planning, 247
internet (1990-2001)
    digital advertising, 17–18
    search engines and SEO, 16–17
    WWW, 16
interoperability, 447
investments
    competitive advantage, 88–89
    customer experience gap, closing of, 86–88
    data-driven decision-making, 91–93
    internal efficiency and collaboration, 89–91
    marketing and communication effectiveness, 84–85
iPhone, 21–22, 74
IT and data teams, 216–217

J

Journey Optimizer, 324

K

Kapost, 210
key performance indicators (KPIs), 230, 341, 369, 373–374, 391, 436, 443
keyword tagging, 218
Klaviyo, 45, 50, 299

L

large language models (LLMs), 25, 107
leadership, marketing
    MarTech, use of, 34–35
    multi-channel attribution platforms, 35
legal and compliance teams, 216
Lei Geral de Proteção de Dados Pessoais (LGPD), 177
level of integration, 441
LinkedIn, 201
live chat
    solutions, 313
    systems, 315
LiveChat, 313
Livepanel, synthetic research platform, 116
loyalty program, 178–179

M

Magento, 324
Mailchimp, 45, 50, 299, 348, 355
mapping findings, 411
marketing
    and analytics connectivity, 268
    automation, 23, 283
        integration, 222
        solutions, 245
        systems, 298
        tools, 220, 352
    and communication effectiveness, 84–85
    leadership, 228–229
    and lead generation, 264
    operations teams, 54–55
    ROI integration, 240
    scalability, 403
    teams (*see* teams, marketing)
marketing automation platforms (MAPs), 244, 283–285, 288, 349, 400; *see also* ActiveCampaign
    evaluation, 290
    example, 289
    features, 288–289
    strengths, 289
    weaknesses, 290
marketing channels, 250

adapting content across channels, 251–253
balancing channel-specific expertise with omnichannel excellence, 255
changing consumer behavior and platform selection, 254—255
complexity of managing multiple channels effectively, 255–256
customer channel switching and evolving preferences, 253
expectation of seamless transitions between channels, 253–254
proliferation of digital touchpoints, 250–251
marketing lifecycle, 246
marketing lifecycle and platform selection, 246
   aligning platforms with, 246–247
   balancing long-term brand-building and short-term performance marketing, 247–248
   fragmentation across teams and workflows, 248–249
   maintaining agility without compromising strategy, 249–250
   planning, creation, management, and delivery, 248
   platform needs across the customer journey, 247
marketing managers, 201–202
marketing operations, 389
   and analytics teams, 230
marketing organizations, 248
marketing platforms, 250, 349
   marketers and, 285
      data-driven decision-making, 286
      efficiency gains, 285
      improved customer journeys, 286
      personalized customer engagement, 285–286
   users of, 286–287
      customer experience teams, 287
      IT and data teams, 288
      marketing teams, 287
      sales teams, 287–288
marketing professionals, 206
marketing qualified leads (MQLs), 59–60
marketing-specific timelines, 234
marketing teams, 214, 317
marketing technology (MarTech) platforms, 3, 182, 184, 186, 190, 199, 228, 237, 246, 250, 255, 409, 448
   AI's role in content creation, 196
   analytics and reporting, 7–8
   categories, 3–5, 189, 243, 339
   content, campaign, and multichannel delivery, 6–7
   customer data, 5–6
   customer needs and expectations of (see customer relationship)
   evaluation, 421
   history (see history of MarTech)
   maturity model, 93, 93–94
   related platforms, 8–9
   stack, ii–iv
   strategy and creative process, 191–192
      aligning creative efforts with marketing objectives, 192–193
      asset management, 192
      automation and AI-driven recommendations, 194
      channel-specific optimization, 193
      collaboration, 195
      content collaboration tools, 195
      content planning and ideation tools, 192
      cross-team coordination, 195–196
      efficiency, 194
      multi-channel needs, 193
      project and task management tools, 194
      remote and global teams, 196

repurposing content and assets, 193
version control and approval
processes, 195
workflow automation, 194–195
VoltAge Motors, 9–11, 197–198
workflow and task automation, 7
Marketo, 6, 23, 36, 45, 55, 60, 289, 321, 324
MarTech Champions, 257
MarTech infrastructure goals framework, 369–370
  business needs, 377
  example KPIs to measure success, 373
  MarTech maturity model, 377–380
    stages, 380–381
  right teams and unique roles, 382–383
  strategic goals, 370–371
    effective governance, compliance, and security, 373
    enhanced automation and personalization, 372–373
    example steps to achieve each, 374–377
    flexible and scalable platform architecture, 372
    seamless system integration and interoperability, 371–372
    unified data management and accessibility, 371
  Voltage Motors' maturity journey, 383–384
MarTech maturity model, 378–380, 411
  advanced stage, 380–381
  foundational stage, 380
  innovative stage, 381
  integrated stage, 380–381
MarTech roadmap, 431–433
  necessity, 432
  steps to create, 432–434
    five: align KPIs with your roadmap, 436
    four: analyze gaps between the current and desired state of technology, 436
    one: start with a static infrastructure map, 434
    six: identify and prioritize current marketing technologies, 436–437
    three: socialize your roadmap, 435–436
    two: determine the best format for your roadmap, 434–435
  up to date, 437
  Voltage Motors, 437–438
    cross-functional collaboration, 438
    personalization and efficiency, 439
    from static map to dynamic strategy, 438
MarTech stack, 284, 317, 321, 359, 369
  deep integration with, 275
MarTech Stack evaluation framework, 385–386
  evaluation, 388–391, 411–413
    automation and personalization capabilities, 402–403
    core infrastructure and architecture, 395–397
    current maturity level, 393–395
    data management and accessibility, 397–399
    gaps and prioritize next steps, 411–413
    governance, compliance, and security, 404–405
    organizational readiness and skills alignment, 407–409
    quantify performance and ROI, 409–411
    scope and purpose, 391–393
    system integration and interoperability, 399–401
    vendor and platform roadmaps, 405–407
  infrastructure map example, 387
  involvement, 388
  overarching gaps, 415

requirements for MarTech infrastructure map, 387–388
specific gaps, 415–417
visualizing, 386–387
MarTech systems, 297; *see also* marketing technology (MarTech) platforms
architecture, 372
ecosystems, 299, 371, 373, 395
infrastructure, 430, 432, 436–437, 446
investments, 248, 254, 395, 410, 445
market, 283
measurement platforms, 347
solutions, 319
Mautic, 329
media library, 265
media mix modelling (MMM), 92
messaging apps and social media chat, 314
Meta Ads Manager, 305
Meta Business Suite, 314
metadata
and tagging features, 222
and tag management, 218
tags, 192
Metaverse, 251
Microsoft Dynamics, 6, 23, 53, 60
Microsoft Excel, 434
Microsoft Office, 8
Microsoft Power BI, 357
Microsoft Visio, 434
mid-sized marketing teams, 232
Miro, 434
Mix Modeler, 324
Mixpanel, 351, 355
mobile emergence (2007-2011)
smartphones, 21–22
SMS marketing, 20–21
mobile-first behavior, 74–75
Monday.com, 194
Moz Pro, 307
Muhney, M., 15
multi-channel campaigns, 227
multi-channel content campaigns, 199

multi-channel content delivery, 275
multi-channel marketing campaigns, 191
multi-channel measurement tools, 352–353
evaluation, 354–355
multichannel personalization platforms, 284–285, 290
evaluations, 292
example, 291
features, 291
strengths, 292
weaknesses, 292
multi-channel tools, 298
multi-factor authentication (MFA), 404
multi-function platforms and suites, 323
Acquia DXP, 328–329
Adobe Experience Platform (AEP), 323–325
evaluation, 333–334
Hubspot Marketing Cloud
current market focus, 330
current state, 330
origins and growth, 329
Optimizely ONE, 325–327
Salesforce Marketing Cloud, 330–331
current market focus, 331
Sitecore DXP, 327–328
strengths of, 332
weaknesses of, 333
lack of interoperability, 333
lack of robust features, 333
vendor lock-in, 333
multiple-channel campaign, 285
multi-touch attribution (MTA), 92
MySpace, 19–20

N

natural language processing (NLP), 313, 320, 362
natural language querying, 361
near-real-time analytics, 360
Neolane, 324

Netflix, 25
net promoter score (NPS), 445
NetSpring, 326
news portals, 264
non-marketing executives, 57–59
Notion, 195

O

omni-channel communication, 316
omnichannel customer experience, 75–76
Omnichannel engagement, 267
Omniture, 324
one-time evaluations, 414
on-prem versions, 267
open source software, 101
    benefits, 103
    drawbacks, 103–104
operational inefficiencies, 227
Optimizely, 268
Optimizely ONE, 325–327
    acquisitions and expansions, 326
    current state, 326–327
    market focus, 325–326
optimizing content distribution, 201
opt-in and consent mechanisms, 178–179
organic and paid social media management, 330
overall return on investment (ROI), 441–442
    adoption rate, 446
        cross-team adoption, 446
        level of usage of specific products, 446
        level of usage of the entire martech ecosystem, 446
    benefit to the business, 447–449
    level of integration, 446–447
    TCO, 442
        customer satisfaction (internal and external), 444–446
        speed to delivery, 443
        speed to insights, 443–444
    Voltage Motors aligns MarTech adoption with business benefits, 449–451
oversee data collection, 230
owned channels, 250

P

paid channels, 250
Pardot, 6, 23, 36, 60, 289, 321, 331
Pathfinder Web site, 18
performance
    marketing team, 236
    measurement, 411
personal blogs, 264
personal digital assistants (PDAs), 21
personal identifiable information (PII), 183
personalization, 200, 372; *see also* AI-powered personalization; customer experience
    engines, 448
    marketing, 84–85
        AI, 25–26, 112–114
        expectation of, 77–78
        first-party data strategies, 26
        hyper-personalization, 26–27
        privacy-compliant platforms, 27
        at scale, 86
        software platforms, 114
personalized customer experiences, 315–316
Podium, 315
PowerPoint, 434
predictive analytics, 109–111
privacy
    and compliance considerations, 343
    concerns, 175
    policies, 175
privacy-first mindset, 186
process optimization, 230, 240
production timelines for branding, 230
productivity gaps, 227

product management teams, 215–216
product roadmap, 447
project and task management tools, 194
project management platforms, 225
    evaluation, 237–238
        automation and workflow efficiency, 239–240
        collaboration and communication features, 239
        customization and scalability, 238–239
        ease of use and adoption, 238
        reporting and performance tracking, 240
    features, 233–234
        analytics and reporting, 237
        approval workflows and version control, 236
        campaign planning and calendar management, 234–235
        integration with other marketing tools, 236
        resource allocation and time tracking, 235
        task and workflow management, 234
        team collaboration and communication tools, 235
    and marketers, 226
        improved collaboration, 226–227
        resource allocation and time management, 227
        scalability and adaptability, 227–228
        streamlined workflows, 226
        visibility and transparency, 227
    users, 228
        campaign and content teams, 229
        creative and design teams, 229–230
        cross-functional teams, 230–231
        enterprise-level marketing departments, 233
        marketing leadership, 228–229
        marketing operations and analytics teams, 230
        mid-sized marketing teams, 232
        small marketing teams and startups, 231–232
        varying needs depending on teams, 231
project management software, 228–229, 237
project managers, 203
project status dashboards, 240
proprietary software, 101
    benefits, 102
    drawbacks, 102
prosumers, 266
protected health information (PHI), 177
public policy regulations, 176–177

Q

quality of content, 200

R

rate of innovation, 447
real-time analytics, 248
real-time campaign performance, 361
real-time collaboration, 223
real-time communication, 311, 371
    platforms, 318, 321
real-time customer data platform (CDP), 324
real-time customer interaction, 321
real-time customer support, 317
real-time dashboards, 227
real-time data personalization, 403
real-time dialogues, 311
real-time interaction, 312
real-time personalization, 263, 266, 372
real-time synchronization of content, 236
real-time tracking of events, 350
remote and global teams, 196
reporting and performance tracking, 240, 330

repurposing content and assets, 193
resource allocation and time management, 227
resource allocation and time tracking, 235
retail brand, 218
retention period, 180–181
return on ad spend (ROAS), 92
return on investment (ROI), 203, 206, 442
role-based access control (RBAC), 222, 399
role-based access structure, 371
Ruler Analytics, 126

S

SaaS, 267
Salesforce, 6, 23, 36, 47, 53, 58, 60, 62–63, 87, 152, 164, 208, 240, 363, 365, 393
Salesforce Commerce Cloud, 50
Salesforce Customer 360, 68
Salesforce Einstein Analytics, 110
Salesforce Interaction Studio, 294
Salesforce Marketing Cloud (SFMC), 101–102, 121, 294, 330–331
    current market focus, 331
Salesforce Pardot, 289
sales qualified leads (SQLs), 59
sales teams, 59–61, 215, 317
scalability, 221, 372
    and adaptability, 227–228
    and future-proofing, 209–210
    and load handling, 396
search engine marketing (SEM), 17
search engine optimization (SEO), 16–17
    platforms, 307–309
    specialists, 202
search marketing, 48–49
search-optimized blog content, 252
second-party data, 135
secure and compliant data deletion, 180–181
secure deletion, 181
self-regulation, 178
self-serve mindset, 78–79

self-serve portal, 217
SEMrush, 307
sensitive data protection, 216
SEO, *see* search engine optimization (SEO)
service-level agreements (SLAs), 406
Simudyne, synthetic research platform, 116
single-channel analytics, 348
single-channel marketing platforms, 297
    examples, 298–299
        digital advertising management, 304–307
        email marketing platforms, 299–302
        search engine optimization (SEO) platforms, 307–309
        social media management, 302–304
    and marketers, 298
single-channel measurement tools, 347–348
    evaluating single-channel analytics platforms, 348–349
single sign-on (SSO), 404
Sitecore, 268
Sitecore DXP, 327–328
    evolution of, 328
    market focus, 327
site-mapping tools, 274
small marketing teams and startups, 231–232
SMS marketing, 283
    and business texting, 314–315
    platforms, 315
social media, 193, 218
    audiences, 252
    customer relationship, 73–74
    engagement tools, 312
    marketing, 42–43
    messaging, 312
    platforms, 298
    snippets, 201
    updates, 193
social media era (2002–2009)
    early social platforms, 19
    marketing, 19

paid social advertising, 20
Web 2.0 and user-generated content, 18–19
social media management, 283, 302–304
    platform, 303
spreadsheets, 225
Sprinklr, 42, 69, 124
Sprout Social, 7, 38, 42, 302, 348
Standard Contractual Clauses (SCCs), 181
Starbucks, 74
startups, 231–232
storage capacity and file type support, 224
storytelling and presentation tools, 362
Sullivan, P., 15
synchronization, 359
synthetic research platforms, 114
    campaign testing, 115
    examples, 116
    product development, 116

T

Tableau, 8, 23, 35, 51, 55, 58, 60, 62, 331, 357
targeted audience segments, 200
task
    assignment and tracking, 234
    automation, 239
    management tools, 229
    and workflow management, 234
TCO, *see* total cost of ownership (TCO)
team collaboration and communication tools, 235
teams, marketing
    analytics, 51–52
    content, 39–40
    CRM, 52–54
    customer acquisition, 36–37
    digital advertising, 43–44
    digital marketing campaigns, 37–38
    e-commerce, 49–51
    email, 45–46
    events, 46–48
    leadership, 34–35
    operations, 54–55
    search, 48–49
    social media, 42–43
    web site, 40–42
technical KPIs, 409
technology teams, 318
television ads, 193
third-party cookies, 179
third-party data, 135–137
Tidio, 313
TikTok, 74
time-to-market, 443
time-to-market for digital campaigns, 450
time tracking, 235
total cost of ownership (TCO), 343, 410, 429, 441–442
    customer satisfaction (internal and external), 444–446
    speed to delivery, 443
    speed to insights, 443–444
tracking performance, 199
Trade Desk, The, 123, 305
transparency in data collection methods, 178
trust gap between consumers and brands, 175–176
Twilio, 314

U

unified data management, 371
user experience (UX) strategies, 309
user-friendly interface, 220
user-generated content (UGC), 18–19, 72–73, 119, 126, 429
UX designers, 352

V

version control, 215, 219
    and approval processes, 195

systems, 236
video-based storytelling, 252
video clips, 215
video content marketing platforms, 211
videographers, 214–215
video producers, 229
virtual reality (VR), 251; *see also* artificial intelligence (AI)
visibility and transparency, 227
visual recognition and image processing, 126–127
voice assistants, 312, 316
voice-controlled interactions, 320
VoltAge Motors, vii–viii, 9–11, 183–185, 197–198, 344, 391
    CDPs, CRMs and DMPs, 172–174
    CMP, 184
    content and campaigns, 10
    customer data, 9, 143–145
    digital marketing capabilities, 437
    e-commerce platform, 401
    internal developers, 407
    marketing campaigns, 9–10
    marketing operations team, 392
    measurement and reporting, 10
    multichannel marketing challenges, 260–262
    and multilayered analytics strategy, 355–356
    personalization framework, 402–403
    roadmap, building of, 10–11
    security standards, 405
    workflow automation, 402

W

W3Catalog, web-based search engine, 16
web analytics platforms, 363
web-based content, 263
Web-based messaging platforms, 314
web content management platforms, 263
Webflow, 266
websites, 193
    analytics, 344
    creation, 264
    marketing, 40–42
WeChat, 312
Welcome, 326
WhatsApp, 312, 316, 320
WhatsApp for Business, 314
What-You-See-Is-What-You-Get (WYSIWYG) editor, 265, 273
Widen, 39, 67, 329
WordPress, 265
workflow automation, 111–112, 156, 194–195, 203, 208–209, 223; *see also* project and task management tools
Workfront, 324
world wide web (www), 16; *see also* websites
Wrike, 194
writers, 202

Y

Yahoo!, 16
YouTube, 19, 135, 211, 252

Z

Zaius, 326
Zapier, 37, 53, 55, 63, 112
ZenDesk, 8, 65, 69, 78, 118, 313
Zendesk Chat, 313
zero-party data, 134

www.ingramcontent.com/pod-product-compliance
Lightning Source LLC
Chambersburg PA
CBHW081102170526
45165CB00008B/2297